T0215043

Lectures, Problems and Solutions
for
Ordinary Differential Equations

Second Edition

Lectures, Problems and Solutions for
Ordinary Differential Equations

Second Edition

Yuefan Deng

Stony Brook University, USA

World Scientific

NEW JERSEY · LONDON · SINGAPORE · BEIJING · SHANGHAI · HONG KONG · TAIPEI · CHENNAI · TOKYO

Published by

World Scientific Publishing Co. Pte. Ltd.
5 Toh Tuck Link, Singapore 596224
USA office: 27 Warren Street, Suite 401-402, Hackensack, NJ 07601
UK office: 57 Shelton Street, Covent Garden, London WC2H 9HE

British Library Cataloguing-in-Publication Data
A catalogue record for this book is available from the British Library.

**LECTURES, PROBLEMS AND SOLUTIONS FOR ORDINARY
DIFFERENTIAL EQUATIONS**
Second Edition

ISBN 978-981-3226-12-8
ISBN 978-981-3226-13-5 (pbk)

Printed in Singapore

Preface

This book, *Lectures, Problems and Solutions for Ordinary Differential Equations*, results from more than 20 revisions of lectures, exams, and homework assignments to approximately 6,000 students in the College of Engineering and Applied Sciences at Stony Brook University over the past 30 semesters. The book contains notes for 25 80-minute lectures and approximately 1,000 problems with solutions. Thanks to a large body of classics for the topics, creating another ODEs book is probably as unnecessary as reinventing the wheel. Yet, I constructed this manuscript *differently* by focusing on ODE examples from sciences, engineering, economics, and other real-world applications of ODEs. These examples, partly adapted from well-known textbooks and partly freshly composed, focus on illustrating the ODE applications. Serendipitously, the examples motivate students to learn theorem-proving while mastering the art of applying them.

While preparing this manuscript, I benefited immensely from many people including undergraduate students who took the class and graduate teaching assistants (TAs) who helped teaching it. C. Han has triple roles: one of the undergraduate students, a TA, and an editor. Students H. Fan, B. Hoefer, Y. Hu and M. Nino proofread the latest versions.

I am proud to announce that I'm the receipt of the 2016 State University of New York Chancellor's Award of Excellence in Teaching. It would not have been possible without our collective efforts.

Yuefan Deng

Stony Brook, New York

January 1, 2017

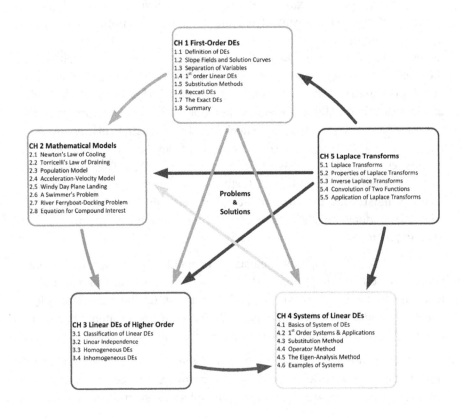

CH 1 First-Order DEs
1.1 Definition of DEs
1.2 Slope Fields and Solution Curves
1.3 Separation of Variables
1.4 1st order Linear DEs
1.5 Substitution Methods
1.6 Reccati DEs
1.7 The Exact DEs
1.8 Summary

CH 2 Mathematical Models
2.1 Newton's Law of Cooling
2.2 Torricelli's Law of Draining
2.3 Population Model
2.4 Acceleration-Velocity Model
2.5 Windy Day Plane Landing
2.6 A Swimmer's Problem
2.7 River Ferryboat-Docking Problem
2.8 Equation for Compound Interest

Problems & Solutions

CH 5 Laplace Transforms
5.1 Laplace Transforms
5.2 Properties of Laplace Transforms
5.3 Inverse Laplace Transforms
5.4 Convolution of Two Functions
5.5 Application of Laplace Transforms

CH 3 Linear DEs of Higher Order
3.1 Classification of Linear DEs
3.2 Linear Independence
3.3 Homogeneous DEs
3.4 Inhomogeneous DEs

CH 4 Systems of Linear DEs
4.1 Basics of System of DEs
4.2 1st Order Systems & Applications
4.3 Substitution Method
4.4 Operator Method
4.5 The Eigen-Analysis Method
4.6 Examples of Systems

Contents

Contents

Chapter 1
First-Order Differential Equations

1.1 Definition of Differential Equations

A differential equation (DE) is a mathematical equation that relates some functions of one or more variables with their derivatives. A DE is used to describe changing quantities and it plays a major role in quantitative studies in many disciplines such as all fields of engineering, physical sciences, life sciences, and economics.

Examples
Are they DEs or not?
$$ax^2 + bx + c = 0 \qquad \text{No}$$

$$ax^2 + bx' + c = 0 \qquad \text{Yes} \qquad\qquad \text{Here } x' = \frac{dx}{dt}$$

$$ax^2 + bx' + cy' = 0 \quad \text{Yes} \qquad\qquad \text{Here } x' = \frac{dx}{dt} \text{ and } y' = \frac{dy}{dt}$$

$$y'' = x^3 \qquad\qquad\quad \text{Yes} \qquad\qquad \text{Here } y' = \frac{dy}{dt}$$

To solve a DE is to express the solution of the unknown function (the dependent variable or DV) in mathematical terms without the derivatives.

Examples

$$ax' + b = 0$$

$$x' = -\frac{b}{a} \qquad \text{is not a solution}$$

$$x = -\int \frac{b}{a} dt \qquad \text{is a solution}$$

In general, there are two common ways in solving DEs: analytically and numerically. Most DEs, difficult to solve by analytical methods, must be "solved" by using numerical methods, although many DEs are too stiff to solve by numerical techniques. Solving DEs by numerical methods is a different subject requiring basic knowledge of computer programming and numerical analysis; this book focuses on introducing analytical methods for solving very small families of DEs that are truly solvable. Although the DEs are quite simple, the solution methods are not and the essential solution steps and terminologies involved are fully applicable to much more complicated DEs.

Classification of DEs

Classification of DEs is itself another subject in studying DEs. We will introduce classifications and terminologies to make the contents of the book flow but one may still need to look up terms undefined here or abbreviations introduced at the end of the book. First, we introduce the terms of dependent variables (DVs) and independent variables (IVs) of functions. A DV represents the output or effect while the IV represents the input or the cause. Truly, a DE is an equation that relates these two variables. A DE may have more than one variable for each and the DE with one IV and one DV is called an ordinary differential equation or ODE. The ODE, or simply referred to as DE, is the object of our book. Continuing, you understand why this 1-to-1 relationship is called *ordinary*. A DE that has one DV and $N \geq 2$ IVs, 1-to-N relationship, is called a partial differential equation or PDE. Studying PDEs, out of the scope of this book, requires solid understanding of partial derivatives and, more desirably, full knowledge of multiple variable calculus. By now, you certainly want to see two other types of DEs. With $N \geq 2$ DVs and 1 IV, you may compose an N-to-1 *system* of ODEs. Similarly, with $N \geq 2$ DVs and $M \geq 2$ IVs, you may compose an N-to-M *system* of PDEs. We have several other classifications to categorize DEs.

Order of DEs

The order of a DE is determined by the highest order derivative of the DV (and sometimes we interchangeably call it unknown function). If you love to generalize things, algebraic equations (AE) may be classified as 0th order DEs as there are no derivatives for the unknowns in the AEs.

Examples
Determine the orders of the following DEs.

1st-order (1st.O) DE:

$$x' + ax = 0 \tag{1.1}$$

$$x' + ax^2 = 0 \tag{1.2}$$

2nd-order (2nd.O) DE:

$$x'' + bx = 0 \tag{1.3}$$

$$x'' + bx^5 = 0 \tag{1.4}$$

nth-order (nth.O) DE:

$$x^{(n)} + bx'' + cx = 0 \tag{1.5}$$

$$x^{(n)} + 5x^{(n-1)} + x = 0 \tag{1.6}$$

	Definition	**Examples**
ODEs vs. PDEs	ODEs contain only one IV. PDEs contain two or more IVs.	ODE: $x'' + \omega^2 x = f(t)$ PDE: $\dfrac{\partial^2 u}{\partial t^2} = \dfrac{\partial^2 u}{\partial x^2} + \dfrac{\partial^2 u}{\partial y^2}$
Order of DEs	The highest order of the derivatives of the DVs determines the order of the DEs	First-order (1st.O): $y' = x + y$ Second-order (2nd.O): $y'' = x + y^2$
Linear vs. Nonlinear	A DE containing nonlinear term(s) for the DVs is a nonlinear DE regardless of the nature of the IVs	Linear: $y' = x^2 + y$ Nonlinear: $y'' = x + y^2$

| Homogeneous vs. Inhomogeneous | A 1st-order (1st.O) DE $M(x,y)dx + N(x,y)dy = 0$ is Homo DE if both $M(x,y)$ and $N(x,y)$ are homogeneous functions. Otherwise, it is InHomo DE.

 In general, a linear DE of order n is a Homo DE if it is of the form $$\sum_{j=0}^{n} a_j(x)y^{(j)} = 0$$ Otherwise, it is InHomo DE. | 1st.O Homo DE: $y^2 y' = x^2 + xy$
 1st.O InHomo DE: $y' = x^2 + y^2$

 Homo DE: $y'' + y^2 \cos x = 0$
 InHomo DE: $y'' + y = \cos x$ |

Linearity of DEs

DEs can also be classified as linear or nonlinear according to the linearity of the DVs regardless of the nature of the IVs. A DE that contains only linear terms for the DV or its derivative(s) is a linear DE. Otherwise, a DE that contains at least one nonlinear term for the DV or its derivative(s) is a nonlinear DE.

Linear	Nonlinear
$y'' + y = 0$	$y'' + y^2 = 0$
$y^{(n)} + x + y = 0$	$y^{(n)} + xy + y^3 = 0$
$y^{(n)} + x^3 + y = 0$	$\left(y^{(n)}\right)^2 + x^3 + y = 0$

Solutions

A function that satisfies the DE is a solution to that DE. Seeking such functions is the main objective of the book while composing the DEs, which may excite engineering majors more, is the secondary objective of this book.

Solutions can also be classified into several categories, for example, general solutions (GS), particular solutions (PS), and singular solutions (SS). A GS, *i.e.*, a complete solution, is the set of all solutions to the DE and it can usually be expressed in a function with arbitrary constant(s). A PS is a subset of the GS whose arbitrary constant(s) are determined by the initial conditions (ICs) or the boundary conditions (BCs) or both. An SS is a solution that is singular. In a less convoluted definition, an SS is one for which the DE, or the initial value problem (IVP) or the Cauchy problem, fails to have a unique solution at some point in the solution. The set in which the solution is singular can be a single point or the entire real line.

Examples
Try to *guess* the solutions of the following DEs:
(1) $x' + \omega^2 x = 0$
(2) $x' + \omega^2 x = t$
(3) $x'' + \omega^2 x = 0$
(4) $x'' + \omega^2 x = \sin(\omega t)$
(5) $x'' + \omega^2 x = \sin(\omega_1 t)$ where, in general, $\omega \neq \omega_1$

As briefly mentioned before, there are several methods to find the solutions of DEs. This book covers the first of the two common methods, *i.e.*, analytical method and numerical method.

1.   To obtain analytical (closed form) solutions.

Only a small percentage of linear DEs and a few special nonlinear DEs are simple enough to allow findings of analytical solutions. There are two major types: the exact methods and the approximation methods. Examples of the exact methods are (1) method of undetermined coefficients (MUC); (2) integrating factor (IF) method; (3) method of separation of variables (SOV); and (4) variation of parameters (VOP). The examples of the approximation but convergent methods are (1) successive approximations; (2) series methods including power series methods and the generalized Fourier series methods; (3) multiple scale analysis; and (4) perturbation methods.

2. To obtain numerical solutions.

Most DEs in science, engineering, and finance are too complicated to allow findings of analytical solutions and numerical methods are the only viable alternatives for *approximate* numerical solutions. Unfortunately, most DEs that are of vital importance for practical purposes belong to this type. Indeed, every rose has its thorn. To solve DEs numerically, one must acquire a different set of skills: numerical analysis and computer programming. The types of numerical methods for DEs are too numerous to name. For ODEs, the examples are the Euler method and the general linear methods such as Runge-Kutta methods and the linear multi-step method. For PDEs, the examples are finite difference methods and finite element methods.

This book does not cover any of the numerical methods.

Problems

Problem 1.1.1 Verify by substitution that each given function is a solution to the given DE. Throughout these problems, primes denote derivatives with respect to (wrt) x.

$$y'' + y = 3\cos 2x, \qquad y_1 = \cos x - \cos 2x, \qquad y_2 = \sin x - \cos 2x$$

Problem 1.1.2 Verify by substitution that each given function is a solution to the given DE. Throughout these problems, primes denote derivatives wrt x.

$$x^2 y'' - xy' + 2y = 0, \qquad y_1 = x\cos(\ln x), \qquad y_2 = x\sin(\ln x)$$

Problem 1.1.3 A function $y = g(x)$ is described by some geometric properties of its graph. Write a DE of the form $dy/dx = f(x, y)$ having the function g as its solution (or as one of its solutions). The graph of g is normal to every curve of the form $y = x^2 + k$ (k is a constant) where they meet.

Problem 1.1.4 Determine by inspection at least one solution to the DE $xy' + y = 3x^2$. That is, use your knowledge of derivatives to make an intelligent guess and, then, test your hypothesis.

Problem 1.1.5 Determine by inspection at least one solution to the DE $y'' + y = 0$. That is, use your knowledge of derivatives to make an intelligent guess and, then, test your hypothesis.

Problem 1.1.6 Verify by substitution that the given function is a solution to the given DE. Primes denote derivatives wrt x.

$$y' + 2xy^2 = 0, \qquad y = \frac{1}{1 + x^2}$$

Problem 1.1.7 Verify that $y(x)$ satisfies the given DE and determine a value of the constant C so that $y(x)$ satisfies the given IC.

$$y' + 3x^2 y = 0, \qquad y(x) = C\exp(-x^3), \qquad y(0) = 7$$

Problem 1.1.8 Verify that $y(x)$ satisfies the given DE and determine a value of the constant C so that $y(x)$ satisfies the given IC.

$$y' + y\tan x = \cos x, \qquad y(x) = (x + C)\cos x, \qquad y(\pi) = 0$$

Problem 1.1.9 Verify that $y(x)$ satisfies the given DE and determine a value of the constant C so that $y(x)$ satisfies the given IC.

$$y' = 3x^2(y^2 + 1), \qquad y(x) = \tan(x^3 + C), \qquad y(0) = 1$$

Problem 1.1.10 Verify that $y(x)$ satisfies the given DE and determine a value of the constant C so that $y(x)$ satisfies the given IC.
$$xy' + 3y = 2x^5, \qquad y(x) = \frac{1}{4}x^5 + \frac{C}{x^3}, \qquad y(2) = 1$$

Problem 1.1.11 Verify by substitution that the given functions are solutions of the given DE. Primes denote derivatives wrt x.
$$y'' = 9y, \qquad y_1 = \exp(3x), \qquad y_2 = \exp(-3x)$$

Problem 1.1.12 Verify by substitution that the given functions are solutions of the given DE. Primes denote derivatives wrt x.
$$\exp(y)\,y' = 1, \qquad y(x) = \ln(x + C), \qquad y(0) = 0$$

Problem 1.1.13 Verify that $y(x)$ satisfies the given DE and determine a value of the constant C so that $y(x)$ satisfies the given IC. In the equation and its solution, n is a given constant.
$$y' + nx^{n-1}y = 0, \qquad y(x) = C \exp(-x^n), \qquad y(0) = 2014$$

1.2   Slope Fields and Solution Curves

Geometrical interpretation is an effective way to help understand the properties of the DEs and their solutions. A solution can be drawn as a curve, which is called a Solution Curve. A series of lines with the same slope (for each line), such as a family of curves $f(x, y) = s$ for DE $y' = f(x, y)$ (Figure 1.1) are called Isocline, meaning the same slope. A line in the xy-plane whose slope is $f(x, y)$ is called a Slope Curve and a collection of such slope curves forms a Slope Field (Figure 1.2), aka, the direction field.

Consider a classical example:

$$y' = 2xy \tag{1.7}$$

whose GS is

$$y = C \exp(x^2) \tag{1.8}$$

In the xy-plane, we can draw the isocline by selecting a few appropriate constant values s for the following:

$$2xy = s \tag{1.9}$$

and the solution curves with the given solution by selecting a few constant values C.

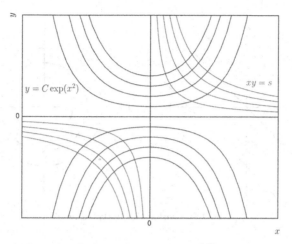

Figure 1.1 Examples of solution curves $y = C \exp(x^2)$ and isocline $xy = s$.

Example 1
Find the isoclines
$$y' = \sin(x - y) \tag{1.10}$$

Solution
$$x - y = \arcsin c$$
$$y = x - \arcsin c$$

Example 2
Find the isoclines
$$y' = x^2 + y^2 = C \tag{1.11}$$

Solution
The isocline equation
$$y' = f(x, y) = x^2 + y^2 = C$$
It is a circle.

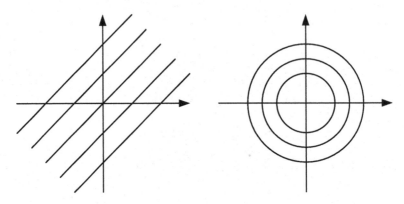

Figure 1.2 The isoclines for the two examples.

Problems

Problem 1.2.1 Plot the solution curves and slope field for the DE for the appropriate ranges for variables x and y
$$y' = x^2 - y$$

Problem 1.2.2 Plot the solution curves and slope field for the DE for the appropriate ranges for variables x and y
$$y' + y = x + 2$$

1.3   Separation of Variables

All 1st.O initial value problem (IVP), aka, the Cauchy problem, can be written as

$$\begin{cases} y' = F(x, y) \\ y(x = a) = b \end{cases} \tag{1.12}$$

If $F(x, y)$ can be written as

$$y' = \frac{g(x)}{f(y)} \tag{1.13}$$

Then, the 1st.O IVP is said to be separable.

Solution

$$\frac{dy}{dx} = \frac{g(x)}{f(y)}$$
$$\int f(y)dy = \int g(x)dx \tag{1.14}$$
$$F(y) = G(x) + C$$

where C is a constant determined by the IC $y(a) = b$.

Example 1A
Find the PS to the following IVP

$$\begin{cases} y' = -6xy \\ y(0) = 7 \end{cases} \tag{1.15}$$

Solution

$$\frac{dy}{y} = -6xdx$$
$$\int \frac{dy}{y} = \int (-6x)dx$$
$$\ln y = -3x^2 + \ln C$$
$$y = C \exp(-3x^2)$$

is the GS to (1.15) as shown in Figure 1.3.

Applying the IC $y(0) = 7$, *i.e.*, $y(0) = C \exp(0) = 7$, we get $C = 7$ and the PS

$$y(x) = 7 \exp(-3x^2)$$

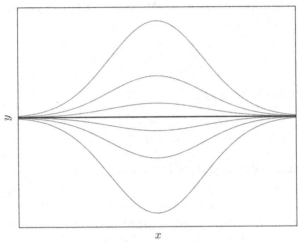

Figure 1.3 The GS $y = C \exp(-3x^2)$ and the SS $y = 0$.

Example 1B
Is $y = 0$ a solution to DE $y' = -6xy$?

Solution
If $y = 0$, for the given DE, we have
$$\text{LHS} = y' = 0$$
and
$$\text{RHS} = -6xy = -6x \cdot 0 = 0$$
Thus,
$$\text{LHS} = \text{RHS}$$
Therefore,
$$y = 0$$
is an SS to $y' = -6xy$.

Example 2
Find the PS to the following IVP

$$\begin{cases} y' = \dfrac{4 - 2x}{3y^2 - 5} \\ y(1) = 3 \end{cases} \tag{1.16}$$

Solution

$$\frac{dy}{dx} = \frac{4 - 2x}{3y^2 - 5}$$

$$\int (3y^2 - 5)dy = \int (4 - 2x)dx$$

Integrating both sides, we get the GS

$$y^3 - 5y = 4x - x^2 + C$$

Applying the IC $y(1) = 3$, *i.e.,* $3^3 - 15 = 4 - 1 + C$, we get $C = 9$ and the PS to the IVP is

$$y^3 - 5y = 4x - x^2 + 9$$

Example 3

Find the GS to the following DE

$$y^2 + x^2 y' = 0 \qquad (1.17)$$

Solution

$$y^2 + x^2 \frac{dy}{dx} = 0$$

$$\frac{dy}{dx} = -\frac{y^2}{x^2}$$

$$\frac{1}{y^2} dy = -\frac{1}{x^2} dx$$

$$\int \frac{1}{y^2} dy = -\int \frac{1}{x^2} dx$$

$$-\frac{1}{y} = \frac{1}{x} - C$$

where we selected $-C$, instead of the more common $+C$, as the integration constant for minor convenience. Therefore,

$$\frac{1}{x} + \frac{1}{y} = C$$

is the GS to the given DE.

Singular Solutions

A singular solution is a solution for which the IVP fails to have a unique solution at some point on the solution. The domain on which a solution is singular is weird as it could be a single point or the entire real line. The SS usually appear(s) when one divides the DE by a term that might be zero while solving the DE.

Suppose we have a DE

$$y' = H(y)G(x) \qquad (1.18)$$

To solve (1.18), we divide it by $H(y)$.

$$\frac{dy}{H(y)} = G(x)dx \qquad (1.19)$$

That gives

$$\int \frac{dy}{H(y)} = \int G(x)dx \qquad (1.20)$$

Such division is mathematically allowed if and only if $H(y) \neq 0$. However,

$$H(y) = 0 \qquad (1.21)$$

is a solution to the DE because it indeed satisfies the original DE (1.18). This solution is peculiar and, thus, it is called SS.

Problems

Problem 1.3.1 Find the GS to the following DE
$$3x(y-2)dx + (x^2+1)dy = 0$$

Problem 1.3.2 Find the GS (implicit if necessary, explicit if convenient) of the following DE. Prime denotes derivatives wrt x.
$$y' = (64xy)^{\frac{1}{3}}$$

Problem 1.3.3 Find the GS (implicit if necessary, explicit if convenient) of the following DE. Prime denotes derivatives wrt x. (*Hint: Factorize the RHS.*)

$$y' = 1 + x + y + xy$$

Problem 1.3.4 Find the explicit PS to the following IVP
$$\begin{cases} \dfrac{dy}{dx} = x\exp(-x) \\ y(0) = 1 \end{cases}$$

Problem 1.3.5 Find the explicit PS to the following IVP
$$\begin{cases} y' = -2\cos 2x \\ y(0) = 2014 \end{cases}$$

Problem 1.3.6 Find the explicit PS to the following IVP
$$\begin{cases} 2\sqrt{x}y' = (\cos y)^2 \\ y(4) = \dfrac{\pi}{4} \end{cases}$$

Problem 1.3.7 Find the explicit PS to the following IVP
$$\begin{cases} y' = 2xy + 3x^2y\exp(x^3) \\ y(0) = 5 \end{cases}$$

Problem 1.3.8 Find the GS (implicit if necessary, explicit if convenient) of the following DE. Primes denote derivatives wrt x.
$$y^3 y' = (y^4 + 1)\cos x$$

Problem 1.3.9 Find the GS to the following DE
$$4xy^2 + y' = 5x^4y^2$$

Problem 1.3.10 Find the explicit PS to the following IVP
$$\begin{cases} y' = 2xy^2 + 3x^2y^2 \\ y(1) = -1 \end{cases}$$

Problem 1.3.11 Find the GS (implicit if necessary, explicit if convenient) of the following DE. Primes denote derivatives wrt x.
$$y' = 2x\sec y$$

Problem 1.3.12 Find the explicit PS to the following IVP
$$\begin{cases} \tan x\, y' = y \\ y\left(\dfrac{\pi}{2}\right) = \dfrac{\pi}{2} \end{cases}$$

Problem 1.3.13 Find the GS (implicit if necessary, explicit if convenient) of the following DE. Primes denote derivatives wrt x.
$$(x^2 + 1)\tan(y)\, y' = x$$

Problem 1.3.14 Find the GS (implicit if necessary, explicit if convenient) of the following DE
$$(x + 3)^3 y' = (y - 2)^2$$

Problem 1.3.15 Find the PS to the following DE
$$\begin{cases} y' - xy = 3y \\ y(1) = 1 \end{cases}$$

Problem 1.3.16 Find the PS to the following DE
$$\begin{cases} (1 + x)y' + y = \cos(x) \\ y(0) = 1 \end{cases}$$

Problem 1.3.17 Find the GS to the following DE
$$\frac{dy}{dt} = \beta y(\alpha - \ln y)$$
where α and β are parameters independent of y and t. If $y(t = 0) = y_0$ and $y(t \to \infty) = y_\infty$, can you express α or β or both in terms of y_0 and y_∞? If so, do it.

1.4   First-Order Linear DEs

All 1st.O linear DEs can be written as

$$A(x)y' + B(x)y = C(x) \qquad (1.22)$$

If $A(x) = 0$, the equation is no longer a DE. Thus, we set $A(x) \neq 0$.

If $B(x) = 0$, DE (1.22) becomes

$$A(x)y' = C(x) \qquad (1.23)$$

which is a separable DE.

If $C(x) = 0$, DE (1.22) becomes

$$A(x)y' + B(x)y = 0 \qquad (1.24)$$

and, after dividing both sides by y, we get

$$\frac{y'}{y} = -\frac{B(x)}{A(x)} \qquad (1.25)$$

which is also a separable DE.

For a general 1st.O linear DE, we divide both sides (1.22) by $A(x)$,

$$y' + \frac{B(x)}{A(x)}y = \frac{C(x)}{A(x)} \qquad (1.26)$$

Let

$$\frac{B(x)}{A(x)} = P(x), \quad \frac{C(x)}{A(x)} = Q(x)$$

Now, we have

$$y' + P(x)y = Q(x) \qquad (1.27)$$

Let's now derive the method to solve the general 1st.O linear DE in the form of (1.27). The key idea is to absorb the $P(x)y$ term to a whole derivative by using an IF $\rho(x)$ whose precise composition will be introduced shortly.

$$\rho(x)(y' + P(x)y) = Q(x)\rho(x)$$

$$\rho(x)y' + \rho(x)P(x)y = Q(x)\rho(x)$$

(1.28)

If we can make the LHS of (1.28)

$$\rho(x)y' + \rho(x)P(x)y = \frac{d}{dx}(\rho(x)y) \qquad (1.29)$$

a whole derivative, (1.28) can be solved easily in terms of $\rho(x)y$. Equation (1.29) can be written as

$$\rho(x)y' + \rho(x)P(x)y = \rho(x)y' + \rho'(x)y \qquad (1.30)$$

Cancelling $\rho(x)y'$ on both sides of DE (1.30), we have

$$\rho(x)P(x)y = \rho'(x)y$$

$$\rho(x)P(x) = \rho'(x)$$

(1.31)

which is a separable DE with unknown function $\rho(x)$. To solve this DE, we have

$$\frac{d\rho}{\rho} = P(x)dx$$

$$\int \frac{d\rho}{\rho} = \int P(x)dx \qquad (1.32)$$

$$\ln \rho = \int P(x)dx$$

Therefore, we have

$$\rho(x) = \exp\left(\int P(x)dx + C_1\right) = C \exp\left(\int P(x)dx\right) \qquad (1.33)$$

Since this function $\rho(x)$ can be multiplied by any non-zero constant to serve as a factor, without loss of generality while gaining simplicity, we set $C = 1$. The resulting factor

$$\rho(x) = \exp\left(\int P(x)dx\right) \tag{1.34}$$

is called the integrating factor (IF) of DE (1.27) and this IF satisfies the condition (1.29). Plugging (1.29) into (1.28), we have

$$\frac{d}{dx}(\rho(x)y) = Q(x)\rho(x) \tag{1.35}$$

i.e.,

$$d(\rho(x)y) = Q(x)\rho(x)dx \tag{1.36}$$

Integrating both sides of DE (1.36), we get

$$\rho(x)y = \int Q(x)\rho(x)dx \tag{1.37}$$

Thus, the GS to the DE (1.27) is

$$y(x) = \frac{1}{\rho(x)}\left(\int Q(x)\rho(x)dx + C\right) \tag{1.38}$$

Replacing the IF $\rho(x)$ by the given function $P(x)$, we find the GS to the general 1st.O linear DE as

$$y(x) = \exp\left(-\int P(x)dx\right)\left(\int Q(x)\exp\left(\int P(x)dx\right)dx + C\right) \tag{1.39}$$

Now, we outline the steps to solve the 1st.O linear DEs without memorizing the formula (1.39).

Step 1: Convert the DE to the standard form $y' + P(x)y = Q(x)$
Step 2: Identify $P(x)$
Step 3: Compute the IF

$$\rho(x) = \exp\left(\int P(x)dx\right) \tag{1.40}$$

Step 4: Multiply both sides of the converted DE by the IF

$$\exp\left(\int P(x)dx\right)y' + \exp\left(\int P(x)dx\right)P(x)y$$
$$= Q(x)\exp\left(\int P(x)dx\right) \tag{1.41}$$

Step 5: Write the LHS as a whole derivative using the product rule,

$$\exp\left(\int P(x)dx\right)y' + \exp\left(\int P(x)dx\right)P(x)y$$
$$= \frac{d}{dx}\left(\exp\left(\int P(x)dx\right)y\right) \tag{1.42}$$

Therefore,

$$\frac{d}{dx}\left(\exp\left(\int P(x)dx\right)y\right) = Q(x)\exp\left(\int P(x)dx\right)$$
$$\exp\left(\int P(x)dx\right)y = \int Q(x)\exp\left(\int P(x)dx\right)dx \tag{1.43}$$

Finally, the solution to the DE is

$$y(x) = \exp\left(-\int P(x)dx\right)\left(\int Q(x)\exp\left(\int P(x)dx\right)dx\right) \tag{1.44}$$

Example 1
Find the GS to the following DE

$$(x^2 + 1)y' + 3xy = 6x \tag{1.45}$$

Solution
Step 1: Converting the DE

$$y' + \frac{3x}{x^2+1}y = \frac{6x}{x^2+1}$$

Step 2: Identify

$$P(x) = \frac{3x}{x^2+1}$$

Step 3: Computing the IF

$$\rho(x) = \exp\left(\int \frac{3x}{x^2+1}dx\right)$$

$$= \exp\left(\frac{3}{2}\int \frac{d(x^2+1)}{x^2+1}\right)$$

$$= \exp\left(\frac{3}{2}\ln(x^2+1)\right)$$

$$= (x^2+1)^{\frac{3}{2}}$$

Step 4: Multiplying the DE by the IF

$$y'(x^2+1)^{\frac{3}{2}} + \frac{3x}{x^2+1}y(x^2+1)^{\frac{3}{2}} = \frac{6x}{x^2+1}(x^2+1)^{\frac{3}{2}}$$

$$y'(x^2+1)^{\frac{3}{2}} + 3xy(x^2+1)^{\frac{1}{2}} = 6x(x^2+1)^{\frac{1}{2}}$$

Step 5: Completing the solution

$$\frac{d}{dx}\left(y(x^2+1)^{\frac{3}{2}}\right) = 6x(x^2+1)^{\frac{1}{2}}$$

$$y(x^2+1)^{\frac{3}{2}} = \int 6x(x^2+1)^{\frac{1}{2}}dx$$

$$y = \frac{1}{(x^2+1)^{\frac{3}{2}}}\int 3(x^2+1)^{\frac{1}{2}}d(x^2+1)$$

$$= (x^2+1)^{-\frac{3}{2}}\left(2(x^2+1)^{\frac{3}{2}} + C\right)$$

$$= 2 + C(x^2+1)^{-\frac{3}{2}}$$

This is the GS to the original DE.

Example 2

Find the PS to the following IVP

$$\begin{cases} x^2y' + xy = \sin x \\ \quad\quad y(1) = y_0 \end{cases} \tag{1.46}$$

Solution

Method 1: The 1st.O linear DE method

$$y' + \frac{y}{x} = \frac{\sin x}{x^2}$$

The IF

$$\rho(x) = \exp\left(\int \frac{1}{x}dx\right) = \exp(\ln x) = x$$

Multiplying the IF yields

$$(xy)' = \frac{\sin x}{x^2}\cdot x$$

$$xy = \int \frac{\sin x}{x} dx$$

$$y(x) = \frac{1}{x} \int_{x_0}^{x} \frac{\sin t}{t} dt + \frac{C}{x}$$

This is the GS to the original DE where x_0 and C are constants.
Using the IC $y(1) = y_0$

$$y(1) = y_0 = \frac{1}{1} \int_{x_0}^{1} \frac{\sin t}{t} dt + \frac{C}{1}$$

$$C = y_0 - \int_{x_0}^{1} \frac{\sin t}{t} dt$$

Now plugging C back into $y(x)$, we have

$$y(x) = \frac{1}{x} \int_{x_0}^{x} \frac{\sin t}{t} dt + \frac{y_0}{x} - \frac{1}{x} \int_{x_0}^{1} \frac{\sin t}{t} dt$$

$$= \frac{y_0}{x} + \frac{1}{x} \int_{1}^{x} \frac{\sin t}{t} dt$$

Method 2: Direct use of the formula (1.35)

$$y(x) = \exp\left(-\int P(x)dx\right)\left(\int Q(x) \exp\left(\int P(x)dx\right) dx\right)$$

$$P(x) = \frac{1}{x}$$

$$Q(x) = \frac{\sin x}{x^2}$$

$$y(x) = \exp\left(-\int \frac{1}{x} dx\right)\left(\int \frac{\sin x}{x^2} \exp\left(\int \frac{1}{x} dx\right) dx + C\right)$$

$$= \frac{1}{x} \int_{x_0}^{x} \frac{\sin t}{t} dt + \frac{C}{x}$$

Using the IC $y(1) = y_0$, we have

$$C = y_0 - \int_{x_0}^{1} \frac{\sin t}{t} dt$$

Therefore,

$$y(x) = \frac{y_0}{x} + \frac{1}{x} \int_{1}^{x} \frac{\sin t}{t} dt$$

Problems

Problem 1.4.1 Find the PS to the following IVP

$$\begin{cases} (x^2 + 1)y' + 3x^3 y = 6x \exp\left(-\frac{3}{2}x^2\right) \\ \qquad\qquad y(0) = 1 \end{cases}$$

Problem 1.4.2 (a) Show that

$$y_C(x) = C \exp\left(-\int P(x)dx\right)$$

is a GS to $y' + P(x)y = 0$.
(b) Show that

$$y_P(x) = \exp\left(-\int P(x)dx\right)\left(\int Q(x)\exp\left(\int P(x)dx\right)dx\right)$$

is a PS to $y' + P(x)y = Q(x)$.
(c) Suppose that $y_C(x)$ is any GS to $y' + P(x)y = 0$ and that $y_P(x)$ is any PS to $y' + P(x)y = Q(x)$. Show that $y(x) = y_C(x) + y_P(x)$ is a GS to $y' + P(x)y = Q(x)$.

Problem 1.4.3 Find the GS to the following IVP
$$\begin{cases} xy' + 2y = 7x^2 \\ \quad y(2) = 5 \end{cases}$$

Problem 1.4.4 Find the GS to the following DE
$$y' + y\cot x = \cos x$$

Problem 1.4.5 Find the GS to the following DE using two different methods
$$y' = 3(y + 7)x^2$$

Problem 1.4.6 Find the GS to the following DE
$$xy' = 2y + x^3 \cos x$$

Problem 1.4.7 Find the GS to the following DE
$$(2x + 1)y' + y = (2x + 1)^{\frac{3}{2}}$$

Problem 1.4.8 Find the GS to the following DE using two different methods
$$y' = \frac{2xy + 2x}{x^2 + 1}$$

Problem 1.4.9 Find the GS to the following DE
$$(1 + 2xy)y' = 1 + y^2$$
(Hint: Regard x as DV and y as IV.)

Problem 1.4.10 Find the GS to the following DE
$$2xy' + y = 10\sqrt{x}$$

Problem 1.4.11 Find the GS to the following DE
$$(x + y\exp(y))y' = 1$$
(Hint: Regard x as DV and y as IV.)

Problem 1.4.12 Find the GS to the following DE
$$2y + (x + 1)y' = 3(x + 1)$$

Problem 1.4.13 Find the GS to the following IVP
$$\begin{cases} (1 + x)y' + y = \sin x \\ \quad\quad y(0) = 1 \end{cases}$$

Problem 1.4.14 Find the GS to the following DE
$$y' = 2(xy' + y)y^3$$
(Hint: Regard x as DV and y as IV.)

Problem 1.4.15 Find the GS to the following 2nd.O DE
$$x^2 y'' + 3xy' = 4x^4$$
(Hint: Use substitution v = y'.)

1.5   The Substitution Methods

Substitution methods (S-methods) are those introducing one or more new variables to represent the variables in the original DE. The procedure will convert the original DE to a separable or other more easily solvable DE, *e.g.*, one can change one variable x to u,

$$F_1(x,y) = 0 \xrightarrow{f(x)\to u} F_2(u,y) \tag{1.47}$$

One may also consider changing both variables.

$$G_1(x,y) = 0 \xrightarrow[h(x,y)\to v]{g(x,y)\to u} G_2(u,v) \tag{1.48}$$

Let's discover the power of the S-methods by solving several families of DEs.

1.5.1   Polynomial Substitution

Solve

$$y' = F(ax + by + c) \tag{1.49}$$

where a, b, and c are constants.

Step 1: Introducing a new variable v.

$$v = ax + by + c \tag{1.50}$$

Step 2: Transform the original DE (1.49) into a new DE of v.

$$\begin{aligned} v' &= a + by' \\ y' &= \frac{1}{b}v' - \frac{a}{b} \end{aligned} \tag{1.51}$$

After the substitution, the original DE (1.49) now becomes

$$y' = F(v) \tag{1.52}$$

Substituting y' with the function containing v', we have

$$\frac{1}{b}v' - \frac{a}{b} = F(v)$$
$$\frac{1}{b}v' = F(v) + \frac{a}{b} \qquad (1.53)$$

which is now separable.

Step 3: Solve DE (1.53).

$$\frac{1}{b}\frac{dv}{dx} = F(v) + \frac{a}{b}$$
$$\frac{\frac{1}{b}dv}{F(v) + \frac{a}{b}} = dx \qquad (1.54)$$
$$\frac{dv}{bF(v) + a} = dx$$
$$x = \int \frac{dv}{bF(v) + a}$$

Example
Find the GS to the following DE

$$y' = (x + y + 3)^2 \qquad (1.55)$$

Solution
Let

$$v = x + y + 3$$
$$v' = 1 + y'$$
$$y' = v' - 1$$

After the substitution, the original DE becomes

$$v' - 1 = v^2$$
$$\frac{dv}{v^2 + 1} = dx$$
$$\int \frac{dv}{v^2 + 1} = \int dx$$
$$\tan^{-1} v = x + C$$
$$v = \tan(x + C)$$

Substituting v back into the above DE, we obtain

$$x + y + 3 = \tan(x + C)$$

Thus,

$$y = \tan(x + C) - x - 3$$

1.5.2   Homogeneous DEs

For the 1st.O homogeneous DEs (Homo DEs)

$$y' = F\left(\frac{y}{x}\right) \tag{1.56}$$

After a different substitution $v = \frac{y}{x}$ with which $y = xv$ and $y' = xv' + v$, we convert the original DE (1.56) into

$$xv' + v = F(v)$$
$$v' = \frac{F(v) - v}{x} \tag{1.57}$$

which is a separable DE. Solving it, we get

$$\int \frac{dv}{F(v) - v} = \int \frac{dx}{x}$$
$$\int \frac{dv}{F(v) - v} = \ln x - C$$
$$x = \exp\left(\int \frac{dv}{F(v) - v} + C_1\right) \tag{1.58}$$
$$x = C \exp\left(\int \frac{dv}{F(v) - v}\right)$$

Example
Find the GS to the following DE

$$2xyy' = 4x^2 + 3y^2 \tag{1.59}$$

Solution
Dividing DE (1.59) by $2xy$

$$y' = \frac{2x}{y} + \frac{3y}{2x}$$

This is a Homo DE.

Let $v = \frac{y}{x}$. Then, $y = vx$ and $y' = v + xv'$.

Substituting this back into the DE, we have

$$v + xv' = \frac{2}{v} + \frac{3}{2}v$$

$$v' = \frac{1}{x}\left(\frac{2}{v} + \frac{v}{2}\right)$$

$$\frac{2v\,dv}{4 + v^2} = \frac{dx}{x}$$

$$\int \frac{d(4 + v^2)}{4 + v^2} = \int \frac{dx}{x}$$

$$\ln(4 + v^2) = \ln x + C$$

$$4 + v^2 = Cx$$

Plugging v back in, we have

$$4 + \left(\frac{y}{x}\right)^2 = Cx$$

$$y^2 + 4x^2 = Cx^3$$

1.5.3   Bernoulli DEs

Given the DE

$$y' + P(x)y = Q(x)y^n \tag{1.60}$$

We have a few cases depending on n.

Case 1 $(n < 0)$:

DE (1.60) is a general 1st.O nonlinear DE.

Case 2 $(n = 0)$:

DE (1.60) is the 1st.O linear DE $y' + P(x)y = Q(x)$ whose solution method was introduced before.

Case 3 $(n = \frac{1}{2})$:

DE (1.60) is a general 1st.O nonlinear DE.

Case 4 $(n = 1)$:

DE (1.60) is 1st.O linear DE $y' + (P(x) - Q(x))y = 0$ which is separable.

Case 5 $(n = 2)$:

DE (1.60) is a special 1st.O nonlinear DE, *i.e.*, the Riccati DE.

Case 6 $(n > 2)$:

DE (1.60) is a general 1st.O nonlinear DE.

To sum up, for $n \neq 0, 1$, we know that DE (1.60) is a 1st.O nonlinear DE and we have not learned any methods to solve the DE for such cases. Now, let's use a substitution to solve the Bernoulli DEs for $n \neq 0, 1$.

Solution steps

Step 1: Select a proper substitution

$$v = y^{1-n} \tag{1.61}$$

Step 2: Find a relationship to replace one variable

$$v' = (1-n)y^{-n}y' \tag{1.62}$$

Step 3: Divide DE (1.60) by y^n

$$y^{-n}y' + P(x)y^{1-n} = Q(x) \tag{1.63}$$

Since $n \neq 1$, we have

$$\frac{1}{1-n}\left((1-n)y^{-n}y'\right) + P(x)y^{1-n} = Q(x) \tag{1.64}$$

Inserting v and v' in (1.61) and (1.62), respectively, into DE (1.64) yields

$$\left(\frac{1}{1-n}\right)v' + P(x)v = Q(x) \tag{1.65}$$

Or

$$v' + (1 - n)P(x)v = (1 - n)Q(x) \qquad (1.66)$$

Step 4: Solve this DE (1.66) in terms of v. Since it is now a 1st.O linear DE, we can solve it by any of the methods that were introduced before.

Step 5: Solve DE (1.66) to find v.

Step 6: Back substituting variable v with y to express the solution in the original variables.

Example

Find the GS to the following DE

$$xy' + 6y = 3xy^{\frac{4}{3}} \qquad (1.67)$$

Solution

Slight reformatting $y' + \frac{6}{x}y = 3y^{\frac{4}{3}}$ leads to an obvious Bernoulli DE for which we have

$$P(x) = \frac{6}{x}$$
$$Q(x) = 3$$
$$n = \frac{4}{3}$$

Let's follow the step-by-step procedure to solve this Bernoulli DE.

Step 1:

$$v = y^{1-n} = y^{-\frac{1}{3}}$$
$$y = v^{-3}$$

Step 2:

$$v' = -\frac{1}{3}y^{-\frac{4}{3}}y'$$
$$y' = -3v'(v^{-3})^{\frac{4}{3}}$$
$$= -3v'v^{-4}$$

Step 3:

$$x(-3v'v^{-4}) + 6v^{-3} = 3x(v^{-3})^{\frac{4}{3}}$$
$$-3xv'v^{-4} + 6v^{-3} = 3xv^{-4}$$
$$v'x - 2v = -x$$

$$v' - \frac{2v}{x} = -1$$

Step 4:   Now, it becomes a 1st.O linear DE that can be solved using IF

$$\rho(x) = \exp\left(\int P(x)dx\right)$$
$$= \exp\left(-\int \frac{2}{x}dx\right)$$
$$= \exp(-2\ln x)$$
$$= x^{-2}$$

where we used $P(x) = -\frac{2}{x}$.

Multiplying the IF on both sides of the DE

$$\frac{1}{x^2}v' - \frac{2v}{x^3} = -\frac{1}{x^2}$$
$$\left(\frac{1}{x^2}v\right)' = -\frac{1}{x^2}$$
$$\frac{1}{x^2}v = -\int \frac{1}{x^2}dx$$
$$\frac{v}{x^2} = \frac{1}{x} + C$$
$$v = x + Cx^2$$

Step 5:   Plug y back in

$$y^{-\frac{1}{3}} = x + Cx^2$$
$$y = \frac{1}{(x + Cx^2)^3}$$
$$y = \frac{1}{x^3(1 + Cx)^3}$$

Problems

Problem 1.5.1 Find the GS to the following DE
$$x^2y' = xy + y^2$$

Problem 1.5.2 Find the GS to the following DE
$$xyy' = y^2 + x\sqrt{4y^2 + x^2}$$

Problem 1.5.3 Show that the substitution $v = \ln y$ transforms the DE $y' + P(x)y = Q(x)(y \ln y)$ into the linear DE $v' + P(x) = Q(x)v(x)$.

Problem 1.5.4 Find the GS to the following DE. Prime denotes derivatives wrt x.

$$5y^4 y' = x^2 y' + 2xy$$

Problem 1.5.5 Find the GS to the following DE

$$xy' - 4x^2 y + 2y \ln y = 0$$

Problem 1.5.6 Find the GS to the following DE

$$tx' - (m+1)t^m x + 2x \ln x = 0$$

Problem 1.5.7 Find the GS to the following DE

$$xy' = 6y + 12x^4 y^{\frac{2}{3}}$$

Problem 1.5.8 Find the GS to the following DE

$$yy'' = 3(y')^2$$

Problem 1.5.9 Find the GS to the following DE

$$y^3 y'' = 3$$

Problem 1.5.10 Find the GS to the following DE using two different methods.

$$y' = xy^3 - xy$$

Problem 1.5.11 Find the GS to the following DE using two different methods.

$$xy' - y = y^2 \sin x$$

Problem 1.5.12 Show that the solution curves of the following DE

$$y' = -\frac{y(2x^3 - y^3)}{x(2y^3 - x^3)}$$

are of the form $x^3 + y^3 = 3Cxy$.
(Hint: Use substitution $u = y/x$.)

Problem 1.5.13 Find the GS to the following DE
$$xy' = y + x \exp\left(\frac{y}{x}\right)$$

Problem 1.5.14 Find the GS to the following DE
$$(x + y)y' = 1$$

Problem 1.5.15 Find the GS to the following DE
$$(2x \sin y \cos y)y' = 4x^2 + \sin^2 y$$

Problem 1.5.16 Find the GS to the following DE
$$y'' = (x + y')^2$$
(Hint: Use substitution $v = y'$.)

Problem 1.5.17 Find the GS to the following DE
$$(x + \exp(y))y' = x \exp(-y) - 1$$

Problem 1.5.18 Find the GS to the following DE
$$(3xy)^2 + x^{\frac{3}{2}}y' = y^2$$

Problem 1.5.19 Find the GS to the following DE
$$y^2(xy' + 1)(1 + x^4)^{\frac{1}{2}} = x$$

Problem 1.5.20 Find the GS to the following DE
$$x \exp(y)\, y' = 2(\exp(y) + x^3 \exp(2x))$$

Problem 1.5.21 Find the GS to the following DE
$$\frac{dy}{dx} - 4xy \ln y + 2\frac{y}{x}(\ln y)^n = 0$$
where $n \geq 2$ is a constant integer.

Problem 1.5.22 Find the GS to the following DE
$$yy'' + (y')^2 = yy'$$

Problem 1.5.23 Find the GS to the following DE
$$6xy^3 + 2y^4 + (9x^2y^2 + 8xy^3)y' = 0$$

Problem 1.5.24 Given a 1st.O nonlinear DE
$$y' + 7yx^{-1} - 3y^2 = 3x^{-2}$$
(1) Prove that after substitution, $y(x) = x^{-1} + u(x)$, one can transform this DE into a Bernoulli DE;
(2) Solve the resulting Bernoulli DE.

Problem 1.5.25 Find the GS to the following DE
$$x^2y'' + 3xy' = 4$$
(Hint: Use substitution $v = y'$.)

Problem 1.5.26 Find the GS to the following DE
$$tx' - 1000t^{1000}x + 2x^2 = 0$$

Problem 1.5.27 Find the GS to the following DE
$$(x^2 + xy)dx - (xy + y^2)dy = 0$$

Problem 1.5.28 Find the GS to the following DE
$$2x^2yy' + 2xy^2 + 1 = 0$$

Problem 1.5.29 Find the GS to the following DE using two different methods
$$(x^2 + 1)y' - 2xy - 2x = 0$$

Problem 1.5.30 Find the PS to the following IVP
$$\begin{cases} y' = \dfrac{y}{x} - b\left(1 + \left(\dfrac{y}{x}\right)^2\right)^{\frac{1}{2}} \\ y(a) = 0 \end{cases}$$
where a and b are constants and $x \in [0, a]$. Additionally, please
(1) sketch the solution for $b < 1$;
(2) sketch the solution for $b = 1$;
(3) sketch the solution for $b > 1$;
(4) identify a case for which $y(x = 0) = 0$ possible?

Problem 1.5.31 Find the GS to the following DE
$$x' = 3x^2 - \frac{8}{t}x + \frac{4}{t^2}$$
(*Hint: Use substitution* $x = \frac{1}{t} + u$.)

Problem 1.5.32 Find the GS to the following DE
$$y' = (K(x) + y + \beta)(y - \beta)$$
When the function $K(x)$ is given as $K(x) = x^{2011}$ and $\beta = 1$, find the specific form of the solution.

Problem 1.5.33 Find the GS to the following DE
$$tx' - x = \beta(x'x + t)$$

Problem 1.5.34 Find the GS to the following DE
$$y' = by^2 + cx^n$$
where $b, c \neq 0$ are constants and $n = 0, -2$. You need to consider both values of n.

1.6   Riccati DEs

DEs that can be expressed as follows are called Riccati DEs, named after the Italian mathematician J. F. Riccati (1676–1754),

$$y' = A_0(x) + A_1(x)y + A_2(x)y^2 \qquad (1.68)$$

If $A_0(x) = 0$, DE (1.68) reduces to a Bernoulli DE.
If $A_2(x) = 0$, DE (1.68) reduces to the 1st.O linear DE.

By intuition or by *guessing*, one may propose a *pilot* solution $y_1(x)$ with which the following solution is proposed,

$$y(x) = y_1(x) + \frac{1}{Z(x)} \qquad (1.69)$$

Thus,

$$y' = y_1' - \frac{Z'}{Z^2}$$
$$y^2 = y_1^2 + \frac{1}{Z^2} + \frac{2y_1}{Z} \qquad (1.70)$$

Plugging (1.69) and its associated terms (1.70) into the original Riccati DE (1.68), we have

$$y_1' - \frac{Z'}{Z^2} = A_0 + A_1\left(y_1 + \frac{1}{Z}\right) + A_2\left(y_1^2 + \frac{1}{Z^2} + \frac{2y_1}{Z}\right) \qquad (1.71)$$

which, after reorganization of the terms, can be written as

$$y_1' - A_0 - A_1 y_1 - A_2 y_1^2$$
$$= \frac{Z'}{Z^2} + A_1\left(\frac{1}{Z}\right) + A_2\left(\frac{1}{Z^2} + \frac{2y_1}{Z}\right) \qquad (1.72)$$

Because y_1 is a solution to the original Riccati DE, the LHS of (1.72) must be zero, *i.e.*,

$$y_1' - A_0 - A_1 y_1 - A_2 y_1^2 = 0 \qquad (1.73)$$

Thus,

$$\frac{Z'}{Z^2} + A_1\left(\frac{1}{Z}\right) + A_2\left(\frac{1}{Z^2} + \frac{2y_1}{Z}\right) = 0 \qquad (1.74)$$

Multiplying DE (1.74) by Z^2, we get

$$Z' + (A_1(x) + 2A_2(x)y_1(x))Z = -A_2(x) \qquad (1.75)$$

which is a 1st.O linear DE that can be solved, conveniently.

If defining

$$P(x) = A_1(x) + 2A_2(x)y_1(x) \qquad (1.76)$$

$$Q(x) = -A_2(x) \qquad (1.77)$$

we get the standard form of a 1st.O linear DE

$$Z' + P(x)Z = Q(x) \qquad (1.78)$$

Example 1
Find the GS to the following DE

$$y' + y^2 = \frac{2}{x^2} \qquad (1.79)$$

Solution
This is a simple Riccati DE and, by inspection, we may assume a PS in the following form $y_1 = \frac{A}{x}$ where A is a constant to be determined. Since it is a PS, it must satisfy the original DE, $i.e.,$

$$-\frac{A}{x^2} + \frac{A^2}{x^2} = \frac{2}{x^2} \qquad (1.80)$$

We can easily find two roots $A = -1, 2$ to satisfy (1.80) for arbitrary values of x, resulting in two PS's: $y_1 = -\frac{1}{x}$ or $y_1 = \frac{2}{x}$.

Next, we may select any one of the PS's to compose the GS as

$$y = -\frac{1}{x} + \frac{1}{Z}$$

where we selected the PS: $y_1 = -\frac{1}{x}$.

Thus,

$$y' = \frac{1}{x^2} - \frac{Z'}{Z^2}$$

$$y^2 = \frac{1}{x^2} - \frac{2}{xZ} + \frac{1}{Z^2}$$

Plugging the above two formulas into the original DE:

$$\left(\frac{1}{x^2} - \frac{Z'}{Z^2}\right) + \left(\frac{1}{x^2} - \frac{2}{xZ} + \frac{1}{Z^2}\right) = \frac{2}{x^2}$$

We get

$$Z' + \frac{2}{x}Z = 1$$

whose solution is

$$Z = \frac{x}{3} + \frac{C}{x^2}$$

Finally, the GS for the DE is

$$y(x) = -\frac{1}{x} + \frac{3x^2}{x^3 + C_1}$$

Example 2

Find the GS to the following DE

$$y' + 2xy = 1 + x^2 + y^2 \tag{1.81}$$

Solution

This is a simple Riccati DE and, by inspection, we found a PS $y_1(x) = x$. The original DE can be written as

$$y' = (1 + x^2) + (-2x)y + y^2$$
$$= A_0(x) + A_1(x)y + A_2(x)y^2$$

where $(x) = 1 + x^2$, $A_1(x) = -2x$, $A_2(x) = 1$. Using substitution $y(x) = y_1(x) + Z(x)^{-1}$, we have

$$Z' + (A_1(x) + 2A_2(x)y_1(x))Z = -A_2(x)$$

i.e.,

$$Z' = -1$$

whose GS is

$$Z(x) = -x + c$$

After back substitution, we get

$$y = x - \frac{1}{x - c}$$

Alternatively, we may use the S-method: $y' = 1 + (x - y)^2$ with substitution $u = x - y$, we get $y' = 1 - u'$ and

$$1 - u' = 1 + u^2$$

Thus, $\frac{1}{u} = x + c$. After back substitution, we get

$$y = x - \frac{1}{x-c}$$

Problems

Problem 1.6.1 The DE
$$y' = A(x)y^2 + B(x)y + C(x)$$
is called a Riccati DE. Suppose that one PS y_1 of this DE is known, show that the substitution $y = y_1 + \frac{1}{v}$ can transform the Riccati DE into a linear DE $v' + (B + 2Ay_1)v = -A$.

Problem 1.6.2 Find the GS to the following DE
$$y' + y^2 = 1 + x^2$$
given that $y_1 = x$ is a solution.

Problem 1.6.3 Find the GS to the following DE
$$y' = 1 + \frac{1}{4}(x-y)^2$$
given that $y_1 = x$ is a solution.

Problem 1.6.4 Find the GS to the following DE
$$y' - 13(x^2 + y^2) + 26xy = 1$$
given that $y_1 = x$ is a solution.

Problem 1.6.5 Find the GS to the following DE
$$\frac{dy}{dx} = (f(x) + y + a)(y - a)$$
given that $y_1 = a$ is a solution.

Problem 1.6.6 Find the GS to the following DE
$$y' = -(a^2 + 4ax^3) + 4x^3y + y^2$$
given that $y_1 = a$ is a solution.

Problem 1.6.7 Find the GS to the following DE
$$y' = \frac{2\cos^2 x - \sin^2 x + y^2}{2\cos x}$$
given that $y_1 = \sin x$ is a solution.

Problem 1.6.8 Find the GS to the following DE
$$y' = y^2 + \alpha(x)(y - x^2) + 2x - x^4$$
and express it in terms of the given function $\alpha(x)$. We know that this DE has one solution $y = x^2$.

Problem 1.6.9 Find the GS to the following DE
$$x^3 y' + x^2 y - y^2 = 2x^4$$
given that $y_1 = x^2$ is a solution.

1.7   The Exact DEs

Consider a DE

$$(6xy - y^3)dx + (4y + 3x^2 - 3xy^2)dy = 0 \qquad (1.82)$$

Is it

- ✓   1st-order?
- ✓   Nonlinear?
- ✓   Homo DE?
- ✓   InHomo DE?
- ✓   Inseparable?
- ✓   Non-Bernoulli?

New methods must be introduced in order to solve DEs of this kind.

The GS to all DEs can be written as

$$F(x, y) = C \qquad (1.83)$$

With this claim, we get

$$d\big(F(x, y)\big) = \frac{\partial F}{\partial x}dx + \frac{\partial F}{\partial y}dy = dC = 0 \qquad (1.84)$$

Thus,

$$\frac{\partial F}{\partial x} + \frac{\partial F}{\partial y}\frac{dy}{dx} = 0 \qquad (1.85)$$

Therefore, we have

$$\left(\frac{\partial F}{\partial x}\right)dx + \left(\frac{\partial F}{\partial y}\right)dy = 0 \qquad (1.86)$$

Let

$$M(x, y) = \frac{\partial F}{\partial x}$$
$$N(x, y) = \frac{\partial F}{\partial y}$$

(1.87)

This gives

$$M(x, y)dx + N(x, y)dy = 0$$

(1.88)

All 1st.O DEs can be written in the above manner.

The condition for the DEs to be exact is

$$\frac{\partial}{\partial x}\left(\frac{\partial F}{\partial y}\right) = \frac{\partial}{\partial y}\left(\frac{\partial F}{\partial x}\right)$$

(1.89)

That is, if $N_x = M_y$, (1.88) is exact. In fact, one prove that the necessary and sufficient condition for (1.88) to be exact is $N_x = M_y$.

Example 1

Is DE

$$(6xy - y^3)dx + (4y + 3x^2 - 3xy^2)dy = 0$$

(1.90)

exact?

Solution

We have

$$M(x, y) = 6xy - y^3$$
$$N(x, y) = 4y + 3x^2 - 3xy^2$$

Here

$$\frac{\partial M}{\partial y} = \frac{\partial}{\partial y}(6xy - y^3) = 6x - 3y^2$$
$$\frac{\partial N}{\partial x} = \frac{\partial}{\partial x}(4y + 3x^2 - 3xy^2) = 6x - 3y^2$$

Since

$$M_y = N_x = 6x - 3y^2$$

The DE is exact.

Example 2

Is DE

$$ydx + 3xdy = 0 \tag{1.91}$$

exact?

Solution
From DE, we get

$$M(x,y) = y$$
$$N(x,y) = 3x$$

Here

$$M_y = \frac{\partial}{\partial y}(y) = 1$$
$$N_x = \frac{\partial}{\partial x}(3x) = 3$$

That means

$$M_y \neq N_x$$

The DE is not exact.

Example 3
Is DE

$$y^3dx + 3xy^2dy = 0 \tag{1.92}$$

exact?

Solution
We get

$$M(x,y) = y^3$$
$$N(x,y) = 3xy^2$$

Here

$$M_y = \frac{\partial}{\partial y}(y^3) = 3y^2$$

$$N_x = \frac{\partial}{\partial x}(3xy^2) = 3y^2$$

That means

$$N_x = M_y = 3y^2$$

The DE is exact.

Let's find its solution. There is a new set of methods for solving the exact DEs. As mentioned in the beginning of this section, the solution to any DE can be written as

$$F(x,y) = C \tag{1.93}$$

From the above example, we have

$$M(x,y) = \frac{\partial F}{\partial x} = y^3 \tag{1.94}$$

$$N(x,y) = \frac{\partial F}{\partial y} = 3xy^2 \tag{1.95}$$

So, from (1.94), we have

$$F(x,y) = \int y^3 dx = y^3 x + C$$

Here we have C as a constant wrt x. Consider the constant be a function of y: $C = g(y)$. Now we have

$$F(x,y) = y^3 x + g(y) \tag{1.96}$$

Plugging (1.96) back into (1.95), we get

$$\frac{\partial}{\partial y}(y^3 x + g(y)) = 3xy^2$$
$$3xy^2 + g'(y) = 3xy^2$$
$$g'(y) = 0$$
$$g(y) = C_1$$

So, we find that $g(y)$ is actually a constant.

Plugging $g(y) = C_1$ into (1.96), we get

$$F(x,y) = y^3 x + C_1 \tag{1.97}$$

Therefore, from (1.93), we have

$$y^3 x + C_1 = C_2 \tag{1.98}$$

Thus, the solution is

$$y^3 x = C \tag{1.99}$$

Steps for solving exact DEs:

Step 1: Write the DE in the following form

$$M(x,y)dx + N(x,y)dy = 0 \qquad (1.100)$$

and get the formulas for

$$M(x,y) = \frac{\partial F}{\partial x} \qquad (1.101)$$

$$N(x,y) = \frac{\partial F}{\partial y} \qquad (1.102)$$

Step 2: Find

$$F(x,y) = \int M(x,y)dx + g(y) \qquad (1.103)$$

Step 3: Plugging $F(x,y)$ into (1.102) obtains $g(y)$.
Step 4: The GS is

$$\int M(x,y)dx + g(y) = C \qquad (1.104)$$

Remarks

In the above steps, we start from (1.101), integrate it and, then, plug it back into (1.102) to get the solution. Similarly, we can start from (1.102), integrate it and plug back into (1.101) to get the *same* solution, different by a constant.

Let's now clarify a few points:

(1) Not all DEs are exact. Some non-exact DEs can be converted to exact DEs while others cannot. So, the question is, under what condition(s) a non-exact DE can be converted to an exact DE.

(2) If a non-exact DE can be converted to an exact DE, is the conversion unique? In other words, is there more than one method to convert a non-exact DE into an exact DE?

Let's now convert a non-exact DE to an exact DE. Suppose we have a DE

$$M(x,y)dx + N(x,y)dy = 0 \qquad (1.105)$$

If we have the IF $I(x)$ such that

$$I(x)M(x,y)dx + I(x)N(x,y)dy = 0 \qquad (1.106)$$

is exact, we should have

$$\frac{\partial}{\partial y}\big(I(x)M(x,y)\big) = \frac{\partial}{\partial x}\big(I(x)N(x,y)\big) \qquad (1.107)$$

That means

$$I(x)\frac{\partial}{\partial y}M(x,y) = N(x,y)\frac{\partial}{\partial x}I(x) + I(x)\frac{\partial}{\partial x}N(x,y)$$

$$I \cdot M_y = N \cdot I' + I \cdot N_x$$

$$I\big(M_y - N_x\big) = I' \cdot N \qquad (1.108)$$

$$\frac{I'}{I} = \frac{M_y - N_x}{N}$$

$$I(x) = \exp\left(\int \frac{M_y - N_x}{N}dx\right)$$

Here the integrand

$$\frac{M_y - N_x}{N} \qquad (1.109)$$

must be a function with a single variable x.

Exact DE Theorem

For DE

$$M(x, y)dx + N(x, y)dy = 0 \tag{1.110}$$

1) If

$$\frac{M_y - N_x}{N} = f(x) \tag{1.111}$$

is a function of one variable x, we have the IF

$$\rho(x) = \exp\left(\int f(x)dx\right)$$

2) If

$$\frac{M_y - N_x}{M} = g(y) \tag{1.112}$$

is a function of one variable y, we have the IF

$$\rho(y) = \exp\left(-\int g(y)dy\right) \tag{1.113}$$

Remarks

(1) For one DE, $\rho(x)$ and $\rho(y)$ may both exist. In this case, use the most convenient form.

(2) It is possible that neither $\rho(x)$ nor $\rho(y)$ exists.

(3) You must have noticed that $\rho(x)$ has no negative sign in the exponential term while $\rho(y)$ has a negative sign in the exponential term. So, given below is the proof of why we need the negative sign for $\rho(y)$ but not for $\rho(x)$.

Proof of the negative sign in $\rho(y)$ in (1.113).

$M(x, y)dx + N(x, y)dy = 0$ is the given DE. The IF is chosen as $\rho(y)$

$$\rho(y)M(x, y)dx + \rho(y)N(x, y)dy = 0 \tag{1.114}$$

Let

$$\bar{M}(x,y) = \rho(y)M(x,y) \tag{1.115}$$

$$\bar{N}(x,y) = \rho(y)N(x,y) \tag{1.116}$$

We get the DE: $\bar{M}(x,y)dx + \bar{N}(x,y)dy = 0$. We want it to be exact and, thus, we set

$$\bar{M}_y = \bar{N}_x \tag{1.117}$$

i.e.,

$$\frac{\partial}{\partial y}\left(\rho(y)M(x,y)\right) = \frac{\partial}{\partial x}\left(\rho(y)N(x,y)\right)$$

$$M(x,y)\frac{\partial}{\partial y}\rho(y) + \rho(y)\frac{\partial}{\partial y}M(x,y) = \rho(y)\frac{\partial}{\partial x}N(x,y) \tag{1.118}$$

$$M\rho' + \rho M_y = \rho N_x$$

$$\frac{\rho'}{\rho} = \frac{N_x - M_y}{M}$$

That is

$$\frac{\rho'}{\rho} = -g(y) \tag{1.119}$$

where

$$g(y) = \frac{M_y - N_x}{M} \tag{1.120}$$

as noted above.

Solving (1.119), we get

$$\rho(y) = \exp\left(-\int g(y)dy\right) \tag{1.121}$$

We sum up the above discussion in the following theorems:

Theorem 1

For a given 1st.O DE $M(x,y)dx + N(x,y)dy = 0$, if

$$\frac{M_y - N_x}{N} = f(x) \tag{1.122}$$

is a function of purely x, the DE can be converted into an exact DE by multiplying the original DE with

$$\rho(x) = \exp\left(\int f(x)dx\right) \tag{1.123}$$

Theorem 2

For a given 1st.O DE $M(x,y)dx + N(x,y)dy = 0$, if

$$\frac{M_y - N_x}{M} = g(y) \tag{1.124}$$

is a function of purely y, the DE can be converted into an exact DE by multiplying the original DE with

$$\rho(y) = \exp\left(-\int g(y)dy\right) \tag{1.125}$$

Next, let's work on one example to enhance the understanding of the theorems expressed by the IF's (1.123) and (1.125). We also make a few remarks afterwards.

Example 4

Determine whether the DE

$$ydx + 3xdy = 0 \tag{1.126}$$

is exact or not. If not, convert it to an exact DE and solve it.

Solution

From the given DE, we get

$$M(x,y) = y$$
$$N(x,y) = 3x$$

Here, we have

$$\frac{\partial M}{\partial y} = \frac{\partial}{\partial y}(y) = 1$$

$$\frac{\partial N}{\partial x} = \frac{\partial}{\partial x}(3x) = 3$$

Since $M_y \neq N_x$, we know that the given DE is not exact. For solving this DE, there are two methods of doing this.

Method 1: Using the IF $\rho(x)$ to covert the DE.

$$\frac{M_y - N_x}{N} = \frac{1-3}{3x} = -\frac{2}{3x} = f(x)$$

It is clear that $f(x)$ is a function of purely x.

$$\rho(x) = \exp\left(\int f(x)dx\right)$$

$$= \exp\left(-\int \frac{2}{3x}dx\right)$$

$$= \exp\left(-\frac{2}{3}\ln x\right)$$

$$= x^{-\frac{2}{3}}$$

Multiplying $\rho(x)$ on both sides of the original DE, we have

$$x^{-\frac{2}{3}}ydx + 3x \cdot x^{-\frac{2}{3}}dy = 0$$

This is now our new DE. Here we have

$$\frac{\partial F}{\partial x} = \bar{M}(x,y) = x^{-\frac{2}{3}}y \tag{1.127}$$

$$\frac{\partial F}{\partial y} = \bar{N}(x,y) = 3x^{\frac{1}{3}} \tag{1.128}$$

Thus, we have

$$\bar{M}_y = \frac{\partial}{\partial y}\left(x^{-\frac{2}{3}}y\right) = x^{-\frac{2}{3}}$$

$$\bar{N}_x = \frac{\partial}{\partial x}\left(3x^{\frac{1}{3}}\right) = x^{-\frac{2}{3}}$$

Since $\bar{M}_y = \bar{N}_x$, the new DE is exact and can be solved by using either (1.127) or (1.128) to find $F(x,y)$. It appears the function in (1.128) is easier to integrate and we opt for (1.128) to find $F(x,y)$. We have

$$\frac{\partial F}{\partial y} = \bar{N}(x,y) = 3x^{\frac{1}{3}}$$

$$F(x,y) = \int 3x^{\frac{1}{3}}dy + g(x) = 3x^{\frac{1}{3}}y + g(x) \tag{1.129}$$

Substituting (1.129) back into (1.127), we have

$$\frac{\partial}{\partial x}\left(3x^{\frac{1}{3}}y + g(x)\right) = x^{-\frac{2}{3}}y + g'(x) = x^{-\frac{2}{3}}y$$

This solves $g'(x) = 0$. That means $g(x) = C_1$. Substituting this back into (1.129), we have

$$F(x,y) = 3x^{\frac{1}{3}}y + C_1$$

That gives the solution to the DE

$$F(x,y) = C_2$$

That is

$$x^{\frac{1}{3}}y = C$$

One may even write the GS as

$$xy^3 = C_3$$

where C_3 is another constant.

Method 2: Using the IF $\rho(y)$ to covert the DE.

We now discuss the second method of converting a non-exact DE to an exact DE.

$$\frac{M_y - N_x}{M} = \frac{1-3}{y} = -\frac{2}{y} = g(y)$$

The IF

$$\rho(y) = \exp\left(-\int g(y)dy\right)$$
$$= \exp\left(\int \frac{2}{y}dy\right)$$
$$= y^2$$

Multiplying $\rho(y)$ on both sides of the original DE, we have

$$y^3 dx + 3xy^2 dy = 0$$

as our new DE. Here we have

$$\frac{\partial F}{\partial x} = \bar{M}(x,y) = y^3 \tag{1.130}$$

$$\frac{\partial F}{\partial y} = \bar{N}(x,y) = 3xy^2 \tag{1.131}$$

and

$$\bar{M}_y = \frac{\partial}{\partial y}(y^3) = 3y^2$$

$$\bar{N}_x = \frac{\partial}{\partial x}(3xy^2) = 3y^2$$

Since $\bar{M}_y = \bar{N}_x$, the new DE formed is now exact.

We can now use either (1.130) or (1.131) to find $F(x,y)$. Since the function in (1.130) is easier to integrate, we choose (1.130) to find $F(x,y)$. We have

$$F(x,y) = \int y^3 dx + g(y) = xy^3 + g(y) \tag{1.132}$$

Substituting (1.132) back into (1.130), we have

$$\frac{\partial}{\partial y}(xy^3 + g(y)) = 3xy^2 + g'(y) = 3xy^2$$

It gives

$$g'(y) = 0$$

which means

$$g(y) = C_1$$

Substituting this back into (1.132), we have

$$F(x, y) = xy^3 + C_1$$

Thus, the GS to the DE is

$$xy^3 + C_1 = C_2$$

That is

$$xy^3 = C$$

This is the same answer as we found in Method 1.

Summary

We have four ways for solving a non-exact DE, which is summarized as follows.

Steps for solving a non-exact DE by converting it into an exact DE

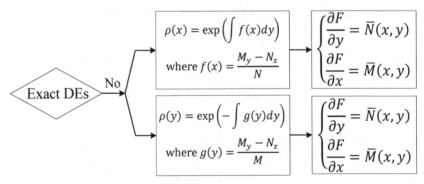

Figure 1.4 Steps for converting a non-exact DE to an exact DE.

Step 1 Determine $M(x, y)$ and $N(x, y)$.

Check if $M_y = N_x$? If so, go to Step 5

Step 2 Check if $\frac{M_y - N_x}{N} = f(x)$ is a function of purely x or if $\frac{M_y - N_x}{M} = g(y)$ is a function of purely y.

Step 3 Find the easiest IF

$$\rho(x) = \exp\left(\int f(x)dx\right) \tag{1.133}$$

Or

$$\rho(y) = \exp\left(-\int g(y)dy\right) \tag{1.134}$$

Step 4 Compute the new $\bar{M}$ and $\bar{N}$ using either $\rho(x)$ or $\rho(y)$

$$\bar{M}(x,y) = \rho\, M(x,y) \tag{1.135}$$

$$\bar{N}(x,y) = \rho\, N(x,y) \tag{1.136}$$

Step 5 Construct the partial DEs

$$\begin{cases} \dfrac{\partial F}{\partial x} = \bar{M}(x,y) \\[2mm] \dfrac{\partial F}{\partial y} = \bar{N}(x,y) \end{cases} \tag{1.137}$$

Step 6 Use either of the DEs (1.137) to find $F(x,y)$

$$F(x,y) = \int \bar{M}(x,y)dx + g(y) \tag{1.138}$$

or equivalently,

$$F(x,y) = \int \bar{N}(x,y)dy + g(x) \tag{1.139}$$

Step 7 Substituting the $F(x,y)$ obtained as (1.138) or (1.139) into one of the DEs (1.137), we have either

$$\frac{\partial}{\partial y}\left(\int \bar{M}(x,y)dx + g(y)\right) = \bar{N}(x,y) \qquad (1.140)$$

Or

$$\frac{\partial}{\partial x}\left(\int \bar{N}(x,y)dy + g(x)\right) = \bar{M}(x,y) \qquad (1.141)$$

Step 8 Solving DE (1.140) or (1.141) produces $g(y)$ or $g(x)$. Inserting $g(y)$ or $g(x)$ into (1.138) or (1.139) forms the final GS to the DE $F(x,y)$ =C.

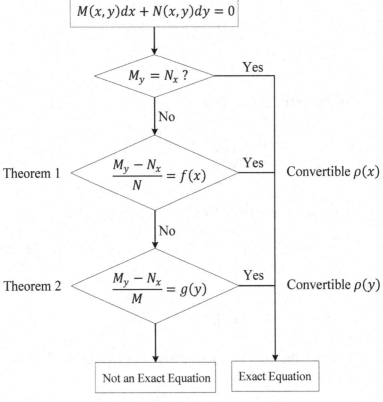

Figure 1.5 Steps for solving a non-exact DE.

Example 5
Find the GS to the following DE
$$M(x,y)dx + N(x,y)dy = 0 \qquad (1.142)$$

Solution
We can compose two simple partial DEs with F as the DV and x and y as the IVs
$$\frac{\partial F}{\partial x} = M(x,y) \qquad (1.143)$$
$$\frac{\partial F}{\partial y} = N(x,y) \qquad (1.144)$$

Solving (1.143), we get
$$F(x,y) = \int M(x,y)dx + g(y)$$

Plugging the above into (1.144), we get
$$\frac{\partial}{\partial y}\left(\int M(x,y)dx\right) + g'(y) = N(x,y)$$
$$g'(y) = N(x,y) - \frac{\partial}{\partial y}\left(\int M(x,y)dx\right)$$

Solving this, we have
$$g(y) = \int N(x,y)dy - \int \left(\frac{\partial}{\partial y}\left(\int M(x,y)dx\right)\right)dy$$

Thus, we get the GS to the DE as
$$\int M(x,y)dx + \int N(x,y)dy - \int \left(\frac{\partial}{\partial y}\left(\int M(x,y)dx\right)\right)dy = C$$

Example 6
Is the following DE exact? Find the GS.
$$(6xy - y^3)dx + (4y + 3x^2 - 3xy^2)dy = 0 \qquad (1.145)$$

Solution
From the given DE, we have
$$\frac{\partial F}{\partial x} = M(x,y) = 6xy - y^3$$
$$\frac{\partial F}{\partial y} = N(x,y) = 4y + 3x^2 - 3xy^2$$

Now we evaluate
$$M_y = \frac{\partial}{\partial y}(6xy - y^3) = 6x - 3y^2$$

$$N_x = \frac{\partial}{\partial x}(4y + 3x^2 - 3xy^2) = 6x - 3y^2$$

Since $M_y = N_x$, the given DE is exact. Now, we find $F(x, y)$

$$\frac{\partial F}{\partial x} = M(x, y) = 6xy - y^3$$

$$F(x, y) = \int (6xy - y^3)dx + g(y)$$

$$= 3x^2 y - y^3 x + g(y)$$

Plugging it into

$$\frac{\partial F}{\partial y} = N(x, y)$$

we have

$$\frac{\partial}{\partial y}\left(3x^2 y - y^3 x + g(y)\right) = 4y + 3x^2 - 3xy^2$$

$$3x^2 - 3xy^2 + g'(y) = 4y + 3x^2 - 3xy^2$$

$$g'(y) = 4y$$

$$g(y) = 2y^2$$

Therefore, the GS to the DE is

$$3x^2 y - xy^3 + 2y^2 = C$$

Let's examine the following to show that not all DEs can be solved this way.

Example 7
Find the GS to the following DE

$$(3y^2 + 5x^2 y)dx + (3xy + 2x^3)dy = 0 \qquad (1.146)$$

Solution
From the given DE, we have

$$M(x, y) = 3y^2 + 5x^2 y$$

$$N(x, y) = 3xy + 2x^3$$

$$M_y = \frac{\partial}{\partial y}(3y^2 + 5x^2 y) = 6y + 5x^2$$

$$N_x = \frac{\partial}{\partial x}(3xy + 2x^3) = 3y + 6x^2$$

Since $M_y \neq N_x$, the given DE is not exact. Let's now check if we can convert it to an exact DE.

$$\frac{M_y - N_x}{N} = \frac{3y - x^2}{3xy + 2x^3}$$

is not a function of purely x, and

$$\frac{M_y - N_x}{M} = \frac{3y - x^2}{3y^2 + 5x^2 y}$$

is not a function of purely y. Therefore, the given DE cannot be converted to an exact DE.

Finally, we show the relationship between the exact DE method and the linear DE we previously studied.

Converting the 1st.O linear DE to an exact DE

Now the question arises whether all linear DEs are exact. If not, are they convertible? Let's now answer these queries. The standard form of the 1st.O linear DE is

$$y' + P(x)y = Q(x) \tag{1.147}$$

That is

$$\frac{dy}{dx} + P(x)y - Q(x) = 0$$
$$(P(x)y - Q(x))dx + dy = 0 \tag{1.148}$$

Now we have

$$M(x, y) = P(x)y - Q(x)$$
$$N(x, y) = 1$$
$$M_y = P(x)$$
$$N_x = 0 \tag{1.149}$$

Since in general $M_y \neq N_x$, (1.147) is not an exact DE.

If a DE is not exact in its original form, we may convert it to an exact DE.

Compare the following two formulas.

$$\frac{M_y - N_x}{N} = P(x) \tag{1.150}$$

$$\frac{M_y - N_x}{M} = \frac{P(x)}{P(x)y - Q(x)}$$

Apparently, working with the first one is much easier than working with the second one. Thus, we have our IF

$$\rho(x) = \exp\left(\int P(x)dx\right) \tag{1.151}$$

Multiplying $\rho(x)$ on both sides of (1.148), we have

$$\exp\left(\int P(x)dx\right)(P(x)y - Q(x))dx$$
$$+ \exp\left(\int P(x)dx\right)dy = 0 \tag{1.152}$$

which is now the new DE. Thus, we have

$$\frac{\partial F}{\partial x} = \bar{M}(x,y) = \exp\left(\int P(x)dx\right)(P(x)y - Q(x)) \tag{1.153}$$

$$\frac{\partial F}{\partial y} = \bar{N}(x,y) = \exp\left(\int P(x)dx\right) \tag{1.154}$$

$$\bar{M}_y = \frac{\partial}{\partial y}\left(\exp\left(\int P(x)dx\right)(P(x)y - Q(x))\right)$$
$$= \exp\left(\int P(x)dx\right)\frac{\partial}{\partial y}(P(x)y - Q(x)) \tag{1.155}$$
$$= P(x)\exp\left(\int P(x)dx\right)$$

$$\bar{N}_x = \frac{\partial}{\partial x}\exp\left(\int P(x)dx\right)$$
$$= P(x)\exp\left(\int P(x)dx\right) \tag{1.156}$$

Since $\bar{M}_y = \bar{N}_x$, the new DE is exact.

Now we can use either (1.153) or (1.154) to find $F(x, y)$. Since the function in (1.154) is easier to integrate, we use it to find:

$$\begin{aligned}
F(x, y) &= \int \exp\left(\int P(x)dx\right) dy \\
&= y \exp\left(\int P(x)dx\right) + g(x)
\end{aligned} \tag{1.157}$$

Substituting this back to (1.153), we have

$$\frac{\partial}{\partial x}\left(y \exp\left(\int P(x)dx\right) + g(x)\right)$$

$$= \exp\left(\int P(x)dx\right)\left(P(x)y - Q(x)\right)$$

$$+ y P(x) \exp(\smallint P(x)dx) + g'(x)$$

$$= y P(x) \exp\left(\int P(x)dx\right) - Q(x) \exp\left(\int P(x)dx\right) \tag{1.158}$$

$$g'(x) = -Q(x) \exp\left(\int P(x)dx\right)$$

$$g(x) = -\int Q(x) \exp\left(\int P(x)dx\right) dx \tag{1.159}$$

Substituting this back to $F(x, y)$, we have

$$F(x, y) = y \exp\left(\int P(x)dx\right) - \int Q(x) \exp\left(\int P(x)dx\right) dx \tag{1.160}$$

Therefore, the GS to the DE is

$$y \exp\left(\int P(x)dx\right) - \int Q(x) \exp\left(\int P(x)dx\right) dx = C \quad (1.161)$$

That is

$$y = \exp\left(-\int P(x)dx\right)\left(\int Q(x) \exp\left(\int P(x)dx\right) dx + C\right) \quad (1.162)$$

Problems

Problem 1.7.1 Find the GS to the following DE
$$(1 + \ln(xy))dx + \left(\frac{x}{y}\right) dy = 0$$

Problem 1.7.2 Given a 1st.O DE
$$A(x,y)dx + B(x,y)dy = 0$$
which is not an exact DE in general, but the function $A(x,y)$ and $B(x,y)$ satisfy the following relationship
$$\left(\frac{\partial A(x,y)}{\partial y} - \frac{\partial B(x,y)}{\partial x}\right) / B(x,y) = P(x)$$
where $P(x)$ is a function of one variable x. Prove that, after multiplying the original DE by an IF $\rho(x) = \exp(\int P(x)\,dx)$, one can transform the original non-exact DE into an exact DE.

Problem 1.7.3 Given 1st.O linear DE
$$\alpha(x)\frac{dy}{dx} + \beta(x)y + \gamma(x) = 0$$
where $\alpha(x) \neq 0$.
(1) Use the exact DE method to solve the DE and express the GS in terms of $\alpha(x), \beta(x),$ and $\gamma(x)$.
(2) Use the 1st.O linear DE method to solve the DE and express the GS in terms of $\alpha(x), \beta(x),$ and $\gamma(x)$.

Problem 1.7.4 The 1st.O linear DE can be expressed alternatively as
$$(P(x)y - Q(x))dx + dy = 0$$
where functions $P(x) \neq 0$ and $Q(x) \neq 0$ are given. Please
(1) Check if this DE is exact.
(2) If not, convert it to an exact DE.
(3) Solve the DE using the exact DE method; your solution may be expressed in terms of the functions $P(x)$ and $Q(x)$.
(4) If $P(x) = \frac{1}{x}$ and $Q(x) = \frac{\cos x}{x}$, get the specific solution.

Problem 1.7.5 Find the GS to the following DE
$$(2x - y^2)dx + xydy = 0$$

Problem 1.7.6 Find the GS to the following DE using two different methods.
$$y' = -\frac{3x^2 + 2y^2}{4xy}$$

Problem 1.7.7 Find the GS to the following DE using two different methods.
$$y' - \frac{x}{x^2 + 1}y = 2x(x^2 + 1)$$

Problem 1.7.8 Find the GS to the following DE using two different methods.
$$y' = \frac{x + 3y}{y - 3x}$$

Problem 1.7.9 Find the GS to the following DE
$$ydx + (2x + y^4)dy = 0$$

Problem 1.7.10 Find the GS to the following DE
$$y'(2xy + 1) = 2x - y^2$$

Problem 1.7.11 Find the GS to the following DE
$$y^3 + xy^2y' - y' = 0$$

Problem 1.7.12 Check whether the following DE is exact

$$(\cos x + \ln y)dx + \left(\frac{x}{y} + \exp(y)\right)dy = 0$$

If it is exact according to your verification above, solve it using the exact DE method. Otherwise, use a different method to solve it.

Problem 1.7.13 Find the GS to the following DE using two different methods.

$$\exp(y) + y \cos x + (x \exp(y) + \sin x)y' = 0$$

1.8   Summary

Most of the DEs we discuss can be summarized in the following Figure 1.6.

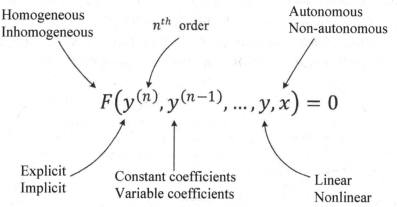

Figure 1.6 This diagram illustrates the classification of DEs by various categories.

Classifying DEs according to homogeneous (Homo) and inhomogeneous (InHomo) will produce two types. Of each type, further classification according to autonomous and non-autonomous will produce two sub-types. Thus, we have four types using these two classifications. Adding linear and nonlinear, explicit and implicit, constant coefficients (c-coeff) and variable coefficients (v-coeff), we have a total of $2^5 = 32$ types. Further, if adding the orders of the DEs, we have 32 types of 1st.O DEs and 32 types of 2nd.O DEs, *etc.*

Based on their intrinsic properties and solution methods, 1st.O DEs (like many other types of DEs) may be classified into several overlapping groups: separable, 1st.O linear, substitutable, the Bernoulli, the Riccati, and the exact DEs.

Commonly, one single DE may belong to multiple groups and, thus, can be solved by multiple solution methods although the algebraic complexity of the methods may differ dramatically. Therefore, when given a DE to solve, the most important first step is to identify the group(s) the DE belongs to and, then, select the most effective solution method(s) to solve it. The following Venn diagram (Figure 1.7) illustrates the logical relationship of all 1st.O DEs and their solution methods.

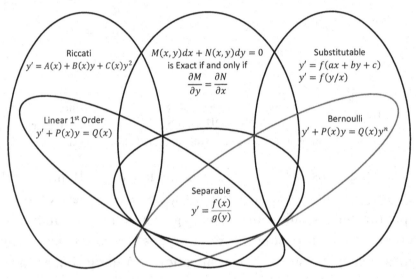

Figure 1.7 This Venn diagram illustrates the logical relationships of 1st.O DEs and their solution methods.

Problems

Problem 1.8.1 Find the GS to the following DE using at least two methods

$$xy' - y = y^2 \sin x$$

Problem 1.8.2 Find the GS to the following DE using at least two methods:

$$y' + 30xy = 1 + 15(x^2 + y^2)$$

Chapter 2
Mathematical Models

2.1   Newton's Law of Cooling

Newton's law of cooling was introduced by Isaac Newton (December 25, 1642 –March 20, 1726) in 1701 to describe convection cooling, in which the heat transfer coefficient is independent or almost independent of the temperature difference between the object and its thermo environment. For cases where the temperature difference is too big to maintain such assumptions, two French Physicists, P. Dulong and A. Petit, introduced corrections to the original Newton's law of cooling in 1817.

The time rate of change of the temperature $T(t)$ of an object is proportional to the difference between $T(t)$ and the temperature A

of the surrounding medium. The medium is assumed to be so big that the heat transfer to the medium from the object, or the reverse, is too insignificant to change the medium's temperature. However, such heat transfer will cause the object's temperature to change, and this is the subject of Newton's law of cooling published anonymously in Philosophical Transactions. Our current discussion is an approximation of the truth that the total energy is conserved for the complete system of the body and the medium.

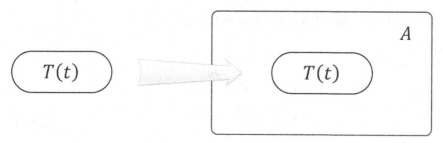

Figure 2.1 Placing an object of temperature T_0 into a large medium of fixed temperature A.

Consider an object at a temperature $T(t)$ that is placed in a medium at a constant temperature A (Figure 2.1). We establish the DE that describes the temperature change of the object while it exchanges energy (heat) with the medium. We define a few quantities:

➢ A is the temperature of the medium, kept at a constant;
➢ $T(t)$ is the temperature of the object at time t;
➢ $T(t = 0) = T_0$ is the initial temperature of the object;
➢ t is the time;
➢ $\frac{dT(t)}{dt} = k(A - T)$ is the rate of temperature variation and k is a constant parameter that characterizes the heat conductivity of the medium with the object.

With this introduction, we establish the following DE with an IC to form an IVP (2.1):

$$\begin{cases} \dfrac{dT(t)}{dt} = k(A - T) \\ T(t = 0) = T_0 \end{cases} \tag{2.1}$$

Solution

$$\frac{dT}{dt} = k(A - T)$$
$$\frac{dT}{T - A} = -kdt \tag{2.2}$$

Integrating both sides of the above, we get

$$\int_{T_0}^{T} \frac{dT}{T - A} = \int_{0}^{t} (-k)dt$$
$$\ln(T - A) - \ln(T_0 - A) = -kt$$
$$\ln\left(\frac{T - A}{T_0 - A}\right) = -kt \tag{2.3}$$
$$\frac{T - A}{T_0 - A} = \exp(-kt)$$
$$T - A = (T_0 - A)\exp(-kt)$$
$$T(t) = A + (T_0 - A)\exp(-kt)$$

Finally, we get the PS to the DE,

$$T(t) = T_0 \exp(-kt) + (1 - \exp(-kt))A \tag{2.4}$$

Thus, we have translated a scientific law into a DE and found its solution. Now, if given the values of A and k, we can predict the temperature of the object at any time.

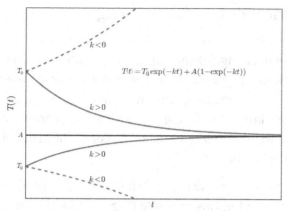

Figure 2.2 The multiple scenarios of the object's temperature $T(t)$ changing with time t.

Remarks

(1) If $A \approx T_0 \Rightarrow \frac{dT}{dt} \approx 0 \Rightarrow T(t)$ does not change with time and the temperature stays approximately constant.

(2) If $A > T_0 \Rightarrow \frac{dT}{dt} > 0 \Rightarrow T(t)$ increases while heat is transferred from the medium to the object. This process will last until the object reaches the same temperature as the medium.

(3) If $A < T_0 \Rightarrow \frac{dT}{dt} < 0 \Rightarrow T(t)$ decreases while heat is transferred to the medium from the object. This process will last until the object reaches the same temperature of the medium.

Problem

Problem 2.1.1 In a warehouse kept at a constant temperature of 60F, a corpse was found at noon with a temperature of 85F. Ninety (90) minutes later, its temperature dropped to 79F. At the time of death, the person's body temperature was at the normal 99F (Actually, its range is 97.7-99.5F). When did the person die?

71

2.2   Torricelli's Law for Draining

The time rate of change of the volume V of water or other liquid in a draining sink is proportional to the square root of the depth y of water in the tank, measured from the sinking point. This DE is called the Torricelli's law for draining or Torricelli's theorem, named after Italian physicist and mathematician E. Torricelli (1608-1647) who discovered the law in 1643.

$V(t)$ is the volume of water in the sink at time t. Although a hemisphere is used in this example (Figure 2.3), the draining model is applicable to containers of any conceivable shapes.

$y(t)$ is the height of the water surface, from the draining point, at time t. The draining model introduced by Torricelli can be written as

$$\frac{dV}{dt} \propto -\sqrt{y} = -k\sqrt{y} \qquad (2.5)$$

where k is the so-called draining constant which can be positive for conventional draining or negative for filling up. It depends on:

(1)   The liquid's properties, such as viscosity. The k value will be different for the following liquids: water, tea, gasoline, honey, or mud water, *etc.*

(2)   Size of the hole. Theoretically, the hole is considered a mathematical dot without size.

(3)   Pressure differential between the inside and the outside of the draining hole.

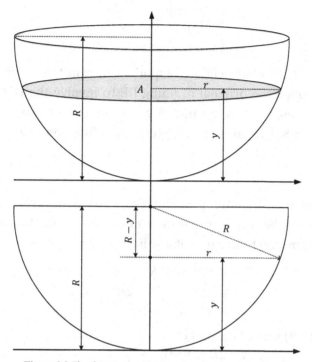

Figure 2.3 The draining model for a hemisphere of radius R.

Let $A(y)$ be the cross-sectional area of the container at height y where the liquid surface lies. The volume element of the liquid at the surface for a thin layer of thickness dy at the height y is

$$dV = A(y)dy \tag{2.6}$$

where we assumed the area remain unchanged from $y + dy$ to y. With the draining DE

$$\frac{dV}{dt} = -k\sqrt{y} \tag{2.7}$$

We can compose an IVP as

$$\begin{cases} \dfrac{A(y)dy}{dt} = -k\sqrt{y} \\ y(t = 0) = y_0 \end{cases} \tag{2.8}$$

where y_0 is the initial height of liquid surface in the container. We have converted the original draining DE to involve the IV t and the DV y, the height. It does not depend on the volume explicitly although one still needs to work out the area function $A(y)$.

Solution

Let's solve the draining problem (2.8) for a peculiar case: the container is a hemisphere whose inner radius is R. We can easily write the cross-sectional area at any height y as

$$\begin{aligned} A(y) = \pi r^2 &= \pi(R^2 - (R-y)^2) \\ &= \pi(R^2 - R^2 + 2Ry - y^2) \\ &= \pi y(2R - y) \end{aligned} \tag{2.9}$$

Plugging (2.9) into (2.8), we get

$$\frac{\pi y(2R - y)dy}{dt} = -k\sqrt{y} \tag{2.10}$$

$$\pi\sqrt{y}(2R - y)dy = -kdt$$

$$\int_R^y \pi\sqrt{y}(2R - y)dy = \int_0^t (-k)dt$$

$$\int_R^y \pi\left(2Ry^{\frac{1}{2}} - y^{\frac{3}{2}}\right)dy = -kt$$

$$\pi\left(2R\int_R^y y^{\frac{1}{2}}dy - \int_R^y y^{\frac{3}{2}}dy\right) = -kt$$

$$\pi\left(\frac{4}{3}Ry^{\frac{3}{2}} - \frac{2}{5}y^{\frac{5}{2}}\right)\Big|_R^y = -kt$$

$$\pi y^{\frac{3}{2}}\left(\frac{4}{3}R - \frac{2}{5}y\right) - \pi R^{\frac{3}{2}}\left(\frac{4}{3}R - \frac{2}{5}R\right) = -kt$$

$$\pi y^{\frac{3}{2}}\left(\frac{4}{3}R - \frac{2}{5}y\right) - \frac{14}{15}\pi R^{\frac{5}{2}} = -kt$$

Therefore, the resulting PS to the draining DE can be written as

$$\pi y^{\frac{3}{2}}(20R - 6y) = 14\pi R^{\frac{5}{2}} - 15kt \qquad (2.11)$$

This equation expresses the height of the liquid surface in the hemisphere container as a function of time, implicitly. The initial state is that the hemisphere container is filled with liquid to the rim.

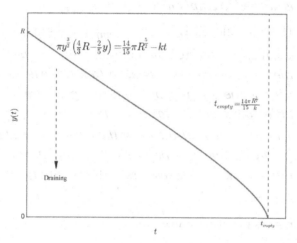

Figure 2.4 The height of the liquid surface in the hemisphere container vs. time.

Remarks

(1) At $t = 0$, the only solution that makes sense is $y = R$ and it is verified by Figure 2.4.

(2) When $y = 0$, the container is emptied. Plugging $y = 0$ into (2.11) finds the time needed to empty the container

$$y = 0 \Rightarrow kt = \frac{14}{15}\pi R^{\frac{5}{2}} \qquad (2.12)$$

$$t_{empty} = \frac{14\,\pi}{15\,k} R^{\frac{5}{2}} = \frac{7}{5k} \frac{V_h}{\sqrt{R}} \tag{2.13}$$

(a) In formula (2.12), $V_h = \frac{1}{2}\left(\frac{4}{3}\pi R^3\right)$ is the volume of the semi-sphere.

(b) The formula (2.12) shows $t_{empty} \propto R^{5/2} \Rightarrow$ the bigger the tank (the larger its radius), the longer the time needed to empty it. *A special case, if the container is tiny, little time is needed to empty it:* $R \to 0 \Rightarrow t_{empty} \to 0$.

(c) The formula (2.12) shows $t_{empty} \propto \frac{1}{k} \Rightarrow$ the bigger the k value, the shorter the empting time. *A special case: if the draining constant is huge (a huge leaking hole, or a very slippery liquid, or a massive pressure differential), little time is needed to empty it:* $k \to \infty \Rightarrow t_{empty} \to 0$.

The opposite is also true: if the draining constant is tiny (a tiny leaking hole, or a very stuffy liquid, or no pressure differential), it takes eternity to empty it: $k \to 0 \Rightarrow t_{empty} \to \infty$.

Everything makes intuitive sense.

Problems

Problem 2.2.1 We drain a fully filled spherical container of inner radius R by drilling a hole at the bottom. We assume the draining process follows Torricelli's law with the draining constant k.
(1) Compute the time needed to empty the container.
(2) If we double the radius of the container which is also fully filled and keep all other conditions unchanged, compute the draining time again.

Problem 2.2.2 A complete spherical container of radius R has two small holes- one at the top for air to come in and one at the bottom for liquid to drain out (think of it as a hollowed bowling ball). At $t = 0$, the container is filled with liquid. Besides the radius R, you are also given the draining constant k. The draining process follows Torricelli's law. Please
(1) compute the time T_1 needed to drain the upper half of the container, *i.e.*, the water level dropping from the top hole to the equator line of the ball;
(2) compute the time T_2 needed to drain the lower half of the container, *i.e.*, the water level dropping from the equator line of the ball to the bottom hole;
(3) identify the correct relationship: $T_1 > T_2$, $T_1 < T_2$, or $T_1 = T_2$?

Problem 2.2.3 A coffee cup from the "Moonbucks" coffee shop is like a cylinder with decreasing horizontal cross-sectional area. The radius of the top circle is R, and that for the bottom is 10% smaller. The cup height is H. Now, a nutty kid drills a pinhole at the half height point of the cup, allowing a full cup of coffee to leak. Compute the time for coffee to leak to the pinhole. Compute the time again if we turn the cup upside down with full cup to start (with the new bottom closed). In both cases, the leak follows the Draining DE with a constant k.

Problem 2.2.4 A spherical container of inner radius R is originally filled up with liquid. We drill one hole at the bottom, and another hole at the top (for air intake) to allow liquid to drain from the bottom hole. The draining process follows the Torricelli's law, and the draining constant depends on the liquid properties, the draining hole size, and the inside-outside pressure differential, *etc.* While draining the upper half, we make the draining constant as k. What draining constant must we set (e.g., by adjusting the bottom hole size) so that the lower half will take the same amount of time to drain as the upper half did?

Problem 2.2.5 A spherical container of inner radius R is filled up with liquid. We drill one hole at the bottom (and another at the top to allow air to come in). The draining process follows the Torricelli's law with the usual draining constant k. We start draining the container, when it is full, until the leftover volume reaches a "magic" value at which the draining time for the top portion is equal to that for the bottom (left-over) portion. Calculate the leftover volume.

2.3   The Population Model

A population model is a DE, or a system of DEs (DE.Syst), that expresses population dynamics. The model describes the change of a population through complex interactions and processes in terms of birth and death rates. The population, extending beyond human beings, can be any beings or things like tigers, insects, viruses, and diseases, as well as radioactive material.

2.3.1   Introduction

The time rate of change of a population $P(t)$ with constant birth and death rates are, in many cases, proportional to the size of the population. Function $P(t)$ = Population count at any time t.

We can establish a simple initial value problem (IVP) to describe the population as a function of time:

$$\begin{cases} \dfrac{dP}{dt} = kP \\ P(t = 0) = P_0 \end{cases} \tag{2.14}$$

Solution

$$\frac{dP}{dt} = kP \tag{2.15}$$

$$\frac{dP}{P} = kdt$$

$$\int_{P_0}^{P} \frac{dP}{P} = \int_{0}^{t} kdt$$

$$\ln P - \ln P_0 = kt$$

$$P(t) = P_0 \exp(kt)$$

For example, if $P(t = 0) = 1000$ and $P(t = 1) = 2000$, we get

$$\begin{cases} P_0 = 1000 \\ k = \ln 2 \end{cases} \qquad (2.16)$$

with which we can construct the model of exponential growth of populations as

$$P(t) = 1000 \exp(t \ln 2) = 1000 \cdot 2^t \qquad (2.17)$$

In this case, the population doubles every time unit.

2.3.2   General Population Equation

It is customary to track the growth or decline of a population in terms of the functions of its birth and death rates, which are defined as follows:

$\beta(t)$: the number of births per unit population per unit time at time t

$\delta(t)$: the number of deaths per unit population per unit time at time t

Then, the numbers of births and deaths that occur during the time interval $(t, t + \Delta t)$ is approximately given by births: $\beta(t)P(t)\Delta t$ and deaths: $\delta(t)P(t)\Delta t$. The change ΔP in the population during the time interval $(t, t + \Delta t)$ of the length Δt is

$$\begin{aligned} \Delta P &= \text{births} - \text{deaths} \\ &= \beta(t)P(t)\Delta t - \delta(t)P(t)\Delta t \\ &= \big(\beta(t) - \delta(t)\big)P(t)\Delta t \end{aligned} \qquad (2.18)$$

Thus,

$$\frac{\Delta P}{\Delta t} = \big(\beta(t) - \delta(t)\big)P(t) \qquad (2.19)$$

Taking the limit $\Delta t \to 0$, we get the DE

$$\frac{dP}{dt} = (\beta - \delta)P \tag{2.20}$$

which is the general population DE where $\beta = \beta(t)$ and $\delta = \delta(t)$ where β and δ can be either pure constants, or functions of t, or they may depend, directly, on the unknown function $P(t)$.

Example 1

Given the following information, find the population after 10 years.

$$\begin{aligned} P(0) = P_0 &= 100 \\ \beta &= 0.0005P \\ \delta &= 0 \end{aligned} \tag{2.21}$$

Solution

Substituting the given data (2.21) into the DE (2.20), we have

$$\frac{dP}{dt} = (\beta - \delta)P$$

$$\frac{dP}{dt} = 0.0005P^2$$

$$\frac{dP}{P^2} = 0.0005dt$$

$$\int \frac{dP}{P^2} = \int 0.0005dt$$

$$-\frac{1}{P} = 0.0005t + C$$

Applying the IC $P(0) = 100$ yields

$$C = -\frac{1}{100}$$

Therefore,

$$P(t) = \frac{2000}{20 - t}$$

and

$$P(10) = 200$$

Thus, the population after 10 years is 200.

2.3.3   The Logistic Equation

The logistic equation (logistic DE) was first introduced by the Belgian mathematician Pierre Verhulst (1804-1849) in the 1840s to model population growth. The logistic DE is, thus, also called the Verhulst model or logistic growth curve. In this model, the time variable is continuous and a modification of the model to discrete quadratic recurrence equation, known as a logistic map, is also very interesting. The model's rediscovery and wider use happened in the 1920s by an American biologist, Raymond Pearl (1879-1940), and by an American biostatistician, Lowell Reed (1886-1966). The logistic DE finds applications in a wide range of fields: artificial neural networks, biomathematics, ecology, chemistry, physics, demography, economics, geosciences, probability, political science, and statistics, *etc.* In fact, the list is still growing.

Let's consider the case of a fruit-fly population growth in a container. It is often observed that the birth rate decreases as the population itself increases due to limited space and food supply. Suppose that the birth rate β is a linear decreasing function of the population size P so that $\beta = \beta_0 - \beta_1 P$ where β_0 and β_1 are positive constants. If the death rate $\delta = \delta_0$ remains constant, the general population DE takes the form

$$\frac{dP}{dt} = (\beta_0 - \beta_1 P - \delta_0)P \qquad (2.22)$$

Let $a = \beta_0 - \delta_0$ and $b = \beta_1$. Then,

$$\frac{dP}{dt} = aP - bP^2 \qquad (2.23)$$

Eq. (2.23) is called the logistic equation for the population model. We can rewrite the logistic DE as

$$\frac{dP}{dt} = kP(M - P) \tag{2.24}$$

where $k = b$ and $M = a/b$ are constants. M is sometimes called the carrying capacity of the environment, *i.e.*, the maximum population that the environment can support on a long-term basis. This DE can be solved by the SOV as follows

$$\begin{cases} \dfrac{dP}{P(M - P)} = kdt \\ P(t = 0) = P_0 \end{cases}$$

$$\int_{P_0}^{P} \frac{dP}{P(M - P)} = \int_0^t kdt$$

$$\frac{1}{M} \int_{P_0}^{P} \frac{M}{P(M - P)} dP = kt$$

$$\int_{P_0}^{P} \left(\frac{1}{M - P} + \frac{1}{P} \right) dP = Mkt \tag{2.25}$$

$$(\ln P - \ln(M - P))\big|_{P_0}^{P} = Mkt$$

$$\left(\ln\left(\frac{P}{P_0}\right) + \ln\left(\frac{M - P_0}{M - P}\right) \right) = Mkt$$

$$\frac{P}{P_0} \frac{M - P_0}{M - P} = \exp(Mkt)$$

This is the PS as an implicit function of P, which can be converted to an explicit form.

$$P = (M - P) \frac{P_0 \exp(Mkt)}{M - P_0} \tag{2.26}$$

$$P\left(1 + \frac{P_0 \exp(Mkt)}{M - P_0} \right) = M \frac{P_0 \exp(Mkt)}{M - P_0}$$

$$P = \frac{MP_0 \exp(Mkt)}{M - P_0 + P_0 \exp(Mkt)}$$

$$P = \frac{M}{1 + \left(\frac{M}{P_0} - 1\right)\exp(-Mkt)}$$

which is the final PS to the logistic DE.

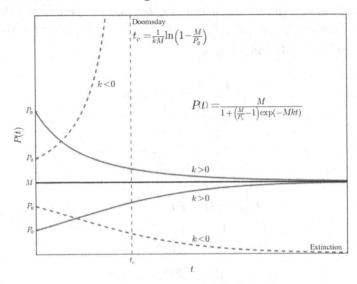

Figure 2.5 This figure is a summary of the solution (2.26) for the population DE for the five cases involving $P_0 > M, P_0 = M, P_0 < M$ coupled with $k > 0$ and $k \leq 0$.

Remarks

(1) At $t = 0$, $P(t = 0) = P_0$

(2) The limiting cases: $\displaystyle\lim_{t\to\infty} P(t) = \begin{cases} P_0, & k = 0 \\ M, & k > 0 \\ 0, & k < 0 \end{cases}$

(3) Limited environment situation:

A certain environment can support at most M populations. It is reasonable to expect the growth rate $\beta - \delta$ (the combined birth and death rates) to be proportional to $M - P$ because we may think of $M - P$ as the potential for further expansion. Thus,

$$\beta - \delta = k(M - P) \tag{2.27}$$

And

$$\frac{dP}{dt} = (\beta - \delta)P = k(M - P)P \tag{2.28}$$

(4) Competition situation:

If the birth rate β is a constant while the death rate δ is proportional to P, *i.e.*, $\delta = \alpha P$, then,

$$\frac{dP}{dt} = (\beta - \alpha P)P = kP(M - P) \tag{2.29}$$

Example 1

Suppose that at time $t = 0$, 10,000 people in a city with population $M = 100{,}000$ are attacked by a virus. After a week the number $P(t)$ of those attacked with the virus has increased to $P(1) = 20{,}000$. Assuming that $P(t)$ satisfies a logistic DE, when will 80% of the city's population have been attacked by that particular virus?

Solution

In order to eliminate the unnecessary manipulation of large numbers, we can consider the population unit of our problem as thousand. Now, substituting $P_0 = 10$ and $M = 100$ into the logistic DE we get

$$P(t) = \frac{MP_0}{P_0 + (M - P_0)\exp(-Mkt)} \tag{2.30}$$
$$= \frac{1000}{10 + 90\exp(-100kt)}$$

Then, substituting $P(t = 1) = 20$ into (2.30), we get

$$20 = \frac{100 \times 10}{10 + (100 - 10)\exp(-100k)}$$
$$\exp(-100k) = \frac{4}{9}$$
$$k = -\frac{1}{100}\ln\left(\frac{4}{9}\right)$$
$$100k = \ln\left(\frac{9}{4}\right)$$

With $P(t) = 80$, we have

$$80 = \frac{1000}{10 + 90\exp(-100kt)}$$

$$\exp(-100kt) = \frac{1}{36}$$

Thus,

$$t = \frac{\ln 36}{100k} = \frac{\ln 36}{\ln(9/4)} \approx 4.42$$

It follows that 80% of the population has been attacked by the virus after 4.42 weeks, *i.e.*, 31 days.

2.3.4   Doomsday vs. Extinction

Consider a population $P(t)$ of unsophisticated animals in which females rely solely on chance of encountering males for reproductive purposes. The rate of which encounters occur is proportional to the product of the number $P/2$ of males and $P/2$ of females, thus, at a rate proportional to P^2. We therefore assume the births occur at the rate of kP^2 (per unit time and k is constant). The birth rate, defined as births per unit time per population, is given by $\beta = kP$. If the death rate δ is a constant, the population DE becomes

$$\frac{dP}{dt} = kP^2 - \delta P \qquad (2.31)$$

Let $M = \delta/k > 0$, we have

$$\frac{dP}{dt} = kP(P - M) \qquad (2.32)$$

Note

The RHS of the DE (2.32) is the negative of the RHS of the logistic DE. The constant M is now called the threshold population. The behavior of the future population depends, critically, on whether the initial population P_0 is less than, or greater than, M.

Let's now solve this DE using a similar method to the logistic DE

$$\frac{dP}{P(P-M)} = kdt$$

$$\int_{P_0}^{P} \frac{dP}{P(P-M)} = \int_0^t kdt$$

$$\int_{P_0}^{P} \left(\frac{1}{P-M} - \frac{1}{P}\right) dP = Mkt \tag{2.33}$$

$$(\ln(P-M) - \ln P)|_{P_0}^{P} = Mkt$$

$$\left(\frac{P_0}{P}\right)\left(\frac{P-M}{P_0-M}\right) = \exp(Mkt)$$

Case 1: $P_0 > M$

Let

$$C_1 = \frac{P_0}{P_0 - M} \tag{2.34}$$

We know that $C_1 > 1$. To solve an explicit DE for P, we have

$$P = (P-M)C_1 \exp(-Mkt)$$

$$P(C_1 \exp(-Mkt) - 1) = MC_1 \exp(-Mkt)$$

$$P(t) = \frac{MC_1 \exp(-Mkt)}{C_1 \exp(-Mkt) - 1} \tag{2.35}$$

$$= \frac{M}{1 - \frac{1}{C_1} \exp(-Mkt)}$$

The population $P(t)$ in (2.35) approaches infinity if

$$1 - \frac{1}{C_1} \exp(-Mkt) = 0$$

or

$$T_{\text{Doom}} = \frac{\ln C_1}{kM} = \frac{1}{kM} \ln \frac{P_0}{P_0 - M} > 0 \qquad (2.36)$$

That is, $\lim\limits_{t \to T} P(t) = \infty \to$ Doomsday.

Case 2: $0 < P_0 < M$

Let

$$C_2 = \frac{P_0}{M - P_0} \qquad (2.37)$$

We know that $C_2 > 0$. A similar solution for P will give

$$P = (M - P)C_2 \exp(-Mkt)$$
$$P(1 + C_2 \exp(-Mkt)) = MC_2 \exp(-Mkt)$$
$$P(t) = \frac{MC_2 \exp(-Mkt)}{C_2 \exp(-Mkt) + 1} \qquad (2.38)$$

Since $C_2 > 0$, it follows that $\lim\limits_{t \to \infty} P(t) = 0 \to$ Extinction.

Problems

Problem 2.3.1 Suppose that when a certain lake is stocked with fish, the birth and death rates β and δ are both inversely proportional to $\sqrt{P}$.
(1) Show that

$$P(t) = \left(\frac{1}{2}kt + \sqrt{P_0}\right)^2$$

where k is constant.
(2) If $P_0 = 100$ and after 6 months there are 169 fish in the lake, how many will there be after 1 year?

Problem 2.3.2 Consider a prolific breed of rabbits whose birth and death rates, β and δ, are each proportional to the rabbit population $P = P(t)$, with $\beta > \delta$ and $P(t = 0) = P_0$.

(1) Show that

$$P(t) = \frac{P_0}{1 - kP_0 t}$$

where k is a constant.

(2) Prove that $P(t) \to \infty$ as $t \to 1/kP_0$. This is doomsday.

(3) Suppose that $P_0 = 6$ and there are 9 rabbits after ten months. When does doomsday occur?

(4) If $\beta < \delta$, what happens to the rabbit population in the long run?

Problem 2.3.3 Consider a population $P(t)$ satisfying the logistic DE

$$\frac{dP}{dt} = aP - bP^2$$

where $B = aP$ is the time rate at which births occur and $D = bP^2$ is the rate at which deaths occur. If initial population is $P(0) = P_0$ and B_0 births per month and D_0 deaths per month are occurring at time $t = 0$, show that the limiting population is

$$M = \frac{B_0 P_0}{D_0}$$

Problem 2.3.4 For the modified logistic population DE $P' = kP(\ln M - \ln P)$, $P(t = 0) = P_0$ where P is the population count as a function of time t; k, P_0 and M are positive constants.

(1) Solve the DE.

(2) Sketch the solution for $P_0 > M$ and show the limit when $t \to \infty$.

(3) Sketch the solution for $P_0 < M$ and show the limit when $t \to \infty$.

Problem 2.3.5 A tumor may be regarded as a population of multiplying cells. It is found empirically that birth rate of the cells in a tumor decreases exponentially with time, so that

$$\beta(t) = \beta_0 \exp(-\alpha t)$$

where α and β_0 are positive constants. Thus, we have the IVP

$$\begin{cases} \frac{dP}{dt} = \beta_0 \exp(-\alpha t) P \\ P(0) = P_0 \end{cases}$$

Solving this IVP yields

$$P(t) = P_0 \exp\left(\frac{\beta_0}{\alpha}(1 - \exp(-\alpha t))\right)$$

leading to the finite limiting population as $t \to \infty$,

$$\lim_{t \to \infty} P(t) = P_0 \exp\left(\frac{\beta_0}{\alpha}\right)$$

Problem 2.3.6 Consider two population functions $P_1(t)$ and $P_2(t)$, both of which satisfy the logistic DE with the limiting population M, but with different values k_1 and k_2 of the constant k in DE

$$\frac{dP}{dt} = kP(M - P)$$

Assume that $k_1 < k_2$. Which population approaches M more rapidly? You can reason geometrically by examining slope fields (especially if appropriate software is available), symbolically by analyzing the solution given in DE

$$P(t) = \frac{MP_0}{P_0 + (M - P_0)\exp(-Mkt)}$$

or numerically by substituting successive values of t.

Problem 2.3.7 Suppose that the fish population $P(t)$ in a lake is attacked by a disease at time $t = 0$, with the result that the fish cease to reproduce (so that the birth rate is $\beta = 0$) and the death rate δ (deaths per week per fish) is thereafter proportional to $1/\sqrt{P}$. If there were initially 900 fish in the lake and 441 were left after 6 weeks, how long did it take all the fish in the lake to die?

Problem 2.3.8 Consider a rabbit population $P(t)$ satisfying the logistic DE $P' = aP - bP^2$, where $B = aP$ is the time rate at which births occur and $D = bP^2$ is the rate at which deaths occur. If the initial population is 120 rabbits and there are 8 births per month and 6 deaths per month occurring at time $t = 0$, how many months does it take for $P(t)$ to reach 95% of the limiting population M? Sketch $P(t)$ with given parameters.

Problem 2.3.9 Consider a prolific animal whose birth and death rates, β and α, are each proportional to its population with $\beta > \alpha$ and $P_0 = P(t = 0)$.
(1) Compute the population as a function of time with the parameters and IC given.
(2) Find the time for doomsday.

(3) Suppose that $P_0 = 2011$ and that there are 4027 animals after 12 time units (days or months or years), when is the doomsday?

(4) If $\beta < \alpha$, compute the population limit when time approaches infinity.

Problem 2.3.10 Fish population $P(t)$ in a lake is attacked by a disease (such as human beings who eats them) at time $t = 0$, with the result that the fish cease to reproduce (so that the birth rate is $\beta = 0$) and the death rate δ (death per week per fish) is thereafter proportional to $P^{-1/2}$. If there were initially 4000 fish in the lake and 2014 were left after 11 weeks, how long did it take all the fish in the lake to die? Can you change the "2014" to a different number such that the fish count never changes with time? To what number if so?

2.4   Acceleration-Velocity Model

2.4.1   Newton's Laws of Motion

Newton's three physical laws describe the motion of a body in response to external forces and these laws laid the foundation for classical mechanics. Newton published these laws in Mathematical Principles of Natural Philosophy in 1687 at the age of 44.

Let the motion of a particle (a mathematical dot) of mass m, along a straight line, be described by its position function $x = f(t)$. Then, the velocity of the particle is

$$v(t) = \frac{dx}{dt} = x'(t) = f'(t) \tag{2.39}$$

and the acceleration is

$$a = \frac{dv}{dt} = \frac{d^2x}{dt^2} = v'(t) = x''(t) = f''(t) \tag{2.40}$$

If a force $F(t)$ acts on a particle and is directed along its line of motion, then,

$$F(t) = ma \tag{2.41}$$

with initial position $x(t = 0) = x_0$ and initial speed $v(t = 0) = x'(t = 0) = v_0$.

We can establish an Equation of Motion (EoM) for the particle as

$$\begin{cases} m\dfrac{d^2x}{dt^2} = F(t) \\ v(t = 0) = x'(t = 0) = v_0 \\ x(t = 0) = x_0 \end{cases} \tag{2.42}$$

Solution

Let's suppose that we are given the acceleration of the particle and that it is a constant.

$$\frac{dv}{dt} = a \tag{2.43}$$

So to find v we integrate both sides of the above DE.

$$v = \int a\,dt$$
$$= at + C_1 \tag{2.44}$$

We know that the initial velocity at time $t = 0$ is v_0. So, after plugging this information into the above equation, we get $C_1 = v_0$ and, thus,

$$v = \frac{dx}{dt} = at + v_0 \tag{2.45}$$

Now, integrating the speed, we get the position of the particle as

$$x(t) = \int v(t)\,dt$$
$$= \int (at + v_0)\,dt \tag{2.46}$$
$$= \frac{1}{2}at^2 + v_0 t + C_2$$

Applying the IC $x(t = 0) = x_0$, we get $C_2 = x_0$.

Finally, we get the position of the particle as a function of time (the IV) and other parameters such as the initial position, speed, and the acceleration

$$x(t) = \frac{1}{2}at^2 + v_0 t + x_0 \tag{2.47}$$

So, if given the acceleration (through the force), the initial position and speed, we can determine the position and speed of the particle at any time.

2.4.2   Velocity and Acceleration Models

We know that

$$v = \frac{dy}{dt} \qquad\qquad y = \int v dt$$

$$a = \frac{dv}{dt} = \frac{d^2 y}{dt^2} \qquad v = \int a dt \qquad y = \int \int a \, dt$$

where

$$\left.\begin{array}{l} a(t) = \text{acceleration} \\ v(t) = \text{velocity} \\ y(t) = \text{position} \end{array}\right\} \text{ of the body}$$

Newton's 2nd law of motion, $F = ma$, implies that to gain an acceleration a for a mass m, one must exert force F.

Example 1

Suppose that a ball is thrown upwards from the ground ($y_0 = 0$) with initial velocity $v_0 = 49$ m/s. Neglecting air resistance, calculate the max-height it can reach and the time required for it to come back to its original position.

Solution

The body, if thrown upwards, reaches its max-height when the velocity is zero, *i.e.*, $y_H = y(v(t) = 0)$. The time required for the body to return back to its original spot is calculated when its position function is zero again, *i.e.*, $y = 0$. From the Newton's Law,

$$m\frac{dv}{dt} = F_G \qquad\qquad (2.48)$$

where $F_G = -mg$ is the (downward-directed) force of gravity, and the gravitational acceleration is $g \approx 9.8 \, m/s^2$. Using the above equation and applying the IC, we have

$$\frac{dv}{dt} = -9.8$$

$$v(t) = -9.8t + 49$$

Thus, the ball's height function $y(t)$ is given by

$$y(t) = \int v dt$$

$$= \int (-9.8t + 49) dt$$

$$= -4.9t^2 + 49t + y_0$$

Since $y_0 = 0$, we have

$$y(t) = -4.9t^2 + 49t$$

So, the ball reaches its max-height when $v = 0$

$$-9.8t + 49 = 0$$

$$t = 5$$

Thus, its max-height is

$$y_H = y(5) = 122.5$$

The time required for the ball to return is

$$y(t) = 0$$

$$-4.9t^2 + 49t = 0$$

$$t = 10$$

2.4.3   Air Resistance Model

Now, let's consider the air resistance in the problems. In order to account for air resistance, the force F_R exerted by air, on the moving mass m must be added. Thus,

$$m\frac{dv}{dt} = F_G + F_R \qquad (2.49)$$

It is found that $F_R = kv^p$ where $1 \le p \le 2$, for most normal air, and k depends on the size and shape of the body as well as the density and viscosity of the air. Usually, $p = 1$ is for motions at relatively low speeds while $p = 2$ is for those at higher speeds.

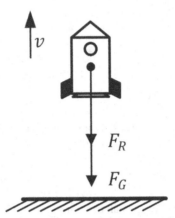

Figure 2.6 A "rocket" model with the consideration of air resistance.

The vertical motion of a projectile of mass m near the surface of the Earth is subject to two forces: a downward gravitational force F_G and a frictional force F_R, due to air resistance, that is directed at the opposite direction of projectile's motion. We have, $F_G = -mg$ and $F_R = -kv$ where k is a positive constant and v is the velocity of the projectile. Note that the negative sign for F_R makes the force positive (an upward force) when the body falls (v is negative) and negative (a downward force) if the projectile moves upwards (v is positive). So, the total force on the projectile is

$$F = F_R + F_G = -kv - mg \qquad (2.50)$$

Therefore,

$$m\frac{dv}{dt} = -kv - mg \qquad (2.51)$$

Let

$$\rho = \frac{k}{m} > 0 \qquad (2.52)$$

The DE becomes

$$\frac{dv}{dt} = -\rho v - g \tag{2.53}$$

This DE is a separable DE that has a solution

$$v(t) = \left(v_0 + \frac{g}{\rho}\right)\exp(-\rho t) - \frac{g}{\rho} \tag{2.54}$$

where v_0 is the initial velocity of the projectile $v(0) = v_0$. For long-time limit, we have

$$v_T = \lim_{t\to\infty} v(t) = -\frac{g}{\rho} = -\frac{mg}{k} \tag{2.55}$$

Thus, the speed of the projectile with air resistance does not increase indefinitely while falling. In fact, it approaches a finite limiting speed or terminal speed

$$|v_T| = \frac{g}{\rho} = \frac{mg}{k} \tag{2.56}$$

The time-varying velocity of the projectile, expressed in the terminal speed, is

$$v(t) = (v_0 - v_T)\exp(-\rho t) + v_T \tag{2.57}$$

The time-varying position of the projectile is the integration of (2.54):

$$\begin{aligned} y(t) &= \int_0^t v(\tau)d\tau \\ &= \int_0^t \left((v_0 - v_T)\exp(-\rho\tau) + v_T\right)d\tau \\ &= \frac{1}{\rho}(v_0 - v_T)(1 - \exp(-\rho t)) + v_T t + y_0 \end{aligned} \tag{2.58}$$

where we used the projectile's initial height $y(0) = y_0$ which can be set to 0 for convenience. Thus, the projectile's position as a function of time is

$$y(t) = v_T t + \frac{1}{\rho}(v_0 - v_T)(1 - \exp(-\rho t)) \qquad (2.59)$$

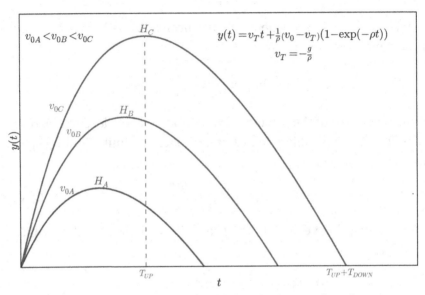

Figure 2.7 The heights as a function of time for three initial speeds.

We can further examine the velocity and the position formulas of the projectile. It is interesting to compute the time T_{UP} for the projectile to reach the highest point at which the projectile reverses its direction of motion and the height H the projectile has reached as shown in Figure 2.7.

First, we know the projectile's velocity at the max-height is 0, *i.e.,*

$$v(T_{UP}) = \left(v_0 + \frac{g}{\rho}\right)\exp(-\rho T_{UP}) - \frac{g}{\rho} = 0 \qquad (2.60)$$

Solving this equation for T_{UP}, we get

$$T_{UP} = \frac{1}{\rho}\ln\left(1 + \frac{\rho v_0}{g}\right) \tag{2.61}$$

Plugging $\rho = \frac{k}{m} = -\frac{g}{v_T}$ into (2.61), we get

$$T_{UP} = \frac{m}{k}\ln\left(1 + \frac{kv_0}{mg}\right) \tag{2.62}$$

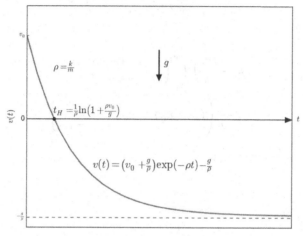

Figure 2.8 The projectile's velocity as a function of time. Velocity changes sign at t_H or T_{UP} when it reaches its highest point.

Second, we can find the max-height. The most straightforward method is to plug the above T_{UP} into the position function. But this is tedious. An alternate method is to re-formulate the original DE in terms of positions and velocity as follows,

$$m\frac{dv}{dt}\frac{dy}{dy} = -kv - mg \tag{2.63}$$

Now, regrouping the LHS and recognizing $v = \frac{dy}{dt}$, we get

$$m\frac{dy}{dt}\frac{dv}{dy} = -kv - mg \tag{2.64}$$

Or,

$$mv\frac{dv}{dy} = -kv - mg \tag{2.65}$$

$$\begin{aligned}
dy &= -\frac{mv}{kv + mg}dv \\
&= -\left(\frac{m}{k}\right)\frac{v}{v + \frac{mg}{k}}dv \\
&= -\left(\frac{m}{k}\right)\left(1 + \frac{v_T}{v - v_T}\right)dv
\end{aligned} \tag{2.66}$$

Thus,

$$\begin{aligned}
y &= -\left(\frac{m}{k}\right)\int_{v_0}^{v}\left(1 + \frac{v_T}{v - v_T}\right)dv \\
&= \frac{v_T}{g}\left((v - v_0) + v_T\ln\frac{v - v_T}{v_0 - v_T}\right)
\end{aligned} \tag{2.67}$$

where $v_T = -\frac{mg}{k}$ was used.

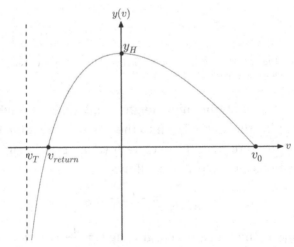

Figure 2.9 The projectile's height as a function of the velocity. The velocity changes sign at t_H or y_H where it reaches the max-height.

Many interesting observations can be made from this formula as shown in Figure 2.9:

(1) We can find the two speeds v at $y = 0$ by solving

$$(v - v_0) + v_T \ln \frac{v - v_T}{v_0 - v_T} = 0 \qquad (2.68)$$

which has two roots for speeds v. The first, $v = v_0$, is the speed at the projectile's launch spot, as expected. The second, impossible to obtain analytically, is the projectile's speed when it returns to the launch spot, labeled as $v = v_{\text{return}}$ in Figure 2.9.

(2) We can find the projectile's max-height H by setting $v = 0$,

$$\begin{aligned}
H &= \frac{v_T}{g}\left((0 - v_0) + v_T \ln \frac{0 - v_T}{v_0 - v_T}\right) \\
&= -\frac{v_T}{g}\left(v_0 + v_T \ln\left(1 - \frac{v_0}{v_T}\right)\right) \qquad (2.69) \\
&= \frac{m}{k}\left(v_0 - \frac{mg}{k}\ln\left(1 + \frac{kv_0}{mg}\right)\right)
\end{aligned}$$

Using (2.62), we get

$$H = \frac{m}{k}(v_0 - gT_{\text{UP}}) \qquad (2.70)$$

(3) We can find the terminal speed $v = v_T$ as we notice

$$y_T = \lim_{v \to v_T}\left(-\frac{m}{k}\left((v - v_0) + v_T \ln \frac{v - v_T}{v_0 - v_T}\right)\right) \qquad (2.71)$$

$$\to -\infty$$

Remarks

(1) The time required to reach the max-height is

$$T_{\text{UP}} = \frac{m}{k}\ln\left(1 + \frac{kv_0}{mg}\right) = \frac{1}{\rho}\ln\left(1 - \frac{v_0}{v_T}\right) \qquad (2.72)$$

(1a) If the resistance is huge, it takes less time to reach the max-height. For an extreme case $k \to \infty$, we have

$$\lim_{k \to \infty} T_{UP} = \lim_{k \to \infty} \frac{m}{k} \ln\left(1 + \frac{kv_0}{mg}\right) = 0 \qquad (2.73)$$

(1b) If the resistance is tiny, the time to reach the max-height must be evaluated with care. For an extreme case $k \to 0$, we have

$$\begin{aligned} \lim_{k \to 0} T_{UP} &= \lim_{k \to 0} \frac{m}{k} \ln\left(\frac{kv_0}{mg} + 1\right) \\ &= \frac{m}{k} \frac{kv_0}{mg} \qquad (2.74) \\ &= \frac{v_0}{g} \end{aligned}$$

which is consistent with the resistance-free case.

(1c) If the initial speed is small, then

$$\begin{aligned} \lim_{v_0 \to 0} T_{UP} &= \lim_{v_0 \to 0} \frac{m}{k} \ln\left(\frac{kv_0}{mg} + 1\right) \\ &= \lim_{v_0 \to 0} \frac{m}{k} \frac{kv_0}{mg} \qquad (2.75) \\ &= \frac{v_0}{g} \end{aligned}$$

which is also consistent with the resistance-free case because the initial speed is so small that the projectile does not have a chance to experience the air resistance.

(1d) If the initial speed is huge, then

$$\lim_{v_0 \to \infty} T_{UP} = \lim_{v_0 \to \infty} \frac{m}{k} \ln\left(\frac{kv_0}{mg} + 1\right)$$

$$= \lim_{v_0 \to \infty} \frac{m}{k} \frac{kv_0}{mg} \qquad (2.76)$$

$$= \frac{v_0}{g}$$

Surprisingly, it is the same formula as that for small initial speed. But, we must not over-claim anything here. Our original assumption for air resistance kv is not applicable at very high speed in the first place.

(2) The max-height is

$$H = \frac{m}{k}\left(v_0 - \frac{mg}{k}\ln\left(1 + \frac{kv_0}{mg}\right)\right) \qquad (2.77)$$

and similar observations can be made as above. For example, if the initial speed $v_0 \to 0$, we find

$$\lim_{v_0 \to 0} H = \lim_{v_0 \to 0} \frac{m}{k}\left(v_0 - \frac{mg}{k}\ln\left(1 + \frac{kv_0}{mg}\right)\right) = 0 \qquad (2.78)$$

which makes good sense: If the initial speed $v_0 = 0$, you do not travel anywhere.

(3) The time to travel down from the max-height to the launch spot is T_{DOWN} which can be calculated by integrating the velocity formula wrt time to travel the following distance, $i.e.,$ height,

$$H = \frac{m}{k}\left(v_0 - \frac{mg}{k}\ln\left(1 + \frac{kv_0}{mg}\right)\right) \qquad (2.79)$$

Alternatively, one can find the total time needed to travel up and down by setting the total travel distance to be 0, $i.e.,$

$$y(t) = v_T t + \frac{1}{\rho}(v_0 - v_T)(1 - \exp(-\rho t)) = 0 \quad (2.80)$$

in which $y_0 = 0$ is set for convenience. The above equation has two roots for $y(t) = 0$. The first root, $t = 0$, is the time the projectile starts to launch. The second root, $T_F = T_{UP} + T_{DOWN}$, is the time when the projectile returns to the launch spot, *i.e.*, the total time the projectile is in the air. Solving

$$v_T T_F + \frac{1}{\rho}(v_0 - v_T)(1 - \exp(-\rho T_F)) = 0 \quad (2.81)$$

produces T_F which is not of an analytical form, in general. Thus, we can't find the formula for T_{DOWN}. One can argue, *e.g.*, by numerical analysis, that $T_{DOWN} > T_{UP}$. The projectile takes longer to travel down than up.

Example 1
We will again consider the ball that is thrown from the ground level with initial velocity $v_0 = 49\, m/s$. But now we will take the air resistance $\rho = 0.04$ into account and find out the max-height and the total time that the ball spent in the air.

Solution
We find out the terminal velocity

$$v_T = -\frac{g}{\rho} = -\frac{9.8}{0.04} = -245$$

Substituting the values of $y_0 = 0$, $v_0 = 49$ and $v_r = 245$ in the equation of velocity and position we get

$$v(t) = (v_0 - v_T)\exp(-\rho t) + v_T$$
$$= \big(49 - (-245)\big)\exp(-0.04t) + (-245)$$
$$= 294\exp\left(-\frac{t}{25}\right) - 245$$

$$y(t) = y_0 + v_T t + \frac{1}{\rho}(v_0 - v_T)(1 - \exp(-\rho t))$$

$$= 0 + (-245)t + \frac{1}{0.04}(49 + 245)(1 - \exp(-0.04t))$$

$$= 7350 - 245t - 7350 \exp\left(-\frac{t}{25}\right)$$

The ball reaches max-height when its velocity is 0. That is

$$v(t) = 0$$

$$294 \exp\left(-\frac{t}{25}\right) - 245 = 0$$

$$t_m = -25 \ln\left(\frac{245}{294}\right)$$

$$\approx 4.558$$

Thus, the time required for the ball to reach the max-height is $t_m = 4.558$ seconds with air resistance as compared to 5 seconds without air resistance. We apply the value of t_m in the position function to compute the max-height

$$y_{max} = y(t_m)$$

$$= 7350 - 245t_m - 7350 \exp\left(-\frac{t_m}{25}\right)$$

$$\approx 108.3 \text{ m}$$

Thus, with air resistance, the max-height is 108.3 m. In contrast, without air resistance, the max-height is 122.5 m.

The time required for the ball to return to its original position is 9.41 seconds, compared to 10.00 seconds from the previous example.

Example 2

A guy dives from a cliff of height H at an initial speed in vertical direction $v_I = 0$ (for simplicity). During the first phase of his fall, he wraps his arms. During the second phase, he stretches his arms to slow himself down so that his touchdown speed at the water's surface is v_F. The air drag coefficients are k and βk during the first and the second phases, respectively. The air drag force is always proportional to the moving speed. At what height should the guy open his arm so that his touchdown speed is v_F? Compute the total falling time.

Solution

For Phase I (wrapped arms), assuming the starting point $y = 0$ and extending downward, we establish the following boundary value problem (BVP):

$$\begin{cases} m\dfrac{dv}{dy}v = mg - kv \\ v(y = 0) = 0 \\ v(y = h_1) = v_1 \end{cases}$$

We illustrate the concept as below:

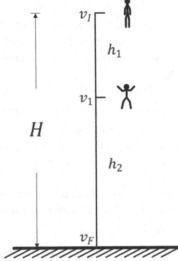

Figure 2.10 A diver leaps from height H at an initial speed v_I with closed arms until reaching speed v_1, at which point the diver opens his arms expecting a touchdown speed v_F.

The BVP can be converted to two definite integrals

$$\int_0^{v_1} m\frac{dv}{mg - kv}v = \int_0^{h_1} dy$$

Or

$$-\frac{m}{k}\int_0^{v_1}\left(1 + \frac{mg}{k}\frac{1}{v - \frac{mg}{k}}\right)dv = \int_0^{h_1} dy$$

Let $v_T = \frac{mg}{k}$, we get

$$-\frac{v_T}{g} \int_0^{v_1} \left(1 + \frac{v_T}{v - v_T}\right) dv = \int_0^{h_1} dy$$

Or

$$v_1 + v_T \ln \left| \frac{v_1}{v_T} - 1 \right| = -\frac{gh_1}{v_T}$$

For Phase II (open arms), we have the following BVP:

$$\begin{cases} m\dfrac{dv}{dy}v = mg - \beta kv \\ v(y = h_1) = v_1 \\ v(y = H) = v_F \end{cases}$$

The BVP can be converted to two definite integrals

$$\int_{v_1}^{v_F} m\frac{dv}{mg - \beta kv}v = \int_{h_1}^{H} dy$$

we get

$$(v_F - v_1) + \frac{v_T}{\beta} \ln \left| \frac{\beta v_F - v_T}{\beta v_1 - v_T} \right| = -\frac{\beta g(H - h_1)}{v_T}$$

Solving these two AEs, one can find h_1 at which (height) the guy should open his arms.

$$\begin{cases} v_1 + v_T \ln \left| \dfrac{v_1}{v_T} - 1 \right| = -\dfrac{gh_1}{v_T} \\ (v_F - v_1) + \dfrac{v_T}{\beta} \ln \left| \dfrac{\beta v_F - v_T}{\beta v_1 - v_T} \right| = -\dfrac{\beta g(H - h_1)}{v_T} \end{cases}$$

Obviously, it's tedious to obtain the explicit expression for h_1. In principle, one can find v_1 from the Phase I solution and plug it into the Phase II solution to find h_1.

Next, we find the expression for finding the total travel time.

For Phase I (wrapped arms), we have the following BVP:

$$\begin{cases} m\dfrac{dv}{dt} = mg - kv \\ v(t = 0) = 0 \\ v(t = t_1) = v_1 \end{cases}$$

The BVP can be converted to two definite integrals

$$-\frac{v_T}{g}\int_0^{v_1} \frac{1}{v - v_T}\, dv = \int_0^{t_1} dt$$

Thus,

$$t_1 = -\frac{v_T}{g}\ln\left|\frac{v_1}{v_T} - 1\right|$$

where v_1 was found previously.

For Phase II (open arms), we have the following BVP:

$$\begin{cases} m\dfrac{dv}{dt} = mg - \beta kv \\ v(t = 0) = v_1 \\ v(t = t_2) = v_F \end{cases}$$

The BVP can be converted to two definite integrals

$$\int_{v_1}^{v_F} \frac{m}{mg - \beta kv}\, dv = \int_0^{t_2} dt$$

Thus,

$$t_2 = \frac{v_T}{\beta g}\ln\left|\frac{\beta v_1 - v_T}{\beta v_F - v_T}\right|$$

Therefore, the total time is

$$T = t_1 + t_2 = -\frac{v_T}{g}\ln\left|\frac{v_1}{v_T} - 1\right| + \frac{v_T}{\beta g}\ln\left|\frac{\beta v_1 - v_T}{\beta v_F - v_T}\right|$$

Example 3

On a straight stretch of road of length L, there is a stop sign at either end at which you must stop, although there is no speed limit between the signs. We further assume the air resistance is μv when your travel speed is v

during the entire stretch. Your car's natural maximum acceleration force is ma where m is the car's mass and its maximum brake force is mb. The values of m, a, b, and L are given constants. Design a travel scheme to minimize the total travel time in this stretch.

Solution

The simplest way to minimize the total travel time is to apply maximum acceleration during Phase 1 and to apply maximum deceleration during Phase 2.

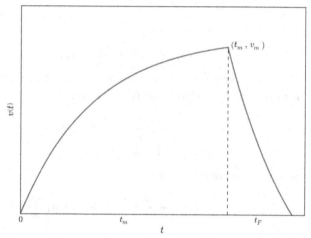

Figure 2.11 The car's speed as a function of time. The car applies maximum acceleration from $t = 0$ to $t = t_m$ and maximum deceleration from $t = t_m$ to $t = t_m+t_F$ when the car comes to a full stop. The car's total travel time is, thus, t_m+t_F.

First, let's consider the problem in (t, v) dimensions.

Phase 1: Accelerating the car to reach some strategic spot (t_m, v_m)

$$\begin{cases} m\dfrac{dv}{dt} = ma - \mu v \\ v(t = 0) = 0 \end{cases}$$

Solving this simple IVP, we get

$$v(t) = \frac{ma}{\mu}\left(1 - \exp\left(\frac{-\mu t}{m}\right)\right)$$

Let $v_a = \frac{ma}{\mu}$, we get

$$v(t) = v_a\left(1 - \exp\left(-\frac{a}{v_a}t\right)\right)$$

a monotonically increasing function of time, *i.e.,* the car continues accelerating.

$$v_m = v_a\left(1 - \exp\left(-\frac{a}{v_a}t_m\right)\right)$$

and

$$t_m = -\frac{v_a}{a}\ln\left(1 - \frac{v_m}{v_a}\right)$$

Phase 2: Decelerating the car from (t_m, v_m) to $(t_F, 0)$. Note: for manipulation conveniences, we reset the clock.

$$\begin{cases} m\dfrac{dv}{dt} = -mb - \mu v \\ v(t = 0) = v_m \end{cases}$$

Solving the above IVP, we get

$$v(t) = -\frac{mb}{\mu} + \left(v_m + \frac{mb}{\mu}\right)\exp\left(\frac{-\mu t}{m}\right)$$

Let $v_b = \frac{mb}{\mu}$, we get

$$v(t) = -v_b + (v_m + v_b)\exp\left(-\frac{b}{v_b}t\right)$$

Decelerating the car for t_F amount of time and, then, reaching at full stop, *i.e.,*

$$0 = -v_b + (v_m + v_b)\exp\left(-\frac{b}{v_b}t_F\right)$$

Thus,

$$t_F = \frac{v_b}{b}\ln\left(1 + \frac{v_m}{v_b}\right)$$

The total travel time is $T = t_m + t_F$.

Second, we consider the problem in (x, v) dimensions to find the condition for the car to stop at the second stop sign, *i.e.*, $(L - x_m, 0)$.

Phase 1 (Accelerating)

$$\begin{cases} mv\dfrac{dv}{dx} = ma - \mu v \\ v(x = 0) = 0 \\ v(x = x_m) = v_m \end{cases}$$

Yielding relationship

$$v_m + v_a \ln\left(\frac{v_m}{v_a} - 1\right) = -\frac{a}{v_a} x_m$$

Phase 2 (Decelerating)

$$\begin{cases} mv\dfrac{dv}{dx} = -mb - \mu v \\ v(x = 0) = v_m \\ v(x = L - x_m) = 0 \end{cases}$$

Solving this simple IVP (ignoring the second condition, for now), we get

$$v - v_b \ln(v + v_b) = -\frac{b}{v_b} x + v_m - v_b \ln(v_m + v_b)$$

Simplifying the above, we get

$$(v - v_m) + v_b \ln\left(\frac{v_m + v_b}{v + v_b}\right) = -\frac{b}{v_b} x$$

At $(L - x_m, 0)$, the car stops at the second stop sign

$$(0 - v_m) + v_b \ln\left(\frac{v_m + v_b}{0 + v_b}\right) = -\frac{b}{v_b}(L - x_m)$$

$$v_m - v_b \ln\left(\frac{v_m}{v_b} + 1\right) = \frac{b}{v_b}(L - x_m)$$

Solving the following two equations, implicitly:

$$\begin{cases} v_m + v_a \ln\left(\dfrac{v_m}{v_a} - 1\right) = -\dfrac{a}{v_a} x_m \\ v_m - v_b \ln\left(\dfrac{v_m}{v_b} + 1\right) = \dfrac{b}{v_b}(L - x_m) \end{cases}$$

by subtracting the two equations, we get

$$\ln\left(\left(\frac{v_m}{v_a}-1\right)^a\left(\frac{v_m}{v_b}+1\right)^b\right) = -\left(\frac{\mu}{m}\right)^2 L$$

$$\left(\frac{v_m}{v_a}-1\right)^a\left(\frac{v_m}{v_b}+1\right)^b = \exp\left(-\left(\frac{\mu}{m}\right)^2 L\right)$$

In principle, we can find v_m. If we want, we can also get x_m.

$$T = t_m + t_F$$
$$= -\frac{m}{\mu}\ln\left(1 - \frac{v_m}{v_a}\right) + \frac{m}{\mu}\ln\left(1 + \frac{v_m}{v_b}\right)$$
$$= \frac{m}{\mu}\ln\left(\frac{1 + \frac{v_m}{v_b}}{1 - \frac{v_m}{v_a}}\right)$$
$$= \frac{m}{\mu}\ln\left(\frac{v_a\, v_b + v_m}{v_b\, v_a - v_m}\right)$$
$$= \frac{m}{\mu}\ln\left(\frac{1 + \frac{\mu}{mb}v_m}{1 - \frac{\mu}{ma}v_m}\right)$$

For a very special case in which the acceleration force is equal to the brake force, i.e., $ma = mb$ and $a = b$, we get v_m by solving

$$a\ln\left(\left(\frac{v_m}{v_a}\right)^2 - 1\right) = -\left(\frac{a}{v_a}\right)^2 L$$

whose root is

$$v_m = v_a\sqrt{1 + \exp\left(-\frac{aL}{v_a^2}\right)}$$

Using the original parameters, we get

$$v_m = \frac{ma}{\mu}\sqrt{1 + \exp\left(-\left(\frac{\mu}{m}\right)^2\frac{L}{a}\right)}$$

One may make numerous analyses on this formula. For example,

$$\lim_{L\to\infty} v_m = \lim_{L\to\infty}\frac{ma}{\mu}\sqrt{1 + \exp\left(-\left(\frac{\mu}{m}\right)^2\frac{L}{a}\right)}$$
$$= \frac{ma}{\mu}$$

This is the max-speed one can achieve, *i.e.*, if the stretch is super long ($L \to \infty$), the strategy for minimizing time is to accelerate your car to the max-speed and cruise until you need to decelerate to allow you to stop at the second stop sign. No surprise!

2.4.4 Gravitational Acceleration

According to Newton's law of universal gravitation first published on July 5, 1687, the gravitational force of attraction between two point-masses M and m located at a distance r apart is given by

$$F = \frac{GMm}{r^2} \tag{2.82}$$

where G is gravitational constant given as $G = 6.67 \times 10^{-11}\ N \cdot (m/kg)^2$ in MKS units.

Also, the initial velocity necessary for a projectile to escape from the Earth is called the Earth's escape velocity and is found to be

$$v_{ESC} = \sqrt{\frac{2GM}{R}} \tag{2.83}$$

where R is the radius of the Earth.

Usually, one can compute the escape speed using two highly related methods. The first is by solving the following IVP that is more relevant

$$\begin{cases} m\dfrac{dv}{dt} = -\dfrac{GMm}{(y+R)^2} \\ v(y=0) = v_0 \end{cases} \tag{2.84}$$

where v is the speed of the projectile at height y from the surface of the Earth. One can convert the DE into

$$v\frac{dv}{dy} = -\frac{GM}{(y+R)^2}$$

$$vdv = -\frac{GM}{(y+R)^2}dy$$

$$\int_{v_0}^{v} vdv = -\int_{0}^{y} \frac{GM}{(y+R)^2}dy \qquad (2.85)$$

$$\frac{1}{2}v^2 - \frac{1}{2}v_0^2 = \frac{GM}{y+R} - \frac{GM}{R}$$

where v_0 is the initial speed of the projectile. The last formula leads to the second method for solving this problem: energy conservation.

The total energy, the sum of the kinetic energy and the potential energy, at launch from Earth's surface

$$\frac{1}{2}mv_0^2 - \frac{GMm}{R} \qquad (2.86)$$

The total energy at any later time, or height, from the Earth's surface

$$\frac{1}{2}mv^2 - \frac{GMm}{y+R} \qquad (2.87)$$

According to the law of conservation of energy, assuming there are no other energies added to, or removed from, the projectile, the above two energies (2.86) and (2.87) must be equal, *i.e.*,

$$\frac{1}{2}mv_0^2 - \frac{GMm}{R} = \frac{1}{2}mv^2 - \frac{GMm}{y+R} \qquad (2.88)$$

which is the solution of the first method with a little cosmetic work.

We can compute the explicit form of the speed as a function of the height

$$v = \left(v_0^2 + \frac{2GM}{y+R} - \frac{2GM}{R}\right)^{\frac{1}{2}} \qquad (2.89)$$

If the projectile approaches $y \to \infty$ and its speed still maintains $v \to 0^+$, it can escape. So, finding the escape speed will require solving

$$\lim_{y \to \infty} \left(v_0^2 + \frac{2GM}{y+R} - \frac{2GM}{R} \right)^{\frac{1}{2}} = 0 \tag{2.90}$$

$$v_0^2 - \frac{2GM}{R} = 0$$

Thus, the escape speed is

$$v_{\text{ESC}} = \sqrt{\frac{2GM}{R}} \tag{2.91}$$

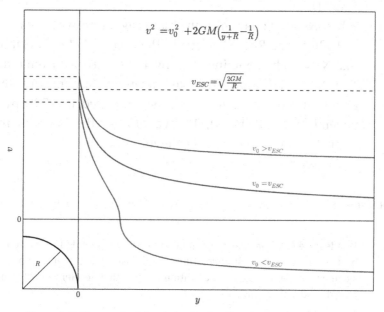

Figure 2.12 The speeds of the projectile after launching at three different initial speeds: $v_0 < v_{\text{ESC}}$, $v_0 = v_{\text{ESC}}$ and $v_0 > v_{\text{ESC}}$.

Figure 2.12 illustrates the speeds of the projectile after launching at three different initial speeds: $v_0 < v_{\text{ESC}}$, $v_0 = v_{\text{ESC}}$ and $v_0 > v_{\text{ESC}}$. At $v_0 < v_{\text{ESC}}$, the projectile does not have enough speed to escape; its speed becomes zero before getting far enough. At $v_0 > v_{ESC}$, the projectile has enough speed to escape; its speed is bigger than zero after getting far enough. At the critical case of $v_0 = v_{ESC}$, the projectile's speed at the most remote point just turns zero, *i.e.*, it can escape if it wishes to as there is no force there.

Remarks

(1) Escape speed wrt Earth's gravity is 11.2 (km/s) or 40,320 (km/h) or 25,054 (mph). Escape speed wrt Mars' gravity is 5.0 (km/s) or 18,000 (km/h) or 11,184 (mph). Escape speed wrt the Moon's gravity is 2.4 (km/s) or 8,640 (km/h) or 5,369 (mph).

(2) When Newton's law of universal gravitation was presented in 1686 to the Royal Society, R. Hooke (1635-1703) claimed that Newton had obtained the inverse square law from him.

(3) The first laboratory test of Newton's law of universal gravitation between masses was conducted in 1798 by H. Cavendish (1731-1810), 111 years after the law was first published and 71 years after Newton's death.

Problems

Problem 2.4.1 A car traveling at 60 mph (88ft/s) skids 176 ft after its brakes are suddenly applied. Under the assumption that the braking system provides constant deceleration, what is that deceleration? For how long does the skid continue?

Problem 2.4.2 Find the position function $x(t)$ of a moving particle with the given acceleration $a(t)$, initial position $x(0) = x_0$ and initial velocity $v(0) = v_0$.

$$\begin{cases} a(t) = 50\sin(5t) \\ \quad v_0 = -10 \\ \quad x_0 = 8 \end{cases}$$

Problem 2.4.3 Find the position function $x(t)$ of a moving particle with the given acceleration $a(t)$, initial position $x(0) = x_0$ and initial velocity $v(0) = v_0$.

$$\begin{cases} a(t) = \dfrac{1}{(t+1)^n} \quad n \geq 3 \\ \quad v_0 = 0 \\ \quad x_0 = 0 \end{cases}$$

Problem 2.4.4 Suppose that a body moves horizontally through a medium whose resistance is proportional to its velocity v, so that $v' = -kv$.
(1) Show that its velocity and position at time t are given by

$$v(t) = v_0 \exp(-kt) \ and \ x(t) = x_0 + \left(\frac{v}{k}\right)(1 - \exp(-kt))$$

(2) Conclude that the body travels only a finite distance, and find that distance.

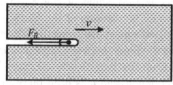

Figure 2.13 The motion model for Problem 2.4.1 and Problem 2.4.5.

Problem 2.4.5 Consider a body that moves horizontally through a medium whose resistance results in the relationship $v' = -kv^{\frac{3}{2}}$. Show that, for given initial velocity v_0 and resistance constant k,

$$v(t) = \frac{4v_0}{\left(kt\sqrt{v_0} + 2\right)^2}$$

$$x(t) = x_0 + \frac{2}{k}\sqrt{v_0}\left(1 - \frac{2}{kt\sqrt{v_0} + 2}\right)$$

Conclude that under the $\frac{3}{2}$-power resistance a body coasts only a finite distance before coming to a stop.

Problem 2.4.6 Consider shooting a bullet of mass m through three different big blocks. The resistance of the first block is proportional to the bullet speed; for the second block, it's proportional to $v^{\frac{3}{2}}$; and for the third block, it's proportional to v^2. If you fire at these blocks one by one with the same initial speed of v_0 for your bullets, compute the farthest the bullet can travel in each of these cases. Note that the gravitational force for this problem is negligible.

Problem 2.4.7 In Jules Verne's original problem, find the minimal launch velocity that suffices for the projectile to make it "From the Earth to the Moon". The projectile launched from the surface of the Earth is attracted by both the Earth and the Moon. The distance r of the projectile from the center of the Earth satisfies the IVP

$$\begin{cases} \dfrac{d^2r}{dt^2} = -\dfrac{GM_e}{r^2} + \dfrac{GM_m}{(S-r)^2} \\ r(0) = R \\ r'(0) = v_0 \end{cases}$$

where M_e and M_m denote the masses of the Earth and the Moon, respectively; R is the radius of the Earth and $S = 384{,}400\,\text{km}$ is the distance between the centers of the Earth and the Moon. To reach the Moon, the projectile must overcome the attraction of the Earth, *i.e.,* it must only just pass the point, between the Earth and the Moon, where the net acceleration from the two objects vanishes. Thereafter it is "under the control" of the Moon, and falls from there to the lunar surface. Find the minimum launch velocity v_0 that suffices for the projectile to make it "From the Earth to the Moon."

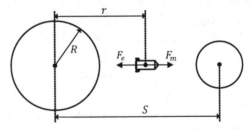

Figure 2.14 The projectile model for Problem 2.4.7.

Problem 2.4.8 One pushes a toy rocket of fixed mass m straight up at an initial velocity of v_0 from the surface of Earth. Assume Earth's gravitational constant to be g and the air frictional force on the rocket $-kv$ where k is the frictional constant and v is the velocity of the rock (Figure 2.6). Ignore the rocket "launcher" height. Please

(1) compute the max height h_{max} the rocket can reach.

(2) compute the time T_{up} for the rocket to reach max height.

Problem 2.4.9 Suppose that a motorboat is moving at 40 ft/s when its motor suddenly quits, and that 10 s later the boat has slowed to 20 ft/s.

(1) Assume that the resistance it encounters while coasting is proportional to its velocity. How far will the boat coast?

(2) Assuming that the resistance is proportional to the square of the velocity, how far does the motorboat coast in the first minute after its motor quits?

Figure 2.15 The motorboat sketch for Problem 2.4.9.

Problem 2.4.10 The resistance of an object of mass m moving in the Earth's gravitational field, with gravitational constant g, is proportional to its speed $v(t)$.

(1) Write down the EoM by Newton's second law, in terms of speed of the object, with assumption that the object was initially thrown upward at speed v_0, and air frictional constant k.

(2) Naturally, the object will move upward in the sky until the point where it reverses its direction to move downward to the Earth. Find the time when the direction of the object has just reversed. Does this time (at which reversing occurs) depend on the initial velocity? How?

(3) Eventually, the object will move at a constant speed, the Terminal Speed. Find this terminal speed. Does this terminal speed depend on the initial velocity? How?

Problem 2.4.11 On a straight stretch of road of length L there is a Stop sign at each end. Assume you follow traffic laws and stop at these signs. Your poor car can accelerate at a constant a_1 and decelerate at another constant a_2. Compute the shortest time to travel from one end of the road

119

to the other if there is no speed limit in between. Do it again where the speed limit between the signs is v_m.

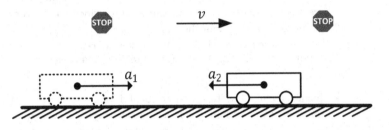

Figure 2.16 The moving car model for Problem 2.4.11.

Figure 2.17 The jumper model for Problem 2.5.2 and Problem 2.5.5.

Problem 2.4.12 If a projectile leaves Earth's surface at velocity v_0, please derivate the formula for the later velocity as a function of the distance r from the projectile to the center of the Earth. The Earth's radius R and gravitational constant G are given. Also, naturally, the velocity has to be a real number. Compute the condition at which the velocity will become imaginary.

Problem 2.4.13 Suppose that a projectile is fired straight upward from the surface of the Earth with initial velocity v_0 and assume its height $y(t)$ above the surface satisfies the IVP

$$\begin{cases} y'' = -\dfrac{GM}{(y+R)^2} - \beta \exp(-y) \\ y(0) = 0 \\ y'(0) = v_0 \end{cases}$$

Compute the max-height the projectile reaches and the time (from launch) for the projectile to reach the max-height. In the above, G, M, v_0 and β are all given constants.

Problem 2.4.14 A bullet of mass m is fired from the barrel of a gun at an initial speed v_0 to a medium whose resistance follows $\alpha v + \beta v^2$ where α and β are constants and v is the bullet's instantaneous speed. Compute the max-distance that the bullet can travel in this medium. If we double the initial speed while keeping the other conditions unchanged, compute the max-distance again. You may ignore gravity.

Problem 2.4.15 Consider shooting a bullet of mass m through three connected (and fixed) media each of equal length L. The resistances are kv, kv^2, kv in Medium-1, -2, -3, respectively. Compute the bullet's entering speed to Medium-1 such that the bullet will pass through Medium-1 and Medium-2 and Medium-3 and, magically, stop precisely at the ending edge of Medium-3. You may neglect gravity.

Problem 2.4.16 Consider shooting a bullet of mass m through a first medium whose resistance is proportional to the speed, through a second medium whose resistance is proportional to $v^{\frac{3}{2}}$, and through a third medium whose resistance is proportional to v^2. Compute in all three cases the farthest the bullet can travel with a given initial speed of v_0. The medium, *e.g.*, water in the ocean, in all three cases is huge.

Problem 2.4.17 Consider shooting, at an initial speed v_0, a bullet of mass m through a medium whose resistance is $-\alpha(\beta + v^2)$ where v is the instantaneous speed, α is the given resistance constant and β is another given constant. You may assume the medium is infinite (for example, the ocean) and you may also neglect the gravity for this problem. Please
(1) compute the farthest distances the bullet can travel in the medium;
(2) compute the time the bullet is in motion.

Problem 2.4.18 An egg is dropped vertically to a fully filled swimming pool of depth H at an initial speed 0 at the surface. Three forces act on the egg: gravity, water buoyancy force, and the friction with water. Assume water density 1, egg density $\rho > 1$, egg mass m. The buoyancy force is given as $\frac{m}{\rho} \times 1 \times g$ where g is the usual gradational constant while the friction force is μv where μ is a friction-related constant and v is the egg's instantaneous speed. Compute the egg's speed of impact at the pool bottom. You may express the solution in an implicit form if necessary.

2.5 Windy Day Plane Landing

A flying plane at $(a, 0)$ approaches an airport whose coordinates are $(0, 0)$, which means the plane is a miles east of the airport. A steady south wind of speed w blows from south to north. While approaching the airport, the plane maintains a constant speed v_0 relative to the wind and maintains its heading directly toward the airport at all times. It is interesting to construct a DE, and find its solution, for the plane's trajectory.

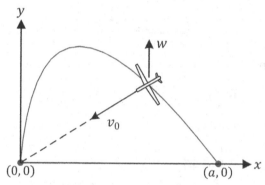

Figure 2.18 A plane at $(a, 0)$ approaches an airport at $(0, 0)$ for landing with wind blowing at a constant speed in the y-direction.

The plane's velocity components relative to the airport tower are

$$
\begin{cases}
\dfrac{dx}{dt} = -v_0 \cos \alpha = -v_0 \dfrac{x}{\sqrt{x^2 + y^2}} & (2.92) \\[3mm]
\dfrac{dy}{dt} = -v_0 \sin \alpha + w = -v_0 \dfrac{y}{\sqrt{x^2 + y^2}} + w & (2.93)
\end{cases}
$$

which is a system of two DEs with x and y as the DVs and t as the IV. We are not interested in time and a wise action is to eliminate the explicit dependence on time to build the following DE between x and y

$$\frac{dy}{dx} = \frac{y}{x} - k\sqrt{1 + \left(\frac{y}{x}\right)^2} \qquad (2.94)$$

where we have defined $k = \frac{w}{v_0}$. Recognizing (2.94) is a Homo DE, we make substitution $u = \frac{y}{x}$ and get

$$\frac{dy}{dx} = u - k\sqrt{1 + u^2} \qquad (2.95)$$

and considering $\frac{dy}{dx} = u + x\frac{du}{dx}$, we have

$$\frac{du}{\sqrt{1 + u^2}} = -\frac{k}{x}dx \qquad (2.96)$$

whose solution is

$$\ln\left(u + \sqrt{1 + u^2}\right) = -k\ln x + C \qquad (2.97)$$

The steps for proving

$$\int \frac{du}{\sqrt{1 + u^2}} = \ln\left(u + \sqrt{1 + u^2}\right) + c \qquad (2.98)$$

Let $u = \sinh z$ and using $u = \sinh z = \frac{\exp(z) - \exp(-z)}{2}$, we get

$$z = \ln\left(u + \sqrt{1 + u^2}\right) \qquad (2.99)$$

Using the following two relationships

$$d(\sinh z) = \cosh z \, dz \qquad (2.100)$$

$$(\cosh z)^2 - (\sinh z)^2 = 1 \qquad (2.101)$$

we get

$$\int \frac{du}{\sqrt{1 + u^2}} = \int \frac{\cosh z \, dz}{\sqrt{1 + (\sinh z)^2}}$$
$$= \int \frac{\cosh z}{\cosh z} dz \qquad (2.102)$$
$$= z + c$$
$$= \ln\left(u + \sqrt{1 + u^2}\right) + c$$

Applying the IC $u(x = a) = \frac{y(a)}{a} = 0$, we get $c = -k \ln a$ and

$$\ln\left(u + \sqrt{1 + u^2}\right) = -k \ln \frac{x}{a}$$
$$u + \sqrt{1 + u^2} = \left(\frac{x}{a}\right)^{-k} \tag{2.103}$$

Thus,

$$u(x) = \frac{1}{2}\left(\left(\frac{x}{a}\right)^{-k} - \left(\frac{x}{a}\right)^{k}\right) \tag{2.104}$$

Back substituting, we get

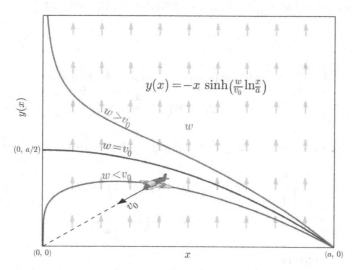

$$y(x) = -x \, \sinh\left(\frac{w}{v_0} \ln \frac{x}{a}\right)$$

Figure 2.19 The plane's trajectories for three cases of the plane speed vs. the wind speed.

$$y(x) = \frac{a}{2}\left(\left(\frac{x}{a}\right)^{1-k} - \left(\frac{x}{a}\right)^{1+k}\right)$$
$$= -x \sinh\left(\frac{w}{v_0} \ln \frac{x}{a}\right) \tag{2.105}$$

which is the trajectory of the plane.

Additionally, we can derive the formula for the total distance the plane would have traveled between $x = a$ and $x = 0$.

The infinitesimal arc length the plane travels during $[t, t + dt]$ is

$$dL = \sqrt{dx^2 + dy^2} = \left(1 + \left(\frac{dy}{dx}\right)^2\right)^{1/2} dx \qquad (2.106)$$

From (2.105), we get

$$\frac{dy}{dx} = -\sinh\left(k \ln\frac{x}{a}\right) - k \cosh\left(k \ln\frac{x}{a}\right) \qquad (2.107)$$

where we have used $\frac{d}{dx}(\sinh x) = \cosh x$ and the chain rule.

Thus,

$$\left(\frac{dy}{dx}\right)^2 = \sinh^2\left(k \ln\frac{x}{a}\right) + k^2 \cosh^2\left(k \ln\frac{x}{a}\right)$$
$$+ 2k \sinh\left(k \ln\frac{x}{a}\right)\cosh\left(k \ln\frac{x}{a}\right) \qquad (2.108)$$

Using $\cosh^2 x - \sinh^2 x = 1$ and $2 \sinh x \cosh x = \sinh 2x$, we get

$$\left(\frac{dy}{dx}\right)^2 = (1 + k^2)\sinh^2\left(k \ln\frac{x}{a}\right) + k^2$$
$$+ k \sinh\left(2k \ln\frac{x}{a}\right) \qquad (2.109)$$

Thus, the length is

$$L = \left| \int_a^0 \left(1 + \left(\frac{dy}{dx}\right)^2\right)^{1/2} dx \right| \qquad (2.110)$$

We can estimate several special cases for $k = 0$ or $k = 1$ or $k > 1$.

At $k = 0$, $\left(\frac{dy}{dx}\right)^2 = 0$ and, thus, $L = \left| \int_a^0 (1 + 0)^{1/2} dx \right| = a$.

At $k = 1$,

$$y(x) = \frac{a}{2} - \frac{x^2}{2a} \tag{2.111}$$

or

$$\frac{dy}{dx} = -\frac{x}{a} \tag{2.112}$$

The integral

$$L = \left| \int_a^0 \left(1 + \left(\frac{dy}{dx}\right)^2\right)^{\frac{1}{2}} dx \right| = \left| \frac{1}{a} \int_a^0 \sqrt{x^2 + a^2}\, dx \right| \tag{2.113}$$

$$= a \frac{\sqrt{2} + \ln(\sqrt{2} + 1)}{2} \approx 1.1478a > a$$

where we used

$$\int \sqrt{x^2 + a^2}\, dx = \frac{x}{2}\sqrt{x^2 + a^2}$$
$$+ \frac{a^2}{2}\ln\left(x + \sqrt{x^2 + a^2}\right) + C \tag{2.114}$$

At $k > 1$, the integral $L = \left| \int_M^0 \left(1 + \left(\frac{dy}{dx}\right)^2\right)^{1/2} dx \right|$ diverges, *i.e.*, $L \to \infty$.

At $k < 1$, we can, in principle, evaluate the integral

$$L = \left| \int_M^0 \left(1 + \left(\frac{dy}{dx}\right)^2\right)^{1/2} dx \right| \tag{2.115}$$

which is also finite. But, we are unable to get the closed form.

Remarks

(1) If $k = \frac{w}{v_0} = 1$, meaning the plane flies at the same speed as the wind, the plane's vertical coordinate is $y(x \to 0) = \frac{a}{2}$ for $x \to 0$, *i.e.*, the plane will reach at $\left(0, \frac{a}{2}\right)$ and miss the airport.

(2) If $k = \frac{w}{v_0} > 1$, meaning the plane flies slower than the wind, the plane's vertical coordinate is $y(x \to 0) \to \infty$ for $x \to 0$, *i.e.*, the plane will get lost.

(3) If $k = \frac{w}{v_0} < 1$, meaning the plane flies faster than the wind, the plane's vertical coordinate is $y(x \to 0) \to 0$ and the plane lands properly.

Problems

Problem 2.5.1 On January 15, 2009, the US Airways Flight 1549 "landed" safely in the Hudson River a couple of minutes after departing from LGA. The air (straight line) distance between LGA and the landing spot on the river is M. We assume (I) the flight's Captain Sullenberger flies the plane at a constant speed v_0 relative to the wind, after realizing the bird problem immediately after taking off; (II) the wind blows at a constant speed w and at a direction perpendicular to the line linking the landing spot and LGA; (III) the plane maintains its heading directly toward the landing spot. Compute the critical wind speed w_c above which the plane would have been blown away. Compute the lengths of the flight trajectories for $w < w_c$ and $w \neq 0$ and $w = 0$.

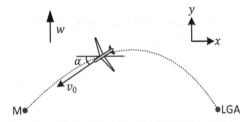

Figure 2.20 The aircraft model for Problem 2.5.1.

Problem 2.5.2 A man with a parachute jumps out of a hovering helicopter at height H. During the fall, the man's drag coefficient is k (with closed parachute) and nk (with open parachute) and air resistance is taken as proportional to velocity. The total weight of the man and his parachute is m. Take the initial velocity when he jumped to be zero. Gravitational constant is g. Find the best time to open his parachute after he leaves the helicopter for the quickest fall and yet "softest" landing at a touchdown speed $\leq v_0$.

Problem 2.5.3 A guy of body mass m jumps out of a horizontally flying plane of speed v_0 at height H. The horizontal component of the jumper's initial velocity is equal to that of the plane's and the jumper's initial vertical speed can be regarded as 0. The air resistance is $-\alpha\vec{v}$ with the parachute open where α is a given constant and $\vec{v}$ is the jumper's velocity. Please compute the total distance, from when the guy was thrown off the plane to the landing spot, if the parachute is opened at the middle height point. For clarification, you compute the length of the trajectory the poor guy draws while in the sky, before hitting ground. You may express your solution implicitly.

(*Hint: Curve length for $x \in [a, b]$ is $L = \int_a^b \sqrt{1 + \left(\frac{dy}{dx}\right)^2}\, dx$*)

Problem 2.5.4 A guy of body mass m jumps out of a horizontally flying plane of speed v_0 at height H. The horizontal component of the jumper's initial velocity is equal to that of the plane's and the jumper's initial vertical speed can be regarded as 0. The air resistance is $\alpha|\vec{v}|$ with the parachute open where α is a given constant and $|\vec{v}| = \left(v_x^2 + v_y^2\right)^{1/2}$ is the magnitude of jumper's velocity. Please compute the times taken for the jumper to reach land if the parachute is opened at heights $\frac{1}{2}H$ and $\frac{1}{4}H$.

Problem 2.5.5 A man with a parachute jumps out of a "frozen" helicopter at height H. During the fall, the man's drag coefficient is k (with closed parachute) and nk (with open parachute) and air resistance is taken as proportional to velocity. The total weight of the man and his parachute is m. Take the initial velocity when he jumped to be zero. Gravitational constant is g. The man opens the parachute t_0 time after he jumps off the helicopter. Please find the total falling time, and the speed he hits the ground. Can you adjust the t_0 to enable the quickest fall, and lightest hit on the ground?

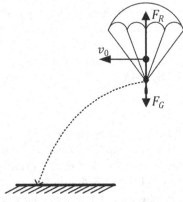

Figure 2.21 The jumper model for Problem 2.5.3 and Problem 2.5.4.

2.6   A Swimmer's Problem

The picture (Figure 2.22) shows a river of width $w = 2a$ that flows north. The lines $x = 0, \pm a$ represent the river's center line and banks.

The river water speed v_R increases as one approaches the river center and the speed can be proven to be

$$v_R(x) = v_0\left(1 - \frac{x^2}{a^2}\right) \qquad (2.116)$$

where x is the distance from the river center line, $v_R(x)$ is the water speed at location x and v_0 is the water speed at the center of the river.

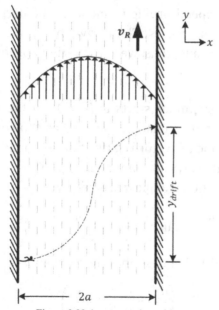

Figure 2.22 A swimmer's problem.

Remarks

(1) Water speed at the center of the river reaches the maximum
$$x = 0, \qquad v_R = v_0(1 - 0) = v_0$$

(2) Water speed at either bank of the river is zero

At left bank where $x = -a$,
$$v_R = v_0(1 - 1) = 0$$
At right bank where $x = a$,
$$v_R = v_0(1 - 1) = 0$$

The two cases where the speed of water is zero at the riverbanks are called the no-slip conditions, a well-known result in fluid mechanics.

(3) The water speed profile formula can be derived from more basic laws of physics. The derivation is out of the scope of this problem and, thus, we use it without proof.

The Swimmer's DE

Suppose that the swimmer starts at $(-a, 0)$ on the west bank and swims heading east at a constant speed v_S, as shown in Figure 2.22. The swimmer's velocity vector relative to the ground had horizontal component v_S and vertical component v_R. Thus, the swimmer's instantaneous directional angle α is given by

$$\frac{dy}{dx} = \tan \alpha = \frac{v_R}{v_S} = \frac{v_0}{v_S}\left(1 - \frac{x^2}{a^2}\right) \qquad (2.117)$$

Thus, we have an IVP that describes the trajectory of the swimmer:

$$\begin{cases} \dfrac{dy}{dx} = \dfrac{v_0}{v_S}\left(1 - \dfrac{x^2}{a^2}\right) \\ y(x = -a) = 0 \end{cases} \qquad (2.118)$$

Solution

Solving this problem, we have

$$dy = \frac{v_0}{v_S}\left(1 - \frac{x^2}{a^2}\right)dx$$

$$\int_0^y dy = \int_{-a}^x \frac{v_0}{v_S}\left(1 - \frac{x^2}{a^2}\right)dx \qquad (2.119)$$

$$y = \frac{v_0}{v_S}\left(x - \frac{x^3}{3a^2}\right)\Bigg|_{-a}^x$$

Finally, we obtain the equation for describing the trajectory of the swimmer as

$$y = \frac{v_0}{v_S}\left(x - \frac{x^3}{3a^2} + \frac{2a}{3}\right) \qquad (2.120)$$

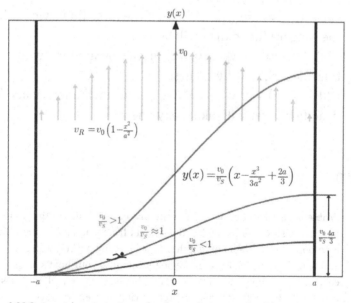

Figure 2.23 Swimmer's trajectories under three relative speeds of the swimmer to water.

Remarks

At the east bank of the river $x = a$, the swimmer's drift is

$$y_D = y(x = a) = \frac{v_0}{v_S}\left(a - \frac{a^3}{3a^2} + \frac{2a}{3}\right) = \frac{v_0}{v_S}\frac{4a}{3} \qquad (2.121)$$

(4) If a increases, the y_D increases proportionally, *i.e.*, the wider the river, the longer the drift.

(5) If the ratio $\frac{v_0}{v_S}$ increases, the y_D increases proportionally, *i.e.*, the faster the river flows, the longer the drift, for a constant swimmer speed v_S. Conversely, the faster the swimmer swims, the shorter the drift, for a constant river water speed v_0.

(2a) $v_0 \to 0 \implies y_D \to 0$, *i.e.*, if the river stays calm, like a lake, drift hardly occurs.

(2b) $v_0 \to \infty \implies y_D \to \infty$, *i.e.*, if the river flows violently, like the Niagara River, drift will be huge.

(2c) $v_S \to \infty \implies y_D \to 0$, *i.e.*, if the swimmer swims like a torpedo, drift hardly occurs.

(2d) $v_S \to 0 \implies y_D \to \infty$, *i.e.*, if the swimmer floats like a dead fish, river will take it anywhere.

Problem

Problem 2.6.1 During the 2014 Stony Brook University Roth Quad Regatta boat race, students propelled their boats from one end to the other of a long pond (not a river) of length L. Since it is not a river, water movement and wind have little effect on the boats. Suppose your boat's mass is M and each of your boaters' mass is m (all boaters are equal in mass). The propelling force from each boater during the entire race is an equal constant f_0. Water resistance is approximated to be proportional to the speed of the boat with a constant μ_0 regardless of the weight of the boat plus boaters. Your boat is always still at the start, and the boaters are the only power source.

(1) Establish the DE, for n boats for relating the boat's speed with time and the given parameters.

(2) Solve the DE established above to find the speed as a function of time.

(3) Which leads to the shortest travel time with all other conditions fixed? One boater or as many as possible?

2.7   River Ferryboat-Docking Problem

Combining the windy day plane landing problem and the swimmer problem, we can design a river ferryboat-docking problem. Study of this problem requires consideration of variable-speed flows that drag the boat at varying *forces* depending on the location of the boat.

In reality, both the direction and magnitude of the flows, e.g., air for the plane landing problem or water for the swimmer's problem or for the ferryboat-docking problem, change with time and location. To be more accurate, we may need to consider secondary effects such as flow changes induced by motions of other neighboring objects such as other passing boats or fellow swimmers.

For mathematical simplifications, we only consider idealistic settings, or a first-order approximation. The following is a simplified ferryboat-docking problem.

We set a 2-dimensional Cartesian coordinate system for a section of a river such that the *y*-axis lies at the river's west bank and points north and the *x*-axis runs from west to east. The water is assumed to flow strictly northward at all times and its speed, denoted by $w(x)$, varies depending on the x-coordinate of the location.

A ferryboat tries to cross the river from a point $(a, 0)$ to dock at $(0,0)$. It can do a lot of good things to cross optimally in time, energy, and safety but it does two simple things: moving at a constant speed v_B relative to water and keeping its head toward the docking point $(0, 0)$ at all times. We attempt to figure out the boat's trajectory.

The boat's velocity components relative to the riverbanks (not to the moving water) are

$$\begin{cases} \dfrac{dx}{dt} = -v_B \cos\alpha = -v_B \dfrac{x}{\sqrt{x^2 + y^2}} & (2.122) \\[4mm] \dfrac{dy}{dt} = -v_B \sin\alpha + w(x) = -v_B \dfrac{y}{\sqrt{x^2 + y^2}} + w(x) & (2.123) \end{cases}$$

which is a system of two DEs with x and y as the DVs and t IV. We can also introduce a dimensionless function $f(x)$ to express the water speed that can be written as $w(x) = v_0 f(x)$. We may build the following DE between x and y

$$\frac{dy}{dx} = \frac{y}{x} - \frac{w(x)}{v_B \left(\dfrac{x}{\sqrt{x^2 + y^2}} \right)} \tag{2.124}$$

The problem becomes the following IVP

$$\begin{cases} \dfrac{dy}{dx} = \dfrac{y}{x} - kf(x)\sqrt{1 + \left(\dfrac{y}{x}\right)^2} \\[4mm] y(x = a) = 0 \end{cases} \tag{2.125}$$

where we have defined a constant $k = \dfrac{v_0}{v_B}$.

Now, let's solve this DE. Let $u = \dfrac{y}{x}$. Then, $\dfrac{dy}{dx} = u + x\dfrac{du}{dx}$ and the DE becomes

$$\begin{aligned} \frac{du}{\sqrt{1 + u^2}} &= -\frac{kf(x)}{x}dx \\[3mm] \ln\left(u + \sqrt{1 + u^2}\right) &= -\int_a^x \frac{kf(z)}{z}dz \end{aligned} \tag{2.126}$$

where we used integration dummy variable z instead of x to avoid potential confusion. For writing convenience, we may define

$$F(x) = -\int_a^x \frac{kf(z)}{z}dz = k\int_x^a \frac{f(z)}{z}dz \tag{2.127}$$

We have

$$u(x) = \frac{1}{2}(\exp(F(x)) - \exp(-F(x))) = \sinh F(x) \qquad (2.128)$$

Back substituting, we get

$$y(x) = x \sinh F(x) \qquad (2.129)$$

We can write $F(x)$ in terms of the given variables and functions

$$F(x) = \frac{1}{v_B} \int_x^a \frac{w(z)}{z} dz \qquad (2.130)$$

Finally, the GS of the trajectory is

$$y(x) = x \sinh\left(\frac{1}{v_B} \int_x^a \frac{w(z)}{z} dz\right) \qquad (2.131)$$

For the windy day plane landing, $w(z) = w$, a constant and the plane speed is another constant v_0, we have the following equation for the trajectory

$$y(x) = x \sinh\left(\frac{1}{v_0} \int_x^a \frac{w}{z} dz\right) = x \sinh\left(\frac{w}{v_0} \ln\frac{a}{x}\right)$$
$$= -x \sinh\left(\frac{w}{v_0} \ln\frac{x}{a}\right) \qquad (2.132)$$

which is exactly what we have found before.

For the ferryboat-docking problem, the water speed is

$$w(x) = 4v_0\left(\frac{x}{a} - \frac{x^2}{a^2}\right) \qquad (2.133)$$

where v_0 is the max-speed at the centerline of the river and the factor "4" is there to enable it. You may check

$$w(x = a) = 0, w(x = 0) = 0, w(x = a/2) = v_0$$

Thus,

$$F(x) = \frac{v_0}{v_B} \int_x^a \frac{1}{z} 4 \left(\frac{z}{a} - \frac{z^2}{a^2} \right) dz$$
$$= k \int_x^a 4 \left(\frac{1}{a} - \frac{z}{a^2} \right) dz = 2k \left(1 - \frac{x}{a} \right)^2$$

(2.134)

Finally, the equation of the ferryboat trajectory is

$$y(x) = x \sinh \left(\frac{2v_0}{v_B} \left(1 - \frac{x}{a} \right)^2 \right)$$

(2.135)

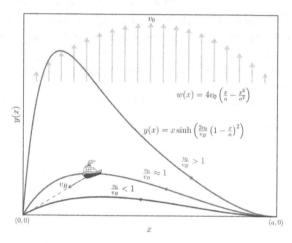

Figure 2.24 The trajectories of a ferryboat.

Problems

Problem 2.7.1 A fighter jet taking off from one aircraft carrier C_1 wishes to land on another carrier C_2, currently located precisely on the northeast of C_1. The distance between the two carriers is L. The carrier C_2 moves east at a constant speed of v_2 just after the jet took off. Wind blows from south to north at a constant speed of v_W and its impact to the carriers is negligible while the impact to the jet must be taken into account fully. Obviously, given the scale of flight distance (usually a thousand miles), the carriers and jet can be treated as mathematical points. Compute the shortest trajectory of the jet to land as it wishes.

Problem 2.7.2 We set a 2-dimensional Cartesian coordinate system for a north running river, so that the y-axis lies on the river's west bank (pointing from south to north), and the x-axis across the river, pointing from west to east. A ferryboat wants to cross the river from a point $(a, 0)$ on the east bank to the west bank, and it hopes to dock at $(0, 0)$. The water speed, depending on the location in the river, is measured by a function $W(x) = \frac{\omega_0 x(a-x)}{a^2}$ where both ω_0 and a are positive constants. Now, we further assume the ferryboat keeps facing the docking point $(0, 0)$ at all times during the trip, and it moves at a constant speed v_0 relative to the water. Please

(1) establish the DE for the boat's trajectory;

(2) solve the DE;

(3) sketch the boat's trajectory if the river flows much slower than the boat.

2.8   Equation for Compound Interest

Consider a situation when one owes a certain amount of money to a bank.

We consider the following terms:

> - $Z(t)$ is the amount owed to the bank at time t;
> - r is the interest rate which is a constant;
> - dt is the period for the time interval $[t, t + dt]$;
> - w is the payment rate which is also a constant;
> - dZ is the change of debt to bank during time dt.

The change of the debt due to interest increment and payments decrement, from t to $t + dt$, can be expressed as

$$(\text{Balance change } dZ) = (\text{Interest}) - (\text{Payment})$$
$$= Z(t)\, rdt - wdt$$

Thus,

$$dZ = (Zr - w)dt$$
$$Z' = Zr - w \tag{2.136}$$

is the loan DE, for lack of a better name.

To solve this loan DE, we use the method of separation-of-variables.

$$\frac{Z'}{r} = Z - \frac{w}{r} \tag{2.137}$$

$$\frac{dZ}{Z - \dfrac{w}{r}} = rdt$$

$$\int_{z_0}^{z} \frac{dZ}{Z - \dfrac{w}{r}} = \int_{t=0}^{t} rdt$$

$$\ln\left(Z - \frac{w}{r}\right)\Big|_{z_0}^{z} = rt$$

$$\ln\left(\frac{Z - \frac{w}{r}}{z_0 - \frac{w}{r}}\right) = rt$$

$$Z - \frac{w}{r} = \left(z_0 - \frac{w}{r}\right)\exp(rt)$$

Thus, the solution to the loan DE is

$$Z(t) = \frac{w}{r} - \left(\frac{w}{r} - z_0\right)\exp(rt) \qquad\qquad (2.138)$$

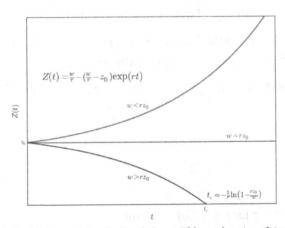

Figure 2.25 This figure shows the loan balance $Z(t)$ as a function of time for three payment schedules: $w < z_0 r$, $w = z_0 r$, and $w > z_0 r$. It also shows the payoff time.

Remarks

(1) At $t = 0$, $Z(t = 0) = z_0$
(2) If $r = 0$, $Z \neq z_0$ (w/r would have become indeterminate)
(3) If the periodic payment is greater than the interest amount,

$$w > z_0 r \qquad\qquad (2.139)$$

then,

$$\frac{w}{r} - z_0 > 0 \tag{2.140}$$

the loan will decrease. Naturally, it will reach a critical time when the loan balance is zero, *i.e.,* the loan is paid off. Now, let's compute such payoff time

$$\frac{w}{r} - \left(\frac{w}{r} - z_0\right)\exp(rt) = 0$$
$$\frac{w}{r} = \left(\frac{w}{r} - z_0\right)\exp(rt)$$
$$\exp(rt) = \frac{\frac{w}{r}}{\frac{w}{r} - z_0} \tag{2.141}$$
$$rt = -\ln\left(1 - \frac{rz_0}{w}\right)$$

We define a function $T(w, r)$ as the payoff time

$$T(w, r) = -\frac{1}{r}\ln\left(1 - \frac{rz_0}{w}\right) \tag{2.142}$$

If one pays more frequently but the total amount is fixed, *i.e.,* paying biweekly, weekly, or daily instead of monthly, the loan balance will decrease faster as shown in Figure 2.25.

If periodic payment amount is doubled while all other terms stay unchanged, the new payoff time will be

$$T(2w, r) = -\frac{1}{r}\ln\left(1 - \frac{rz_0}{2w}\right) \tag{2.143}$$

Obviously,

$$\frac{T(2w, r)}{T(w, r)} = \frac{\ln\left(1 - \frac{rz_0}{2w}\right)}{\ln\left(1 - \frac{rz_0}{w}\right)} \neq \frac{1}{2} \tag{2.144}$$

Thus, doubling the payment amount will not half the payoff time.

Next, we use two figures to make even more elaborate observations of the payoff time $T(w, r)$ as a function of the periodic payment amount w and the interest rate r for a loan of, *e.g.*, $z_0 = 500,000$ (name your favorite currency).

Figure 2.26 shows the relationships of payoff times vs. the periodic payment amount for three different interest rates: $r = 2.0\%, 3.0\%, 4.0\%$. Obviously, the higher the interest rate, the longer the payoff time for the same periodic payment amount w. For each interest rate, there is always a critical w value at which $T(w, r) \to \infty$, which means you will never pay off at a given interest rate, if your periodic payment amount is below a threshold.

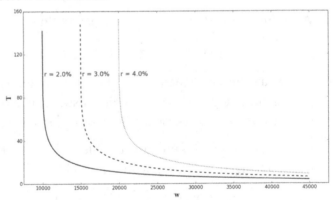

Figure 2.26 Payoff times vs. the periodic payment amount for three different interest rates: $r = 2.0\%, 3.0\%, 4.0\%$.

Figure 2.27 shows the relationships of payoff times vs. the interest rate for three different periodic payment amounts: $w = 10,000, 20,000, 30,000$. Obviously, the higher the w value, the shorter the payoff time for the same interest rate. For each w value, there is always a critical interest rate at which $T(w, r) \to \infty$, which means you will never pay off at a given periodic payment amount, if your interest rate is above a threshold.

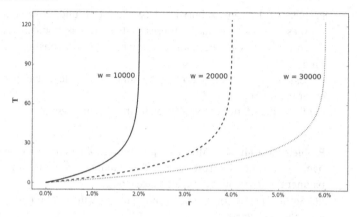

Figure 2.27 Payoff times vs. the interest rate for three different periodic payment amounts: $w = 10,000, 20,000, 30,000$.

(4) If the periodic payment is precisely equal to the interest amount, *i.e.,*

$$\frac{w}{r} - z_0 = 0 \qquad (2.145)$$

the loan balance is

$$z(t) = \frac{w}{r} = z_0 \qquad (2.146)$$

which remains constant forever.

(5) If the periodic payment is smaller than the interest amount, then

$$\left(\frac{w}{r} - z_0\right) < 0 \qquad (2.147)$$

the loan balance will blow up and the loan will never be paid off.

Problems

Problem 2.8.1 A person borrowed Z_0 amount of funds from a bank which charges a fixed (constant) interest rate r. The periodical payment the borrower makes to the bank is $Q_0 t$, where Q_0 is a constant, and t is the time.

(1) Establish the DE governing the time-varying loan balance $Z(t)$ with other parameters.

(2) Solve the DE to express $Z(t)$ explicitly as a function of t.

Problem 2.8.2 Mr. and Mrs. Young borrowed Z amount of money from a bank at a daily interest rate r. The Youngs make the same "micropayment" of D amount to the bank each day.

(1) Set up the DE for amount x owed to the bank at the end of each day after the loan starts.

(2) Solve the above DE for x as a function of t, Z, and r.

(3) Obviously, if D is too small (paid too little), the Youngs will never pay off the loan. Find the critical D at which the loan will not change.

(4) The Youngs wish to pay off the loan in N days. What do they have to pay each day? Write it as a function of Z, N, and r.

(5) The Youngs wish to pay off the loan in $N/2$ days. Find the daily payment D as a function of Z, N, and r.

Problem 2.8.3 Mr. A borrowed Z_0 from a bank at a fixed interest rate r. Mr. A pays the bank fixed amount W periodically so that he can pay off the loan at time T. Please

(1) derive a formula for T in terms of Z_0, W, and r. In other words, find the exact form for function $T = T(Z_0, W, r)$, and compute

   (a) $T_1 = T\left(Z_0, W, r \to 0\right)$

   (b) $T_2 = T\left(Z_0, W, r = \dfrac{W}{2Z_0}\right)$

(*Hint:* $\lim_{x \to 0} \ln(1 + x) = x$)

(2) compute the amount Mr. A must add to his current payment W to pay off at $T/2$ instead of T?

Problem 2.8.4 Mr. Wyze and Mr. Fulesch each borrows z_0 from a bank at the same time. Wyze got it at a fixed interest rate r and will pay it off at time T with a fixed periodic payment W_0. Fulesch's rate changes with time according to $r_F = \frac{1}{5}r(1 + t)$ so Fulesch's rate is only $1/5$ of Wyze's at the start. For example, if Wyze's rate is 6.0%, Fulesch's starting rate is only 1.2%. If Fulesch also pays W_0 periodically, his debt drops faster than Wyze's initially before it drops slower and, then, grows. Compute the time when both loaners owe the bank the same amount, again. The first time when they owe the same is when they just got the loans.

Problem 2.8.5 Given that a loan of Z_0 with a fixed interest rate r would be paid off at time T with a fixed periodic payment W_0. Now, one pays a new periodic payment $(1 + \alpha)W_0$, compute the new time to pay off the loan. Compute total interests paid to the bank in both cases.

Problem 2.8.6 One borrows Z_0 from a bank at a fixed rate r. Now, we assume the time to pay off the loan with fixed periodic payment W is T_1. Please
(1) derive a formula for the new payoff time if the periodic payment is doubled while all other terms remain unchanged;
(2) derive a formula for the new periodic payment if the interest rate is doubled while all other terms remain unchanged.

Problem 2.8.7 One borrows Z_0 dollars from a bank at a fixed rate r. Now, we assume the time to pay off the loan with fixed periodic payment W is T. Please
(1) derive a formula for the new payoff time if the periodic payment is nW while all other terms remain unchanged;
(2) derive a formula for the new periodic payment if the interest rate is br while all other terms remain unchanged.

Problem 2.8.8 One borrows Z_0 from a bank at a fixed interest rate r. We assume the time to pay off the loan with a fixed periodical payment W_0 is T_0. Now, if the payment is changed to αW_0 while all other terms remain unchanged, derive a formula for the new payoff time T_1 in terms of the given parameters.
(1) If $\alpha > 1$, which of the following is true?
    (a) $T_1 > T_0$
    (b) $T_1 = T_0$

      (c) $T_1 < T_0$

(2) If $0 < \alpha < 1$, which of the following is true?

      (a) $T_1 > T_0$

      (b) $T_1 = T_0$

      (c) $T_1 < T_0$

(3) If $\alpha < 0$, compute T_1.

Chapter 3
Linear DEs of Higher Order

3.1 Classifications of Linear DEs

In classifications of DEs, we have categorized them into 32 different groups by their properties. For higher-order DEs, we will consider linear Homo and InHomo DEs with constant coefficients and variable coefficients. Thus, we have four different groups. The solution method for each group is usually applicable only to this particular group since a general method for solving all types of DEs does not exist. In fact, for higher-order nonlinear DEs, only a few peculiar cases may be solvable, analytically. Many DEs resort to numerical means for their solutions.

Linearity Coefficients	Yes	No
Constant	$y'' + y = 0$	$y'' + y^2 = 0$
Variable	$y'' + P(x)y = 0$	$y'' + P(x)y^2 = 0$

Based on the coefficients and the "external" term, we will study four types of linear DEs, namely,

Type 1A. The general 2nd.O c-coeff linear homogeneous DE (Homo DE)

$$y'' + ay' + by = 0 \qquad (3.1)$$

Type 1B. The general 2nd.O c-coeff linear inhomogeneous DE (InHomo DE)

$$y'' + ay' + by = c \neq 0 \qquad (3.2)$$

Type 2A. The general 2nd.O v-coeff linear Homo DE

$$y'' + P(x)y' + Q(x)y = 0 \qquad (3.3)$$

Type 2B. The general 2nd.O v-coeff linear InHomo DE

$$y'' + P(x)y' + Q(x)y = R(x) \qquad (3.4)$$

Generally, Type 1A is the easiest to solve while Type 2B is the hardest because one has to deal with v-coeff and the InHomo term.

For linear Homo DEs, the superposition principle, widely used in physics and systems theory, is a powerful tool to search for the GS's of linear DEs. The superposition principle states that, for all linear systems, the net response at a given space and time caused by two or more stimuli is the sum of the responses that would have been caused by each stimulus individually.

Superposition principle contains two key elements: the additivity and homogeneity.

The additivity can be expressed as

$$F(x + y) = F(x) + F(y) \tag{3.5}$$

The homogeneity can be expressed as

$$F(cx) = cF(x) \tag{3.6}$$

A linear function satisfies the superposition properties (3.5) and (3.6).

Theorem 1

For a general 2nd.O linear Homo DE

$$y'' + P(x)y' + Q(x)y = 0 \tag{3.7}$$

if $y_1(x)$ is a solution and $y_2(x)$ is another solution, then, the linear combination of y_1 and y_2

$$y(x) = C_1 y_1(x) + C_2 y_2(x) \tag{3.8}$$

is also a solution. Further, if $y_1(x)$ and $y_2(x)$ are linearly independent, then, the linear combination is the GS.

Corollaries of Theorem 1

If $y_1, y_2, \ldots, y_n$ are the n solutions to the DE

$$y^{(n)} + P_1(x)y^{(n-1)} + \cdots + P_n(x)y^{(0)} = 0 \tag{3.9}$$

then, the linear combination of $y_1, y_2, \ldots, y_n$ is also a solution to the DE. If $y_1, y_2, \ldots, y_n$ are n linearly independent (LI) solutions of the DE, then, their linear combination is the GS to the DE.

Proof of Theorem 1

Let

$$y_G = C_1 y_1 + C_2 y_2 \tag{3.10}$$

be the linear combination of y_1 and y_2 where C_1 and C_2 are constants. Since y_1 and y_2 are the solutions to the DE (3.7), they satisfy

$$y_1'' + P(x)y_1' + Q(x)y_1 = 0 \tag{3.11}$$

$$y_2'' + P(x)y_2' + Q(x)y_2 = 0 \tag{3.12}$$

$C_1 \times$ (3.11) gives

$$C_1 y_1'' + C_1 P(x)y_1' + C_1 Q(x)y_1 = 0 \tag{3.13}$$

Since

$$\begin{aligned} C_1 y_1' &= (C_1 y_1)' \\ C_1 y_1'' &= (C_1 y_1)'' \end{aligned} \tag{3.14}$$

we have

$$(C_1 y_1)'' + P(x)(C_1 y_1)' + Q(x)(C_1 y_1) = 0 \tag{3.15}$$

Similarly, $C_2 \times$ (3.12) gives

$$(C_2 y_2)'' + P(x)(C_2 y_2)' + Q(x)(C_2 y_2) = 0 \tag{3.16}$$

Adding DEs (3.15) and (3.16) generates

$$\begin{aligned} ((C_1 y_1)'' + (C_2 y_2)'') &+ P(x)((C_1 y_1)' + (C_2 y_2)') \\ &+ Q(x)(C_1 y_1 + C_2 y_2) = 0 \\ (C_1 y_1 + C_2 y_2)'' &+ P(x)(C_1 y_1 + C_2 y_2)' \\ &+ Q(x)(C_1 y_1 + C_2 y_2) = 0 \end{aligned} \tag{3.17}$$

That is

$$y_G'' + P(x)y_G' + Q(x)y_G = 0 \tag{3.18}$$

Therefore, $y_G = C_1 y_1 + C_2 y_2$ is also a solution to the original DE.

Example 1

Check whether $y_1 = \cos \omega x$ and $y_2 = \sin \omega x$ are solutions of the DE

$$y'' + \omega^2 y = 0 \tag{3.19}$$

where ω is a constant. Verify whether $y_G = C_1 y_1 + C_2 y_2$ is also a solution.

Solution

Check y_1:

$$y_1' = -\omega \sin \omega x$$
$$y_1'' = -\omega^2 \cos \omega x$$
$$y_1'' + \omega^2 y_1 = 0$$

Thus, y_1 is a solution to the given DE.

Check y_2:

$$y_2' = \omega \cos \omega x$$
$$y_2'' = -\omega^2 \sin \omega x$$
$$y_2'' + \omega^2 y_2 = 0$$

Thus, y_2 is a solution to the given DE.

Check y_G:

$$y_G = C_1 y_1 + C_2 y_2 = C_1 \cos \omega x + C_2 \sin \omega x$$

Thus,

$$y_G' = -\omega C_1 \sin \omega x + \omega C_2 \cos \omega x$$
$$y_G'' = -\omega^2 C_1 \cos \omega x - \omega^2 C_2 \sin \omega x$$

Thus,

$$y_G'' + \omega^2 y_G = 0$$

Therefore, y_G is also a solution to the DE $y'' + \omega^2 y = 0$.

Example 2

Given $y_1 = \sin \omega x$, $y_2 = 5 \sin \omega x$. Both are solutions to the DE
$$y'' + \omega^2 y = 0$$
Find the third solution.

Solution

According to Theorem 1

$$y_3 = C_1 y_1 + C_2 y_2$$
$$= C_1 \sin \omega x + 5 C_2 \sin \omega x$$
$$= (C_1 + 5 C_2) \sin \omega x$$
$$= C_3 \sin \omega x$$

where $C_3 = C_1 + 5 C_2$

Problems

Problem 3.1.1 Show that $y = \frac{c}{x}$ is a solution to $y' + y^2 = 0$ only if $c = 1$.

Problem 3.1.2 Show that $y = cx^3$ is a solution to $yy'' = 6x^4$ only if $c = 1$.

Problem 3.1.3 Show that $y_1 = 1$ and $y_2 = \sqrt{x}$ are solutions to $yy'' + (y')^2 = 0$, but their sum $y = y_1 + y_2$ is not a solution.

Problem 3.1.4 Let y_p be a PS to the InHomo DE $y'' + py' + qy = f(x)$ and let y_c be a solution to its associated Homo DE $y'' + py' + qy = 0$. Show that $y = y_c + y_p$ is a solution to the given InHomo DE. With $y_p = 1$ and $y_c = C_1 \cos x + C_2 \sin x$ in the above notation, find a solution to $y'' + y = 1$ satisfying the IC $y(0) = y'(0) = -1$.

3.2 Linear Independence

Definition

For functions y_1 and y_2, if y_1 is not a constant multiple of y_2, then, y_1 is linearly independent (LI) of y_2. That is, if one can find two constants c_1 and c_2 that are not zero simultaneously to make $c_1 y_1 + c_2 y_2 = 0$, then, y_1 and y_2 are called linear dependent (LD). O.W., they are LI.

For the two-function case, if $\frac{y_1}{y_2} \neq$ constant, then, they are LI.

Example
Determine if y_1, y_2 are LI

y_1	y_2	$\dfrac{y_1}{y_2}$	LI
$\sin x$	$\cos x$	$\tan x$	Y
$\sin 2x$	$\sin x$	$2\cos x$	Y
$2\sin x$	$\sin x$	2	N
$2\exp(x)$	$\exp(x)$	2	N
$\exp(2x)$	$\exp(x)$	$\exp(x)$	Y
$\exp(ax)$	$\exp(bx)$	$\exp\big((a-b)x\big)$	Y $(a \neq b)$

More generally, a set of functions $\{f_1(x), f_2(x), \dots, f_n(x)\}$, defined on some given interval, is said to be LI if

$$\sum_{i=1}^{n} \big(c_i f_i(x)\big) = 0 \tag{3.20}$$

implying that $c_i = 0 \ \forall \ i \leq n$. Otherwise, the set is linearly dependent.

Let's now define the differential operator $(\mathcal{D})$

$$\mathcal{D} = \frac{d}{dx} \tag{3.21}$$

Example 1

$$\mathcal{D}y = \left(\frac{d}{dx}\right)y = \frac{dy}{dx} = y'$$

$$\mathcal{D}^2 y = \left(\frac{d}{dx}\right)^2 y = \frac{d^2 y}{dx^2} = y''$$

$$\cdots$$

$$\mathcal{D}^n y = \left(\frac{d}{dx}\right)^n y = \frac{d^n y}{dx^n} = y^{(n)}$$

(3.22)

So now, our n^{th} order DE

$$y^{(n)} + P_1(x)y^{(n-1)} + \cdots + P_n(x)y = 0 \qquad (3.23)$$

can be written as

$$D^n y + P_1(x)D^{n-1}y + \cdots + P_n(x)y = 0$$

$$\left(D^n + P_1(x)D^{n-1} + \cdots + P_n(x)\right)y = 0$$

(3.24)

Definition

A linear differential operator L is defined as

$$L = \mathcal{D}^n + P_1(x)\mathcal{D}^{n-1} + \cdots + P_n(x) \qquad (3.25)$$

so that (3.24) can now be written as $Ly = 0$.

Example 2

Find the linear differential operator L in $y'' + y = 0$

Solution

$$\frac{d^2 y}{dx^2} + y = 0$$

$$\left(\frac{d^2}{dx^2} + 1\right)y = 0$$

$$L = \mathcal{D}^2 + 1$$

Example 3

Find the linear differential operator L in $y'' - 5y' + 6y = 0$

Solution

$$\frac{d^2y}{dx^2} - 5\frac{dy}{dx} + 6y = 0$$

$$\left(\frac{d^2}{dx^2} - 5\frac{d}{dx} + 6\right)y = 0$$

$$L = \mathcal{D}^2 - 5\mathcal{D} + 6$$

Superposition properties of linear differential operator

$$(L_1 + L_2)y = L_1y + L_2y \tag{3.26}$$

$$C(Ly) = L(Cy) \tag{3.27}$$

The above two DEs (3.26) and (3.27) imply that the differential operator L are truly linear, satisfying superposition properties.

Theorem 2

For an nth order linear DE $Ly = 0$, if $y_1, y_2, \ldots, y_n$ are n LI solutions to the DE, then, the linear combination

$$y_G = \sum_{i=1}^{n} C_i y_i \tag{3.28}$$

is the GS to the DE.

Proof

Since $\forall\, 1 \leq i \leq n$, y_i is a solution to $Ly = 0$,

$$Ly_i = 0 \tag{3.29}$$

Multiplying both sides of (3.29) by C_i, we have

$$C_i Ly_i = 0 \tag{3.30}$$

By the linearity of L, we get

$$L(C_i y_i) = 0 \quad \forall\, 1 \leq i \leq n \tag{3.31}$$

Summing over i, we have

$$\sum_{i=1}^{n} L(C_i y_i) = 0 \tag{3.32}$$

i.e.,

$$L\left(\sum_{i=1}^{n} C_i y_i\right) = 0 \tag{3.33}$$

By the definition, we reach $L y_G = 0$.

LI for More than Two Functions

For more than two functions, we have defined the relationship of LI. The methods at the beginning of the section for determining the LD of two functions are valid only for two functions. However, a set of functions that are mutually pair-wise LI does not necessarily indicate they are LI as a whole. We need to determine the dependence properties of more than two functions.

Example 4

Given $y_1 = \sin x$, $y_2 = \cos x$ and $y_3 = \sin x + \cos x$, prove that they are mutually pairwise LI

Solution

It is easy to verify that

$$\frac{y_1}{y_2} = \tan x$$
$$\frac{y_3}{y_1} = 1 + \cot x$$
$$\frac{y_3}{y_2} = \tan x + 1$$

These prove that they are mutually LI

Remarks

Despite the fact they are mutually pairwise LI, intuitively we think they are not LI as a whole since we can easily find that $y_3 = y_1 + y_2$. From the definition to be introduced below, we find that they are indeed LD.

For a formal definition of the LI for more than two functions, we have to expand the definition for two functions from a new direction. For any two LI functions y_1 and y_2, we know that we cannot find a constant c such that

$$\frac{y_1}{y_2} = c \tag{3.34}$$

for all x where y_1 and y_2 are defined. Let's now write this definition in a balanced fashion. Consider the equation

$$C_1 y_1 + C_2 y_2 = 0 \tag{3.35}$$

If we can find a non-zero vector $[C_1, C_2]^T$ that satisfies (3.35), we know we can find a constant C that satisfies (3.34). Note that the zero in (3.35) means the zero function that is 0 everywhere y_1 and y_2 are defined. Thus, we know that the LI of y_1 and y_2 means that the only solution to (3.35) is $[C_1, C_2]^T = [0,0]^T$. Now we can generalize this definition to n functions for a formal definition for arbitrary number of functions.

Definition

A set of n functions $y_1, y_2, \ldots, y_n$, defined on a given interval, is said to be LI if

$$C_1 y_1 + C_2 y_2 + \cdots + C_n y_n = 0 \tag{3.36}$$

has only one solution: $[C_1, C_2, \ldots, C_n]^T = [0,0,\ldots,0]^T$. The zero on the RHS of (3.36) is the zero function defined where $y_1, \ldots, y_n$ are defined.

It is very difficult to verify directly from the above original definition if a set of functions is truly LI. In practice, several other approaches can be derived. One of them is that of the Wronskian method.

Let's now introduce the Wronskian of n functions.

Definition

The Wronskian of n functions $y_1, y_2, \ldots, y_n$ is the determinant (3.37)

$$W(y_1, \ldots, y_n) = \begin{vmatrix} y_1 & y_2 & \cdots & y_n \\ y_1' & y_2' & \cdots & y_n' \\ \vdots & \vdots & \ddots & \vdots \\ y_1^{(n-1)} & y_2^{(n-1)} & \cdots & y_n^{(n-1)} \end{vmatrix}_{n \times n} \tag{3.37}$$

The Wronskian was introduced by a Polish mathematician Jozef Maria Hoene-Wronski (1776-1853) in 1812.

Theorem 3

For a set of n functions $y_1, y_2, \ldots, y_n$, if the Wronskian of the set

$$W(y_1, \ldots, y_n) \neq 0 \tag{3.38}$$

then, they are LI. Conversely, if $y_1, y_2, \ldots, y_n$ are LD, then, the Wronskian of the set

$$W(y_1, \ldots, y_n) = 0 \tag{3.39}$$

In summary, if the Wronskian of a set of functions is nonzero for a given interval, the set is LI on that interval. However, if a set of functions is LD, then, the Wronskian of the set of functions is zero.

Example 5
Check if $y_1 = \exp(r_1 x)$ and $y_2 = \exp(r_2 x)$ ($r_1 \neq r_2$) are LI.

Solution

$$\begin{aligned} W(y_1, y_2) &= \begin{vmatrix} \exp(r_1 x) & \exp(r_2 x) \\ r_1 \exp(r_1 x) & r_2 \exp(r_2 x) \end{vmatrix} \\ &= \exp(r_1 x) \cdot r_2 \exp(r_2 x) - \exp(r_2 x) \cdot r_1 \exp(r_1 x) \\ &= (r_2 - r_1) \exp\big((r_2 - r_1)x\big) \end{aligned}$$

Since $r_1 \neq r_2$, we know that $W(y_1, y_2) \neq 0$. Thus, y_1 and y_2 are LI.

Example 6
Check if $y_1 = \exp(rx)$ and $y_2 = x \exp(rx)$ are LI.

Solution

$$W(y_1, y_2) = \begin{vmatrix} \exp(rx) & x\exp(rx) \\ r\exp(rx) & \exp(rx) + rx\exp(rx) \end{vmatrix}$$
$$= (1 + rx)\exp(2rx) - rx\exp(2rx)$$
$$= \exp(2rx) \neq 0$$

Thus, y_1 and y_2 are LI.

Example 7

Check if $y_1 = \sin \omega x$ and $y_2 = \cos \omega x$ are LI, where $\omega \neq 0$ is a constant.

Solution

$$W(y_1, y_2) = \begin{vmatrix} \sin \omega x & \cos \omega x \\ \omega \cos \omega x & -\omega \sin \omega x \end{vmatrix}$$
$$= -\omega \sin^2 \omega x - \omega \cos^2 \omega x$$
$$= -\omega(\sin^2 \omega x + \cos^2 \omega x)$$
$$= -\omega \neq 0$$

Thus, y_1 and y_2 are LI.

Remarks

If the Wronskian is non-zero, the functions are LI. But, if the Wronskian is zero, it does not necessarily imply that the functions are LD. For example, for $y_1 = x^2$ and $y_2 = x|x|$, though they are LD on either $[0, +\infty)$ or $(-\infty, 0]$, we cannot find a set of non-zero $[C_1, C_2]^T$ such that (3.35) holds for the entire real axis. This means y_1 and y_2 are LI when considered as a function on $(-\infty, +\infty)$, but the Wronskian gives for $x \in [0, +\infty)$

$$W(y_1, y_2) = \begin{vmatrix} x^2 & x^2 \\ 2x & 2x \end{vmatrix} = 0 \tag{3.40}$$

and for $x \in (-\infty, 0)$, we have

$$W(y_1, y_2) = \begin{vmatrix} x^2 & -x^2 \\ 2x & -2x \end{vmatrix} = 0 \tag{3.41}$$

These two mean $W(y_1, y_2) = 0 \; \forall x$. This example indicates that the Wronskian being zero does not give sufficient information about the LI of the functions.

Problems

Problem 3.2.1 Define two operators with given constants $\alpha_1, \alpha_2, \beta_1, \beta_2$:
$$L_1 \equiv D^2 + \alpha_1 D + \beta_1$$
$$L_2 \equiv D^2 + \alpha_2 D + \beta_2$$
Applying these two operators to a function $x(t)$, we get $L_1 L_2 x(t)$ and $L_2 L_1 x(t)$. Check if $L_1 L_2 x(t) = L_2 L_1 x(t)$. Check it again if we define the two operators as:
$$L_1 x(t) \equiv Dx(t) + tx(t)$$
$$L_2 x(t) \equiv tDx(t) + x(t)$$

Problem 3.2.2 Check the LI of the following three functions
$$\exp(x), \ \cosh x, \ \sinh x$$

Problem 3.2.3 Check the LI of the following three functions
$$0, \ \sin x, \exp(x)$$

Problem 3.2.4 Check the LI of the following three functions
$$2x, \ 3x^2, \ 5x - 8x^2$$

Problem 3.2.5 Check the LI of the following three functions
$$x^2, \ x^3, \ x^3 - x^2$$

Problem 3.2.6 Determine if $f(x)$ and $g(x)$ are LI.
$$f(x) = \pi \qquad g(x) = \cos^2 x + \sin^2 x$$

Problem 3.2.7 Let y_1 and y_2 be two solutions of $A(x)y'' + B(x)y' + C(x)y = 0$ on an open interval I where, A, B and C are continuous and $A(x)$ is never zero.
(1) Let $W = W(y_1, y_2)$, show that
$$A(x)\frac{dW}{dx} = y_1(Ay_2'') - y_2(Ay_1'')$$
Then, substitute for Ay_1'' and Ay_2'' from the original DE to show that
$$A(x)\frac{dW}{dx} = -B(x)W(x)$$

(2) Solve this 1st.O DE to deduce Abel's formula

$$W(x) = K \exp\left(-\int \frac{B(x)}{A(x)} dx\right)$$

Problem 3.2.8 Prove directly that the functions $f_0(x) = 1$, $f_1(x) = x$, $f_2(x) = x^2, \ldots, f_n(x) = x^n$ are LI.

Problem 3.2.9 Show that the Wronskian of the following nth.O InHomo DE
$$x^{(n)} + P_1(t)x^{(n-1)} + \cdots + P_n(t)x = f(t)$$
is a function of $P_1(t)$ only, and find the expression of the Wronskian in terms of this $P_1(t)$.

Problem 3.2.10 Suppose that the three numbers r_1, r_2 and r_3 are distinct. Show that the three functions $\exp(r_1 x), \exp(r_2 x)$ and $\exp(r_3 x)$ are LI by showing that their Wronskian

$$W = \exp\left((r_1 + r_2 + r_3)x\right) \begin{vmatrix} 1 & 1 & 1 \\ r_1 & r_2 & r_3 \\ r_1^2 & r_2^2 & r_3^2 \end{vmatrix} \neq 0, \forall x$$

Problem 3.2.11 Compute the Vandermonde determinant

$$V = \begin{vmatrix} 1 & 1 & \cdots & 1 \\ r_1 & r_2 & \cdots & r_n \\ \vdots & \vdots & \ddots & \vdots \\ r_1^{n-1} & r_2^{n-1} & \cdots & r_n^{n-1} \end{vmatrix}$$

Problem 3.2.12 Given that the Vandermonde determinant

$$\begin{vmatrix} 1 & 1 & \cdots & 1 \\ r_1 & r_2 & \cdots & r_n \\ \vdots & \vdots & \ddots & \vdots \\ r_1^{n-1} & r_2^{n-1} & \cdots & r_n^{n-1} \end{vmatrix}$$

is non-zero if the numbers $r_1, r_2, \ldots, r_n$ are distinct. Prove that the functions $f_i(x) = \exp(r_i x), 1 \leq i \leq n$ are LI.

3.3   Homogeneous DEs

The word "homogeneous" is over-used, even in mathematics. There are a number of distinct cases: homogeneous functions, homogeneous type of 1st.O DEs, and homogeneous DEs. To get its precise meaning, we must put it in mathematical context.

A function $f(x)$ is said to be Homo of degree n if, for any constant c,

$$f(cx) = c^n f(x) \qquad (3.42)$$

A linear DE is Homo if the following condition is satisfied: if $y(x)$ is a solution, so is $cy(x)$ where $c \neq 0$ is a constant.

3.3.1   DEs with Constant Coefficients

Consider a 2nd.O c-coeff DE,

$$ay'' + by' + cy = 0 \qquad (3.43)$$

where $a \neq 0$, b and c are constants. Let's try out a solution. Without proper preparation, we propose a trial solution (TS) to the DE (3.43) as $\exp(rx)$. Thus,

$$\begin{aligned} y(x) &= \exp(rx) \\ y'(x) &= r \exp(rx) \\ y''(x) &= r^2 \exp(rx) \end{aligned} \qquad (3.44)$$

Plugging these into (3.43), we have

$$ar^2 \exp(rx) + br \exp(rx) + c \exp(rx) = 0 \qquad (3.45)$$

That is

$$(ar^2 + br + c) \exp(rx) = 0 \qquad (3.46)$$

an AE. As the proposed solution, it cannot be trivial, *i.e.,* $\exp(rx) \neq 0$. To satisfy the AE (3.46), we must set

$$ar^2 + br + c = 0 \qquad (3.47)$$

This is called the characteristic equation (C-Eq) of the original DE (3.43). Solving this C-Eq, we obtain two roots

$$r_{1,2} = \frac{-b \pm \sqrt{b^2 - 4ac}}{2a} \qquad (3.48)$$

So, through "back substitution", we get the solutions of the DE,

$$y_1(x) = \exp(r_1 x) \\ y_2(x) = \exp(r_2 x) \qquad (3.49)$$

Since $a \neq 0$, there are three possible cases for r_1 and r_2 depending on the discriminant $b^2 - 4ac$:

(1) $b^2 - 4ac > 0$
(2) $b^2 - 4ac = 0$
(3) $b^2 - 4ac < 0$

Case 1: $\Delta = b^2 - 4ac > 0$

The C-Eq (3.47) has two distinct real roots $r_1 \neq r_2$ and the two solutions are

$$y_1(x) = \exp(r_1 x) \\ y_2(x) = \exp(r_2 x)$$

which are LI and, thus, the GS for the DE is

$$y(x) = C_1 \exp(r_1 x) + C_2 \exp(r_2 x) \qquad (3.50)$$

Check

Since r_1 and r_2 are the roots of the AE (3.47), we have

$$\begin{cases} ar_1^2 + br_1 + c = 0 \\ ar_2^2 + br_2 + c = 0 \end{cases} \tag{3.51}$$

Plugging the TS's $\exp(r_1 x)$ and $\exp(r_2 x)$ into the DE (3.43), we have

$$\begin{aligned} a(\exp(r_1 x))'' &+ b(\exp(r_1 x))' + c\exp(r_1 x) \\ &= ar_1^2 \exp(r_1 x) + br_1 \exp(r_1 x) + c\exp(r_1 x) \\ &= (ar_1^2 + br_1 + c)\exp(r_1 x) \end{aligned} \tag{3.52}$$

From (3.51), we know formula (3.52) yields 0. Thus, $\exp(r_1 x)$ is a solution to the DE. Similarly, $\exp(r_2 x)$ can be proven to be another solution.

Example 1
Find the GS to the following DE

$$y'' - 5y' + 6y = 0 \tag{3.53}$$

Solution
Step 1: Selecting the TS $y = \exp(rx)$ and we get
$$y' = r\exp(rx)$$
$$y'' = r^2 \exp(rx)$$
Step 2: Plugging the TS into the DE, we have
$$r^2 \exp(rx) - 5r\exp(rx) + 6\exp(rx) = 0$$
$$(r^2 - 5r + 6)\exp(rx) = 0$$
We get the C-Eq
$$r^2 - 5r + 6 = 0$$
$$(r - 2)(r - 3) = 0$$
$$r_{1,2} = 2, 3$$
Step 3: Because $r_1 \neq r_2$ and both are real, we conclude it is Case 1
Step 4: Compose the GS
$$y = C_1 \exp(2x) + C_2 \exp(3x)$$
Step 5: Check
$$y = C_1 \exp(2x) + C_2 \exp(3x)$$
$$y' = 2C_1 \exp(2x) + 3C_2 \exp(3x)$$
$$y'' = 4C_1 \exp(2x) + 9C_2 \exp(3x)$$

$$y'' - 5y' + 6y = 4C_1 \exp(2x) + 9C_2 \exp(3x) - 5(2C_1 \exp(2x) + 3C_2 \exp(3x))$$
$$+6(C_1 \exp(2x) + C_2 \exp(3x))$$
$$= C_1(4 - 10 + 6) \exp(2x) + C_2(9 - 15 + 6) \exp(3x)$$
$$= 0$$

Thus, the GS is

$$y(x) = C_1 \exp(2x) + C_2 \exp(3x)$$

Case 2: $\Delta = b^2 - 4ac = 0$

The C-Eq (3.47) has two equal real roots

$$r_1 = r_2 = r = -\frac{b}{2a} \qquad (3.54)$$

It is clear that $\exp(r_1 x) = \exp(r_2 x)$ and they are not LI. In order to get the GS, we must recover the missing LI solution using the so-called Reduction of Order method.

A 2nd.O DE whose C-Eq has two identical roots r must be of the form **(3.55)** which is slightly simplified to save writing.

$$\boldsymbol{y'' - 2ry' + r^2 y = 0} \qquad (3.55)$$

After finding one solution $y_1(x) = \exp(rx)$, we may try a LI solution to be of the form $y_2(x) = v(x)y_1(x)$. If $v(x)$ is not a constant, $y_1(x)$ and $y_2(x)$ are LI. All we need to do is to find a non-consistent $v(x)$ to form $y_2(x)$ for the original DE.

$$\begin{aligned}
y_2(x) &= v(x) \exp(rx) \\
y_2'(x) &= v'(x) \exp(rx) + v(x)r \exp(rx) \\
y_2''(x) &= v''(x) \exp(rx) + 2v'(x)r \exp(rx) \\
&\quad + v(x)r^2 \exp(rx)
\end{aligned} \qquad (3.56)$$

Plugging these terms into the DE, we get

$$y_2''(x) - 2ry_2'(x) + r^2 y_2(x) = v''(x) \exp(rx) = 0 \qquad (3.57)$$

Thus,

$$v''(x) = 0$$

whose solution should be

$$v(x) = A_1 + A_2 x$$

Thus,

$$y_2(x) = v(x)y_1(x) = (A_1 + A_2 x)y_1(x) = A_1 y_1(x) + A_2 x y_1(x)$$

Since $A_1 y_1(x)$ is a multiple of $y_1(x)$, we just need to write

$$y_2(x) = x y_1(x) \tag{3.58}$$

Therefore, our second solution may take the form of (3.59):

$$y_2(x) = x \exp(rx) \tag{3.59}$$

Thus, the GS is

$$y = C_1 \exp(rx) + C_2 x \exp(rx) \tag{3.60}$$

Sometimes, we also write it in the form of (3.61):

$$y = (C_1 + C_2 x) \exp(rx) \tag{3.61}$$

Example 2
Find the GS to the following DE

$$y'' - 2y' + y = 0 \tag{3.62}$$

Solution
Selecting a TS $y = \exp(rx)$, we get
$$y' = r \exp(rx)$$
$$y'' = r^2 \exp(rx)$$
Plugging the TS into the DE, we have
$$r^2 \exp(rx) - 2r \exp(rx) + \exp(rx) = 0$$
$$(r^2 - 2r + 1) \exp(rx) = 0$$
Thus, we get the C-Eq
$$r^2 - 2r + 1 = 0$$
$$(r - 1)^2 = 0$$
$$r_{1,2} = 1, 1$$

$$y_1 = \exp(x)$$

and we have

$$y_2 = x \exp(x)$$

Thus, the GS is

$$y = C_1 \exp(x) + C_2 x \exp(x)$$

Example 3

Suppose the C-Eq of the given DE is

$$(r - a)^3 = 0 \tag{3.63}$$

Find the GS to the DE.

Solution

$(r - a)^3 = 0$ means that the C-Eq has three identical roots, thus, we have the following

$$y_1 = \exp(ax)$$
$$y_2 = x \exp(ax)$$
$$y_3 = x^2 \exp(ax)$$

and the GS is

$$y = (C_1 + C_2 x + C_3 x^2) \exp(ax)$$

Case 3: $\Delta = b^2 - 4ac < 0$

The C-Eq (3.47) has two complex roots

$$r_{1,2} = \alpha \pm i\beta \tag{3.64}$$

where $\alpha = -\dfrac{b}{2a}$ and $\beta = \dfrac{\sqrt{|b^2 - 4ac|}}{2a}$ are both real numbers. Now we can form our solutions

$$y_{1,2} = \exp\big((\alpha \pm i\beta)x\big) \tag{3.65}$$

It is easy to check that, for $\beta \neq 0$, y_1 and y_2 are LI. There is no need to bother using the Wronskian for two functions.

$$\frac{y_1}{y_2} = \frac{\exp\big((\alpha + i\beta)x\big)}{\exp\big((\alpha - i\beta)x\big)} \tag{3.66}$$
$$= \exp(2i\beta x) \neq \text{constant}$$

So the GS is

$$y = C_1 \exp\big((\alpha + i\beta)x\big) + C_2 \exp\big((\alpha - i\beta)x\big) \qquad (3.67)$$

Since the DE is raised in a real domain, we also want to write the solutions using real coefficients. The trick here is to manipulate the constants in (3.67). We can rewrite (3.67) as

$$y = \exp(\alpha x)\,(C_1 \exp(i\beta x) + C_2 \exp(-i\beta x)) \qquad (3.68)$$

From the most well-known Euler's formula

$$\exp(\pm i\beta x) = \cos \beta x \pm i \sin \beta x \qquad (3.69)$$

we have

$$\begin{aligned}
y &= \exp(\alpha x)\,\big(C_1(\cos \beta x + i \sin \beta x) \\
&\qquad + C_2(\cos \beta x - i \sin \beta x)\big) \\
&= \exp(\alpha x)\,\big((C_1 + C_2)\cos \beta x \\
&\qquad + (iC_1 - iC_2)\sin \beta x\big)
\end{aligned} \qquad (3.70)$$

Let

$$\begin{aligned}
c_1 &= C_1 + C_2 \\
c_2 &= iC_1 - iC_2
\end{aligned} \qquad (3.71)$$

we write the GS in real form

$$y = \exp(\alpha x)\,(c_1 \cos \beta x + c_2 \sin \beta x) \qquad (3.72)$$

Example 4
Find the GS to the following IVP

$$\begin{cases} y'' - 4y' + 5y = 0 \\ \qquad\qquad y(0) = 1 \\ \qquad\qquad y'(0) = 5 \end{cases} \qquad (3.73)$$

Solution
After discussing many examples, we are now able to find the C-Eq directly from the DE.

$$r^2 - 4r + 5 = 0$$

$$(r-2)^2 + 1 = 0$$

This C-Eq has two roots $2 \pm i$. Thus, the GS is

$$y(x) = \exp(2x)\,(c_1 \cos x + c_2 \sin x)$$

To find the PS, we apply the IC. First, we have

$$y(0) = c_1 = 1$$

For y', we have

$$y'(x) = 2\exp(2x)\,(c_1 \cos x + c_2 \sin x) + \exp(2x)\,(-c_1 \sin x + c_2 \cos x)$$

and

$$y'(0) = 2c_1 + c_2 = 5$$

Plugging $c_1 = 1$ into the above equation, we get

$$c_2 = 3$$

So, the PS to the IVP is

$$y(x) = \exp(2x)\,(\cos x + 3\sin x)$$

Example 5

Find the GS to the following DE

$$(\mathcal{D}^2 + 6\mathcal{D} + 13)^2 y = 0 \tag{3.74}$$

Solution

The C-Eq of the given DE is

$$(r^2 + 6r + 13)^2 = 0$$

That is

$$((r+3)^2 + 4)^2 = 0$$

and gives

$$r_1 = r_2 = -3 + 2i$$
$$r_3 = r_4 = -3 - 2i$$

Thus, the GS to the given DE is

$$y(x) = \exp(-3x)\,(c_1 \cos 2x + d_1 \sin 2x) + x\exp(-3x)\,(c_2 \cos 2x \\ + d_2 \sin 2x)$$

Summary

For 2nd.O c-coeff Homo DEs

$$ay'' + by' + cy = 0 \tag{3.75}$$

we select a TS

$$ar^2 + br + c = 0$$

$$r_{1,2} = \frac{-b \pm \sqrt{b^2 - 4ac}}{2a} \qquad (3.76)$$

and the C-Eq

$\Delta = b^2 - 4ac$	r_1, r_2	LI Solution	GS
$\Delta > 0$	$r_1 \neq r_2$ Real roots	$\exp(r_1 x), \exp(r_2 x)$	$C_1 \exp(r_1 x) + C_2 \exp(r_2 x)$
$\Delta = 0$	$r_1 = r_2 = r$ Real root	$\exp(rx), x \exp(rx)$	$(C_1 + C_2 x) \exp(rx)$
$\Delta < 0$	$r_1 = \alpha + i\beta$ $r_2 = \alpha - i\beta$ Complex roots	$\exp\big((\alpha + i\beta)x\big)$ $\exp\big((\alpha - i\beta)x\big)$	$C_1 \exp\big((\alpha + i\beta)x\big)$ $+C_2 \exp\big((\alpha - i\beta)x\big)$ $= \exp(\alpha x)\,(A\cos\beta x + B\sin\beta x)$

Theorem 4

If a c-coeff Homo DE has n repeated and identical roots $r_1 = r_2 = \cdots = r_n = r$, then, there are n LI solutions

$$\exp(rx), x \exp(rx), \ldots, x^{n-1} \exp(rx) \qquad (3.77)$$

The GS for such DE is

$$y = C_1 \exp(rx) + C_2 x \exp(rx) + \cdots + C_n x^{n-1} \exp(rx) \qquad (3.78)$$

Equivalently,

$$y = (C_1 + C_2 x + \cdots + C_n x^{n-1}) \exp(rx) \qquad (3.79)$$

Proof

Suppose we have a DE in the form of

$$(\mathcal{D} - r_1)^n y = 0 \qquad (3.80)$$

We prove that the functions in (3.77) are the solutions to the DE (3.80) by mathematical induction.

First, we have that for $y_1 = \exp(r_1 x)$

$$
\begin{aligned}
(\mathcal{D} - r_1)y_1 &= \mathcal{D}y_1 - r_1 y_1 \\
&= \frac{d}{dx} \exp(r_1 x) - r_1 \exp(r_1 x) \\
&= r_1 \exp(r_1 x) - r_1 \exp(r_1 x) \\
&= 0
\end{aligned}
\tag{3.81}
$$

This means that y_1 is a solution to $(\mathcal{D} - r_1)y = 0$ and, also, a solution to $(\mathcal{D} - r_1)^n y = 0$.

Now supposed for $k < n$, $y_k = x^{k-1} \exp(r_1 x)$ is a solution to

$$
(\mathcal{D} - r_1)^k y = 0
\tag{3.82}
$$

and it is a solution to (3.80). For $k + 1$, we prove

$$
y_{k+1} = x^k \exp(r_1 x)
\tag{3.83}
$$

is also a solution.

$$
\begin{aligned}
(\mathcal{D} - r_1)^{k+1} & y_{k+1} \\
&= (\mathcal{D} - r_1)^k (\mathcal{D}y_{k+1} - r_1 y_{k+1}) \\
&= (\mathcal{D} - r_1)^k \big(x^k r_1 \exp(r_1 x) + k x^{k-1} \exp(r_1 x) \\
&\qquad - r_1 x^k \exp(r_1 x) \big) \\
&= (\mathcal{D} - r_1)^k k x^{k-1} \exp(r_1 x) \\
&= k(\mathcal{D} - r_1)^k y_k \\
&= 0
\end{aligned}
\tag{3.84}
$$

This means y_{k+1} is a solution to

$$
(\mathcal{D} - r_1)^{k+1} y = 0
\tag{3.85}
$$

To prove that the functions in (3.77) are LI, we can use the conclusion of Problem 3.3.17 and directly apply the definition of LI on them. The detail of this part of proof is left as a homework problem for you.

Example 6
Find the GS to the following DE

$$y''' + y'' + y' + y = 0 \qquad (3.86)$$

Solution
The DE can be written in terms of differential operator as
$$(\mathcal{D}^3 + \mathcal{D}^2 + \mathcal{D} + 1)y = 0$$
Thus, the C-Eq is
$$r^3 + r^2 + r + 1 = r^2(r+1) + (r+1)$$
$$= (r^2 + 1)(r+1)$$
$$= 0$$
whose roots are $r_{1,2,3} = i, -i, -1$. Thus,
$$\exp(-x),\ \exp(ix),\ \exp(-ix)$$
are solutions to the original DE and they are LI. Thus, the GS is
$$y(x) = C_1 \exp(-x) + C_2 \cos x + C_3 \sin x$$

Example 7
Find the GS to the following DE
$$9y^{(5)} - 6y^{(4)} + y^{(3)} = 0$$

Solution
The DE can be written in terms of differential operators
$$(9\mathcal{D}^5 - 6\mathcal{D}^4 + \mathcal{D}^3)y = 0$$
So, the C-Eq is
$$9r^5 - 6r^4 + r^3 = r^3(9r^2 - 6r + 1)$$
$$= r^3(3r - 1)^2$$
$$= 0$$
whose roots are $r_{1,2,3,4,5} = 0, 0, 0, \frac{1}{3}, \frac{1}{3}$. Thus, the GS is

$$y(x) = C_1 + C_2 x + C_3 x^2 + (C_4 + C_5 x)\exp\left(\frac{1}{3}x\right)$$

Example 8
Find the GS for the following DE

$$(\mathcal{D}^2 - 5\mathcal{D} + 6)(\mathcal{D} - 1)^2(\mathcal{D}^2 + 1)y = 0 \qquad (3.87)$$

Solution

The C-Eq is

$$(r^2 - 5r + 6)(r - 1)^2(r^2 + 1) = (r - 2)(r - 3)(r - 1)^2(r^2 + 1)$$
$$= 0$$

whose roots are $r_{1,2,3,4,5,6} = 2, 3, 1, 1, i, -i$. Thus, the GS is

$$y(x) = C_1 \exp(2x) + C_2 \exp(3x) + (C_3 + C_4 x) \exp(x) + C_5 \cos x + C_6 \sin x$$

3.3.2   DEs with Variable Coefficients

The general 2nd.O linear v-coeff DE is

$$a_2(x)y'' + a_1(x)y' + a_0(x)y = f(x) \tag{3.88}$$

where $a_2(x) \neq 0$ and the coefficients $a_0(x)$, $a_1(x)$, $a_2(x)$ are, in general, functions of x. If the RHS $f(x) \neq 0$, this is a 2nd.O linear v-coeff InHomo DE and will be studied in the following section.

Usually, linear DEs of higher orders with v-coefficients are not solvable analytically even though they are linear. There are, however, several special types of v-coeff DEs that are solvable and the following is an incomplete list

Example 1

The nth.O Cauchy-Euler DEs

$$\sum_{n=0}^{N} a_n x^n y^{(n)} = 0 \tag{3.89}$$

These DEs are called the Cauchy-Euler DEs or, simply, called the Euler's DEs. Sometimes, they are also known as Equidimensional DEs because of the particular equidimensional structure of the DEs. This class of DEs is attached to two greatest mathematicians of the 18th century, Augustin-Louis Cauchy (1789-1857) and Leonhard Euler (1707-1783). Cauchy is a French mathematician with more than 800 research articles while Euler is a Swiss mathematician and physicist who published more during his lifetime.

Example 2

Consider the Euler's hypergeometric DE

$$x(1-x)\frac{d^2y}{dx^2} + (c-(a+b+1)x)\frac{dy}{dx} + aby = 0 \qquad (3.90)$$

where a, b, and c are constants. The Euler's hypergeometric DE can be converted into the so-called Q-form as

$$\frac{d^2u}{dz^2} + Q(z)u(z) = 0 \qquad (3.91)$$

where

$$Q(z) = \frac{z^2(1-(a-b)^2) + z(2c(a+b+1)-4ab) + c(2-c)}{4z^2(1-z)^2} \qquad (3.92)$$

Example 3

Bessel's DE

$$x^2\frac{d^2y}{dx^2} + x\frac{dy}{dx} + (x^2-\alpha^2)y = 0 \qquad (3.93)$$

where α is an arbitrary complex number. The canonical solutions of the DE are the Bessel functions, defined initially by the Swiss mathematician D. Bernoulli (1700-1782) and generalized by the German mathematician F. W. Bessel (1784-1846). When the complex constant α is an integer or half integer, the Bessel functions are called cylinder functions or spherical Bessel functions, respectively.

All of these DEs have been thoroughly studied and the Cauchy-Euler DEs are much more manageable. A special case is the 2nd.O Cauchy-Euler DEs

$$a_2x^2y'' + a_1xy' + a_0y = 0 \qquad (3.94)$$

We may use this special case to demonstrate the solution methods and, in fact, higher order DEs can also be solved in a similar fashion. The real *geist* of the solution methods is to transform the coefficients of the DEs from variables into constants.

First, we propose a trial solution,

$$y(x) = x^\lambda \tag{3.95}$$

Thus,

$$y'(x) = \lambda x^{\lambda-1}$$
$$y''(x) = \lambda(\lambda - 1)x^{\lambda-2} \tag{3.96}$$

Then, the original DE becomes

$$a_2\lambda(\lambda - 1)x^\lambda + a_1\lambda x^\lambda + a_0 x^\lambda = 0 \tag{3.97}$$

Or

$$(a_2\lambda(\lambda - 1) + a_1\lambda + a_0)x^\lambda = 0 \tag{3.98}$$

Since $x^\lambda \neq 0$, we must have

$$a_2\lambda(\lambda - 1) + a_1\lambda + a_0 = 0 \tag{3.99}$$

i.e.,

$$a_2\lambda^2 + (a_1 - a_2)\lambda + a_0 = 0 \tag{3.100}$$

which is the C-Eq whose roots are used to compose the GS to the original Cauchy-Euler DE.

As usual, there are three cases for the roots of this AE (3.100) and, as such, there are three cases for the GS to the DE.

Case 1: $\Delta = (a_1 - a_2)^2 - 4a_2a_0 > 0$:

The AE (3.100) has two distinct real roots $\lambda_1 \neq \lambda_2 \in R$ and the DE has two LI solutions x^{λ_1} and x^{λ_2}. The GS is

$$y(x) = c_1 x^{\lambda_1} + c_2 x^{\lambda_2} \tag{3.101}$$

Case 2: $\Delta = (a_1 - a_2)^2 - 4a_2a_0 = 0$:

The AE (3.100) has two identical real roots

$$\lambda_1 = \lambda_2 = \frac{a_2 - a_1}{2a_2} \equiv r \tag{3.102}$$

and the DE has two linearly dependent (actually, identical) solutions x^r and x^r. We must recover the missing LI solution using the Reduction of Order method.

Using the one solution $y_1(x) = x^r$, we try a second solution

$$y_2(x) = v(x)y_1(x) = v(x)x^r \tag{3.103}$$

Thus,

$$
\begin{aligned}
y_2'(x) &= v'(x)\, x^r + v(x)rx^{r-1} \\
y_2''(x) &= v''(x)x^r + 2v'(x)\, rx^{r-1} \\
&\quad + v(x)r(r-1)x^{r-2}
\end{aligned}
\tag{3.104}
$$

Plugging these terms into the DE, we get

$$a_2 x^2 y_2''(x) + a_1 x y_2'(x) + a_0 y_2(x) = 0 \tag{3.105}$$

Thus,

$$v(x) = \ln x \tag{3.106}$$

Thus,

$$y_2(x) = x^r \ln x \tag{3.107}$$

The GS to the DE is

$$y(x) = c_1 x^{\lambda_1} + c_2 x^{\lambda_1} \ln x = (c_1 + c_2 \ln x) x^{\lambda_1} \tag{3.108}$$

Case 3: $\Delta = (a_1 - a_2)^2 - 4a_2 a_0 < 0$:

The AE (3.100) has two complex roots

$$
\begin{aligned}
\lambda_1 &= \alpha + i\beta \\
\lambda_2 &= \alpha - i\beta
\end{aligned}
\tag{3.109}
$$

One can write the solutions as

$$
\begin{aligned}
y_1(x) &= x^{\alpha+i\beta} \\
&= x^\alpha x^{i\beta} \\
&= x^\alpha \exp(i\beta \ln|x|)
\end{aligned}
\tag{3.110}
$$

Thus,

$$y_1(x) = x^\alpha(\cos(\beta \ln|x|) + i \sin(\beta \ln|x|)) \tag{3.111}$$

Similarly,

$$y_2(x) = x^{\alpha-i\beta} = x^\alpha(\cos(\beta \ln|x|) - i \sin(\beta \ln|x|)) \tag{3.112}$$

Thus, the GS to the DE is

$$y(x) = x^\alpha(c_1 \cos(\beta \ln|x|) + c_2 \sin(\beta \ln|x|)) \tag{3.113}$$

where c_1 and c_2 are two complex constants.

A more general method, substitution, is to transform the v-coeff DEs into c-coeff DEs whose solution methods are well known.

Using substitution

$$x = \exp(t) \tag{3.114}$$

we have

$$\frac{dy}{dx} = \frac{dt}{dt}\frac{dy}{dx} = \frac{dt}{dx}\frac{dy}{dt} = \frac{1}{x}\frac{dy}{dt} \tag{3.115}$$

$$x\frac{dy}{dx} = \frac{dy}{dt} \equiv \dot{y} \tag{3.116}$$

Similarly,

$$\begin{aligned}
\frac{d^2y}{dx^2} &= \frac{d}{dx}\left(\frac{dy}{dx}\right) \\
&= \frac{d}{dx}\left(\frac{1}{x}\frac{dy}{dt}\right) \\
&= -\frac{1}{x^2}\frac{dy}{dt} + \frac{1}{x}\frac{d}{dx}\left(\frac{dy}{dt}\right) \\
&= -\frac{1}{x^2}\dot{y} + \left(\frac{1}{x}\right)\frac{1}{x}\frac{d}{dt}\frac{dy}{dt} \\
&= -\frac{1}{x^2}\dot{y} + \frac{1}{x^2}\ddot{y}
\end{aligned} \tag{3.117}$$

where we used $\frac{d^2y}{dt^2} \equiv \ddot{y}$.

Thus,

$$x^2 y'' = -\dot{y} + \ddot{y} \tag{3.118}$$

This approach can be further generalized to higher orders so that

$$\begin{cases} xy' = \dot{y} \\ x^2 y'' = \ddot{y} - \dot{y} \\ x^3 y''' = \dddot{y} - 3\ddot{y} + 2\dot{y} \end{cases} \tag{3.119}$$

With the formulas in (3.119), we transform our original DE into

$$a_2(-\dot{y} + \ddot{y}) + a_1\dot{y} + a_0 y = 0$$
$$a_2\ddot{y} + (a_1 - a_2)\dot{y} + a_0 y = 0 \tag{3.120}$$

which is now a c-coeff DE and all methods discussed before for c-coeff DEs can be adopted.

Example 4
Find the GS to the following DE

$$x^2 y'' + xy' - y = 0 \tag{3.121}$$

Solution
Method 1: TS method.
$$y(x) = x^\lambda$$

we have
$$y'(x) = \lambda x^{\lambda-1}$$
$$y''(x) = \lambda(\lambda - 1)x^{\lambda-2}$$
The DE becomes $(\lambda(\lambda - 1) + \lambda - 1)x^\lambda = 0$ with the following C-Eq
$$\lambda(\lambda - 1) + \lambda - 1 = 0$$
whose roots are $\lambda_{1,2} = \pm 1$ and the GS is
$$y(x) = c_1 x^1 + c_2 x^{-1}$$
Method 2: The S-method to replace IV.
$$x = \exp(t)$$
We generate a new DE
$$y'' - y = 0$$
whose GS is
$$y(t) = c_1 \exp(t) + c_2 \exp(-t)$$
Back substitution leads to
$$y(x) = c_1 x^1 + c_2 x^{-1}$$

Example 5

Find the GS to the following DE with one given PS $y_1(x) = x$

$$x^2 y'' - x(x+2)y' + (x+2)y = 0 \qquad (3.122)$$

Solution

Given the PS, we assume the GS with an undetermined function $u(x)$

$$y(x) = u(x)y_1(x) = u(x)x$$

Thus,

$$y'(x) = u + u'x$$
$$y''(x) = 2u' + xu''$$

Plugging into the original DE, we have

$$x^2(2u' + xu'') - x(x+2)(u + u'x) + (x+2)ux = 0$$

Simplifying the above, we have

$$u'' = u'$$

This DE is now solvable by the SOV method,

$$u' = C_1 \exp(x)$$

Integrating the above, we have

$$u = C_1 \exp(x) + C_2$$

Back substituting leads to the GS

$$y(x) = (C_1 \exp(x) + C_2)x$$

In general, one can obtain the GS to the following v-coeff DE

$$y'' + a_1(x)y' + a_0(x)y = 0 \qquad (3.123)$$

if one solution $y_1(x)$ is known, by the reduction of order method. One may propose the second solution as

$$y(x) = y_1(x)u(x) \qquad (3.124)$$

Thus,

$$\begin{aligned} y' &= y_1 u' + y_1' u \\ y'' &= y_1' u' + y_1 u'' + y_1'' u + y_1' u' \\ &= 2y_1' u' + y_1 u'' + y_1'' u \end{aligned} \qquad (3.125)$$

where we omitted the (x) in $y_1(x)$, $y'(x)$ and $u(x)$ for writing convenience. Plugging the above into the original DE, we have

$$2y_1' u' + y_1 u'' + y_1'' u + a_1(x)(y_1 u' + y_1' u)$$
$$+ a_0(x)y_1 u = 0$$
$$y_1 u'' + (2y_1' + a_1(x)y_1)u' \qquad (3.126)$$
$$+ (y_1'' + a_1(x)y_1' + a_0(x)y_1)u = 0$$

Since $y_1(x)$ is a solution, we have

$$y_1'' + a_1(x)y_1' + a_0(x)y_1 = 0 \qquad (3.127)$$

Thus,

$$y_1 u'' + (2y_1' + a_1(x)y_1)u' = 0$$

$$\frac{u''}{u'} = -\left(a_1(x) + 2\frac{y_1'}{y_1}\right)$$

$$\ln(u') = -\int \left(a_1(x) + 2\frac{y_1'}{y_1}\right) dx + c_1 \qquad (3.128)$$

$$u' = C_1 \exp\left(-\int \left(a_1(x) + 2\frac{y_1'}{y_1}\right) dx\right)$$

where C_1 and c_1 are integrating constants. Integrating the above, one obtains $u(x)$ and, then, the second solution $y_1(x)u(x)$.

Example 6
Find the GS to the following DE with one given PS $y_1(x) = \frac{\sin x}{x}$

$$xy'' + 2y' + xy = 0 \qquad (3.129)$$

Solution
We first convert the DE into

$$y'' + \left(\frac{2}{x}\right)y' + y = 0$$

and obtain the coefficients $a_1(x) = \frac{2}{x}$. Using $y_1(x) = \frac{\sin x}{x}$, we get

$$a_1(x) + 2\frac{y_1'}{y_1} = 2\frac{\cos x}{\sin x}$$

Thus, using $u' = C_1 \exp\left(-\int \left(a_1(x) + 2\frac{y_1'}{y_1}\right) dx\right)$, we get

$$u' = C_1 \exp\left(-2\int \frac{\cos x}{\sin x} dx\right)$$

$$= C_1 \exp(\ln(\sin x)^{-2})$$

$$= C_1(\sin x)^{-2}$$

Thus,

$$u = C_1 \int \frac{1}{(\sin x)^2} dx + C_2$$

$$= C_1 \frac{\cos x}{\sin x} + C_2$$

The GS is

$$y(x) = \left(C_1 \frac{\cos x}{\sin x} + C_2\right)\frac{\sin x}{x}$$

$$= \frac{1}{x}(C_1 \cos x + C_2 \sin x)$$

Problems

Problem 3.3.1 Use the quadratic formula to solve the following equations:
(1) $x^2 + ix + 2 = 0$
(2) $x^2 - 2ix + 3 = 0$

Problem 3.3.2 (a) Use Euler's formula to show that every complex number can be written in the form $r \exp(i\theta)$, where $r \geq 0$ and $-\pi < \theta \leq \pi$.
(b) Express the numbers 4, -2, $3i$, $1+i$ and $-1+i\sqrt{3}$ in the form of $r \exp(i\theta)$.
(c) The two square roots of $r \exp(i\theta)$ are $\pm\sqrt{r} \exp\left(\frac{i\theta}{2}\right)$. Find the square roots of the numbers $2 - 2i\sqrt{3}$ and $-2 + 2i\sqrt{3}$.

Problem 3.3.3 Compute the Wronskian of the three LI solutions (x_1, x_2, x_3) of the following 3rd-order DE.
$$t^3 x''' + 6t^2 x'' + 7tx' + x = 0$$

Problem 3.3.4 Find the highest point of the solution curve in
$$\begin{cases} y'' + 3y' + 2y = 0 \\ \qquad\qquad y(0) = 1 \\ \qquad\qquad y'(0) = 6 \end{cases}$$

Problem 3.3.5 Find the third quadrant point of intersection of the solution curves with different IC
$$\begin{cases} y'' + 3y' + 2y = 0 \\ y_1(0) = 3, y_1'(0) = 1 \\ y_2(0) = 0, y_2'(0) = 1 \end{cases}$$

Problem 3.3.6 Find the PS to the following IVP

$$\begin{cases} 3y''' + 2y'' = 0 \\ \quad y(0) = -1 \\ \quad y'(0) = 0 \\ \quad y''(0) = 1 \end{cases}$$

Problem 3.3.7 Find a linear c-coeff InHomo DE with the given GS
$$y(x) = (A + Bx + Cx^2)\cos 2x + (D + Ex + Fx^2)\sin 2x$$

Problem 3.3.8 Find the PS to the following IVP
$$\begin{cases} y^{(3)} = y \\ y(0) = 1 \\ y'(0) = y''(0) = 0 \end{cases}$$

Problem 3.3.9 Find the PS to the following IVP
$$\begin{cases} y^{(4)} = y''' + y'' + y' + 2y \\ y(0) = y'(0) = y''(0) = 0 \\ y'''(0) = 30 \end{cases}$$

Problem 3.3.10 The DE
$$y'' + (\text{sign } x)\, y = 0$$
has the discontinuous coefficient function
$$\text{sign } x = \begin{cases} 1, & x > 0 \\ -1, & x < 0 \end{cases}$$
Show that this DE nevertheless has two LI solutions $y_1(x)$ and $y_2(x)$ defined for all x such that
- Each satisfies the DE at each point $x \neq 0$
- Each has a continuous derivative at $x = 0$
- $y_1(0) = y_2(0) = 1$ and $y_1'(0) = y_2'(0) = 0$

Problem 3.3.11 Find the GS to the following DE
$$D^3(D - 2)(D + 3)(D^2 + 1)y = 0$$
where $D = d/dx$

Problem 3.3.12 Find the PS to the following IVP

$$\begin{cases} y'' - 2y' + 2y = 0 \\ \quad y(0) = 0 \\ \quad y'(0) = 5 \end{cases}$$

Problem 3.3.13 Find the PS to the following IVP

$$\begin{cases} y''' + 9y' = 0 \\ \quad y(0) = 3 \\ \quad y'(0) = -1 \\ \quad y''(0) = 2 \end{cases}$$

Problem 3.3.14 Find the GS to the following DE

$$(D - 1)^3 (D - 2)^2 (D - 3)(D^2 + 9)y(x) = 0$$

where $D = d/dx$.

Problem 3.3.15 Find the GS to the following DE

$$y'' - y' - 15y = 0$$

Problem 3.3.16 Find the GS to the following DE

$$9y'' - 12y' + 4y = 0$$

Problem 3.3.17 For a Homo DE with repeated real roots, please
(1) find the GS to the DE $(D - r_1)^{k_1} y(x) = 0$ where $D = \frac{d}{dx}$ is the usual derivative operator, r_1 is a real constant and k_1 is a positive integer;
(2) compose the GS to the following Homo DE

$$(D - r_1)^{k_1}(D - r_2)^{k_2} \cdots (D - r_n)^{k_n} = 0$$

where $r_1, r_2, \ldots, r_n$ are known distinct real constants while $k_1, k_2, \ldots, k_n$ are known positive integers.

Problem 3.3.18 Find the GS to the following DE

$$\left(x\frac{d}{dx} - \alpha \right)^n y(x) = x$$

where $x > 0, \alpha =$ constant and $n =$ positive integer.

Problem 3.3.19 Find the GS to the following DE

$$(ax^2 D^2 + bxD + c)y = 0$$

where $D = d/dx$, and a, b and c are constants.

Problem 3.3.20 Find the GS to the following DE
$$x^2 y'' + xy' - 9y = 0$$

Problem 3.3.21 Find the GS to the following DE
$$ax^2 y'' + bxy' + cy = 0$$

Problem 3.3.22 Find the PS to the following IVP
$$\begin{cases} x^2 y'' - 2xy' + 2y = 0 \\ y(1) = 3, \ \ y'(1) = 1 \end{cases}$$

Problem 3.3.23 Find the GS to the following DE
$$(x + 2)^2 y'' - (x + 2)y' + y = 0$$

Problem 3.3.24 Find the GS to the following DE
$$x^2 y'' - 2xy' - 10y = 0$$

Problem 3.3.25 Find the GS to the following DE
$$x^3 y''' + x^2 y'' - xy' + y = 0$$
where $x > 0$.

3.4   Inhomogeneous Linear DEs

Theorem 1

If y_c is a GS to $Ly = 0$ and y_p is a PS to $Ly = f(x)$, $y = y_C + y_P$ is the GS to the InHomo DE $Ly = f(x)$.

Proof

Since y_C is a GS to $Ly = 0$, we have

$$Ly_C = 0 \tag{3.130}$$

and y_P is a PS to $Ly = f(x)$ gives

$$Ly_P = f(x) \tag{3.131}$$

Then,

$$\begin{aligned} L(y_C + y_P) &= Ly_C + Ly_P \\ &= 0 + f(x) \\ &= f(x) \end{aligned} \tag{3.132}$$

Therefore $y_C + y_P$ is a GS to $Ly = f(x)$.

There are many methods for finding the PS and each has its own merits and drawbacks, depending on the properties of the DEs in hand. These methods include the MUC and VOP that we will discuss immediately, as well as the method of Laplace transforms that we will discuss in a later chapter.

3.4.1   Method of Undetermined Coefficients

If the RHS $f(x) = a \cos kx + b \sin kx$, it is reasonable to expect a PS of the same form

$$y_P = A \cos kx + B \sin kx \qquad (3.133)$$

which is a linear combination with undetermined coefficients A and B. The reason is that any derivative of such a linear combination of $\cos kx$ and $\sin kx$ has the same form. Therefore, we may substitute this form of y_P in the given DE to determine the coefficients A and B by equating coefficients of $\cos kx$ and $\sin kx$ on both sides. The technique of finding a PS to an InHomo DE is rather tricky and tedious; one must remember a few rules.

Let's consider a few examples.

Example 1
Find the PS to the following DE

$$y'' + 3y' + 4y = 3x + 2 \qquad (3.134)$$

Solution
Since $f(x) = 3x + 2$, let's guess that
$$y_P = Ax + B$$
Therefore
$$y_P' = A, \quad y_P'' = 0$$
Substituting these in the given DE, we have
$$y_P'' + 3y_P' + 4y_P = 0 + 3A + 4(Ax + B)$$
$$= 3x + 2$$
$$A = \frac{3}{4}, \quad B = -\frac{1}{16}$$
Therefore, we have a PS
$$y_P = \frac{3}{4}x - \frac{1}{16}$$

Example 2
Find the PS to the following DE

$$y'' - 4y = 2\exp(3x) \qquad (3.135)$$

Solution
Since any derivative of $\exp(3x)$ is a constant multiple of $\exp(3x)$, we select the TS

$$y_P = A \exp(3x)$$

Thus,

$$y_P'' = 9A \exp(3x)$$

Substituting it into the given DE, we have

$$9A \exp(3x) - 4A \exp(3x) = 2 \exp(3x)$$

$$A = \frac{2}{5}$$

Therefore, we find a PS to the given DE

$$y_P = \frac{2}{5} \exp(3x)$$

Example 3

Find the PS to the following DE

$$3y'' + y' - 2y = 2 \cos x \tag{3.136}$$

Solution

Though our first guess might be $y_P = A \cos x$, the presence of y' on the LHS shows us that we will have to include a term of $\sin x$ as well. Selecting the TS

$$y_P = A \cos x + B \sin x$$

we get

$$y_P' = -A \sin x + B \cos x$$
$$y_P'' = -A \cos x - B \sin x$$

Substituting these into the given DE, we have

$$3(-A \cos x - B \sin x) + (-A \sin x + B \cos x) - 2(A \cos x + B \sin x)$$
$$= 2 \cos x$$

Equating the terms of $\cos x$ and $\sin x$, we have

$$\begin{cases} -5A + B = 2 \\ -A - 5B = 0 \end{cases}$$

Solving the above two simultaneous equations we have

$$A = -\frac{5}{13}$$

$$B = \frac{1}{13}$$

Thus, the PS is

$$y_P = -\frac{5}{13} \cos x + \frac{1}{13} \sin x$$

Next, let's introduce a few rules for a few common cases.

Rule 1

If $f(x), f'(x)$ and $f''(x)$ do not satisfy $Ly = 0$, *i.e.*, $f(x)$ is not the solution to $Ly = 0$, the TS y_P can be a linear combination of all terms in $f(x)$.

Meaning of the Rule

Suppose that no term appearing in either $f(x)$ or any of its derivatives satisfies the associated Homo DE $Ly = 0$. We select a TS y_P using the linear combination of all such LI terms and their derivatives. Then, we determine the coefficients by substitution of this TS into the InHomo DE $Ly = f(x)$.

Necessary Conditions

We check the supposition made in Rule 1 by first using the C-Eq to find the complementary function y_c and, then, write a list of all terms appearing in $f(x)$ and its successive derivatives. If none of the terms in this list duplicates a term in y_c, we proceed with Rule 1.

Example 4

Find the GS to the DE by the MUC

$$y'' - 3y' - 4y = 15\exp(4x) \tag{3.137}$$

Solution

Let's find the GS to the Homo DE
$$y'' - 3y' - 4y = 0$$
whose C-Eq is
$$r^2 - 3r - 4 = 0$$
$$r_{1,2} = 4, -1$$
Therefore,
$$y_C = C_1 \exp(4x) + C_2 \exp(-x)$$
Given $f(x) = 15\exp(4x)$, we select a TS $y_P = A\exp(4x)$ so that
$$y_P' = 4A\exp(4x)$$
$$y_P'' = 16A\exp(4x)$$
Plugging this back into the DE, we have

$$\text{LHS} = y_P'' - 3y_P' - 4y$$
$$= 16A\exp(4x) - 3(4A\exp(4x)) - 4(A\exp(4x))$$
$$= (16 - 12 - 4)A\exp(4x)$$
$$= 0$$

While

$$\text{RHS} = 15\exp(4x)$$

Thus, LHS $\neq$ RHS $\forall A$.

No matter what value we opt for A, we will never be able to find the solution. So, after reading the Rule mentioned above and the fact that $\exp(4x)$ term in $y_C(y_C = C_1\exp(4x) + C_2\exp(-x))$ also appears in $f(x)(15\exp(4x))$, we cannot use the term $A\exp(4x)$ as the TS, as the PS should be LI with all terms present in $f(x)$ (that is the PS should not be a constant multiple of the terms present in $f(x)$). So, now, the question is how to solve the DE. Consider Rule 1A.

Rule 1A

If $f(x)$ is a solution to $Ly = 0$, one may compose the TS by multiplying the solution with x.

Example 4a (continued)

Selecting a new TS to the DE $y'' - 3y' - 4y = 15\exp(4x)$

$$y_P = Ax\exp(4x) \tag{3.138}$$

we get

$$y_P' = A\exp(4x) + 4Ax\exp(4x)$$
$$y_P'' = 8A\exp(4x) + 16Ax\exp(4x)$$

Plugging these into the DE, we have

$$\text{LHS} = y_P'' - 3y_P' - 4y_P$$
$$= 8A\exp(4x) + 16Ax\exp(4x) - 3(A\exp(4x) + 4Ax\exp(4x))$$
$$\quad - 4Ax\exp(4x)$$
$$= 5A\exp(4x)$$
$$= \text{RHS} = 15\exp(4x)$$

This gives

$$A = 3$$

Therefore, we have a PS

$$y_P = 3x\exp(4x)$$

and the GS

$$y = C_1 \exp(4x) + C_2 \exp(-x) + 3x \exp(4x)$$

Example 5
Find the PS to the following DE

$$y'' + 4y = 3x^3 \tag{3.139}$$

Solution
We can find the solution to the Homo portion of the above DE as
$$y_C = c_1 \cos 2x + c_2 \sin 2x$$
The function $f(x) = 3x^3$ and its derivatives are constant multiples of the LI functions x^3, x^2, x and 1. Because none of these appears in y_C, we select a TS
$$y_P = Ax^3 + Bx^2 + Cx + D$$
Therefore
$$y_P' = 3Ax^2 + 2Bx + C, \qquad y_P'' = 6Ax + 2B$$
Substituting the above in the DE, we have
$$(6Ax + 2B) + 4(Ax^3 + Bx^2 + Cx + D) = 3x^3$$
Equating the coefficients of the like terms, we have
$$\begin{cases} 4A = 3 \\ 4B = 0 \\ 6A + 4C = 0 \\ 2B + 4D = 0 \end{cases}$$
These give
$$A = \frac{3}{4}, \qquad B = 0, \qquad C = -\frac{9}{8}, \qquad D = 0$$
Therefore, we have the PS
$$y_P = \frac{3}{4}x^3 - \frac{9}{8}x$$

Example 6
Find the general form of a PS to the following DE

$$y''' + 9y' = x \sin x + x^2 \exp(2x) \tag{3.140}$$

Solution
The C-Eq is
$$r^3 + 9r = 0$$
$$r_1 = 0, \qquad r_2 = 3i, \qquad r_3 = -3i$$
So, the GS to the Homo DE is
$$y_C = C_1 + C_2 \cos 3x + C_3 \sin 3x$$

The derivatives of the RHS involves the terms

$$\cos x, \sin x, x \cos x, x \sin x$$

and

$$\exp(2x), x \exp(2x), x^2 \exp(2x)$$

Since there is no duplication with the terms of the GS, the TS takes the form

$$y_P = A \cos x + B \sin x + Cx \cos x + Dx \sin x + E \exp(2x) + Fx \exp(2x) + Gx^2 \exp(2x)$$

Upon substituting y_P in the DE and equating the coefficient terms we get seven equations determining the seven coefficients A, B, C, D, E, F and G

Rule 2 (The case of duplication)

If $f(x)$ contains $P_m(x) \exp(rx) \cos kx$ or $P_m(x) \exp(rx) \sin kx$, the TS is

$$y_P(x) = x^s \big((A_0 + A_1 x + \cdots + A_m x^m) \exp(rx) \cos kx + (B_0 + B_1 x + \cdots + B_m x^m) \exp(rx) \sin kx \big)$$

where P_m is a polynomial of order m and s is the smallest non-negative integer which does not allow duplication of the above solution as the solution to the Homo DE. We, then, determine the coefficients for y_p by substituting y_p into the InHomo DE.

$f(x)$	$y_p(x)$
$P_m(x) = b_0 + b_1 x + \cdots + b_m x^m$	$x^s(A_0 + A_1 x + A_2 x^2 + \cdots + A_m x^m)$
$a \cos kx + b \sin kx$	$x^s(A \cos kx + B \sin kx)$
$\exp(rx)$	$x^s(A \exp(rx))$
$\exp(rx)(a \cos kx + b \sin kx)$	$\exp(rx)(A \cos kx + B \sin kx)$
$P_m(x) \exp(rx)$	$\exp(rx)(A_0 + A_1 x + \cdots + A_m x^m)$
$P_m(x)(a \cos kx + b \sin kx)$	$(A \cos kx + B \sin kx)(A_0 + A_1 x + \cdots + A_m x^m)$

The above table lists the functions that can be chosen as the TSs for a given RHS.

Example 7

Find the PS and GS to the following DE

$$y''' + y'' = 3\exp(x) + 4x^2 \tag{3.141}$$

Solution

We have the C-Eq

$$r^3 + r^2 = 0$$

$$r_1 = r_2 = 0, \qquad r_3 = -1$$

The complementary solution is

$$y_C(x) = C_1 + C_2 x + C_3 \exp(-x)$$

We now select our TS

$$A\exp(x) + B + Cx + Dx^2$$

$f(x)$	y_P
$3\exp(x)$	$\exp(x)$
$4x^2$	$x^s(B + Cx + Dx^2)$

Since we have C_1 and $C_2 x$ in complementary solution, we will have duplication in $B + Cx + Dx^2$ terms in y_C. Therefore $s \neq 0$. If we choose $s = 1$, $Bx + Cx^2 + Dx^3$ will still duplicate the x term. Now let $s = 2$, $Bx^2 + Cx^3 + Dx^4$ will not duplicate any terms in y_C. Therefore, we have $s = 2$. The part $A\exp(x)$ corresponding to $3\exp(x)$ does not duplicate any part of the complementary function, but the part $B + Cx + Dx^2$ must be multiplied by x^2 to eliminate duplication. Thus, we have

$$y_P = A\exp(x) + Bx^2 + Cx^3 + Dx^4$$
$$y_P' = A\exp(x) + 2Bx + 3Cx^2 + 4Dx^3$$
$$y_P'' = A\exp(x) + 2B + 6Cx + 12Dx^2$$
$$y_P''' = A\exp(x) + 6C + 24Dx$$

Substituting these back to the given DE gives

$$A\exp(x) + 6C + 24Dx + A\exp(x) + 2B + 6Cx + 12Dx^2 = 3\exp(x) + 4x^2$$
$$2A\exp(x) + (2B + 6C) + (6C + 24D)x + 12Dx^2 = 3\exp(x) + 4x^2$$

Equating the like terms, we have

$$\begin{cases} 2A = 3 \\ 2B + 6C = 0 \\ 6C + 24D = 0 \\ 12D = 4 \end{cases}$$

Solving the above simultaneous equations gives us

$$A = \frac{3}{2}, \qquad B = 4, \qquad C = -\frac{4}{3}, \qquad D = \frac{1}{3}$$

A PS is

$$y_P = \frac{3}{2}\exp(x) + 4x^2 - \frac{4}{3}x^3 + \frac{1}{3}x^4$$

So, the GS is

$$y = C_1 + C_2 x + C_3 \exp(-x) + \frac{3}{2}\exp(x) + 4x^2 - \frac{4}{3}x^3 + \frac{1}{3}x^4$$

Example 8

Determine the appropriate form for the complementary solution and the PS if given the roots of the C-Eq and $f(x)$. The roots are $r = 2, 2, 2$ and $-2 \pm 3i$ and $f(x) = x^2 \exp(2x) + x \sin 3x$.

Solution

Being given the roots of the C-Eq, we can compose the complementary function as

$$y_C(x) = (C_1 + C_2 x + C_3 x^2) \exp(2x) + \exp(-2x)(C_4 \cos 3x + C_5 \sin 3x)$$

To form the PS, we examine the sum

$$(A + Bx + Cx^2) \exp(2x) + \exp(-2x)\big((D + Ex)\cos 3x + (F + Gx)\sin 3x\big)$$

To eliminate duplications with terms of $y(x)$, we multiply the first part corresponding to $x^2 \exp(2x)$ by x^3, and the second part corresponding to $x \sin 3x$ by x. Thus, the TS is

$$y_P(x) = (Ax^3 + Bx^4 + Cx^5)\exp(2x)$$
$$+ \exp(-2x)\big((Dx + Ex^2)\cos 3x + (Fx + Gx^2)\sin 3x\big)$$

Example 9

Find the GS to the following c-coeff InHomo DE

$$y''' + 9y' = x \sin x + x^2 \exp(2x) \tag{3.142}$$

Solution

First, let's find the GS. We have the C-Eq

$$r^3 + 9r = 0$$
$$r_1 = 0, \qquad r_2 = 3i, \qquad r_3 = -3i$$

The GS is

$$y_C = C_1 + C_2 \cos 3x + C_3 \sin 3x$$

Let's now find the PS from observing the RHS. We compose our trial.

$$y_P = A_1 \sin x + A_2 \cos x + A_3 x \sin x + A_4 x \cos x + A_5 \exp(2x)$$
$$+ A_6 x \exp(2x) + A_7 x^2 \exp(2x)$$

Example 10
Solve the following InHomo DE

$$x^2 y'' + xy' - 16y = x^4 + x^{-4} \qquad \forall x > 0 \qquad (3.143)$$

Solution
Let $x = \exp(t)$. Then,

$$xy' = \frac{dy}{dt}$$
$$x^2 y'' = \frac{d^2 y}{dt^2} - \frac{dy}{dt}$$

The original DE becomes

$$\frac{d^2 y}{dt^2} - \frac{dy}{dt} + \frac{dy}{dt} - 16y = \frac{d^2 y}{dt^2} - 16y$$
$$= \exp(4t) + \exp(-4t)$$

The GS for the new Homo DE is
$$y_C(t) = C_1 \exp(4t) + C_2 \exp(-4t)$$

The TS is
$$y_P(t) = At \exp(4t) + Bt \exp(-4t)$$
$$y_P'(t) = (A + 4At) \exp(4t) + (B - 4Bt) \exp(-4t)$$
$$y_P''(t) = (8A + 16At) \exp(4t) - (8B - 16Bt) \exp(-4t)$$

Thus, plugging into the DE, we get
$$y_P''(t) - 16y_P(t) = 8A \exp(4t) - 8B \exp(-4t)$$
$$= \exp(4t) + \exp(-4t)$$

Thus, $A = \frac{1}{8}$ and $B = -\frac{1}{8}$ and

$$y_P(t) = \frac{1}{8} t(\exp(4t) - \exp(-4t))$$

The GS to the DE is

$$y(t) = C_1 \exp(4t) + C_2 \exp(-4t) + \frac{1}{8} t(\exp(4t) - \exp(-4t))$$

Back substituting, we get

$$y(x) = C_1 x^4 + C_2 x^{-4} + \frac{1}{8} \ln x \, (x^4 - x^{-4})$$

3.4.2 Variation of Parameters

Consider an example $y'' + y = \tan x$. Since $f(x) = \tan x$ has infinitely many LI derivatives $\sec^2 x$, $2 \sec^2 x \tan x$, $4 \sec^2 x \tan^2 x + 2 \sec^4 x$, ..., we do not have a finite linear combination to use as a TS and we do not have corresponding rules to follow for solving DEs of this type. We have to introduce a new method, *i.e.*, VOP.

For $Ly = f(x)$, the Homo DE $Ly = 0$ has n solutions $y_1, y_2, ..., y_n$.

(1) The GS to Homo DE is $y_C = C_1 y_1 + C_2 y_2 + \cdots + C_n y_n$

(2) Let's propose to change the $C_1, C_2, ..., C_n$ in the complementary function to variables $u_1(x), u_2(x), ..., u_n(x)$ for composing a TS to the InHomo DE

$$
\begin{aligned}
y_P(x) &= u_1(x)y_1 + u_2(x)y_2 + \cdots + u_n(x)y_n \\
&= \sum_{i=1}^{n} u_i(x)y_i(x) \\
&= (u_1 \quad u_2 \quad \cdots \quad u_n) \begin{pmatrix} y_1 \\ y_2 \\ \vdots \\ y_n \end{pmatrix} \\
&= u^T Y
\end{aligned}
\tag{3.144}
$$

Example 1

Consider a 2nd.O InHomo DE

$$y'' + P(x)y' + Q(x)y = f(x) \tag{3.145}$$

Compose a formula for the PS for the above DE.

Solution

1) Assume the solutions to the Homo DE are y_1 and y_2. This gives us

$$
\begin{cases} y_1'' + P(x)y_1' + Q(x)y_1 = 0 \\ y_2'' + P(x)y_2' + Q(X)y_2 = 0 \end{cases}
\tag{3.146}
$$

2) Compose a TS y_p as

$$y_P = u_1(x)y_1 + u_2(x)y_2 \tag{3.147}$$

Therefore we have

$$y_P' = (u_1 y_1 + u_2 y_2)'$$
$$= u_1' y_1 + u_1 y_1' + u_2' y_2 + u_2 y_2'$$
$$= (u_1' y_1 + u_2' y_2) + (u_1 y_1' + u_2 y_2')$$

In the above process, we introduced two "free" parameters $u_1(x)$ and $u_2(x)$. With such, we can impose two constraints. Let's force the following condition (although one may impose many other conditions).

$$u_1' y_1 + u_2' y_2 = 0 \tag{3.148}$$

Therefore, we have

$$y_P' = u_1 y_1' + u_2 y_2' \tag{3.149}$$

$$y_P'' = (u_1 y_1' + u_2 y_2')'$$
$$= u_1' y_1' + u_1 y_1'' + u_2' y_2' + u_2 y_2''$$

$$y_P'' = (u_1 y_1'' + u_2 y_2'') + (u_1' y_1' + u_2' y_2') \tag{3.150}$$

From (3.146) we have

$$y_1'' = -P(x) y_1' - Q(x) y_1$$
$$y_2'' = -P(x) y_2' - Q(x) y_2 \tag{3.151}$$

Substitute these into $u_1 y_1'' + u_2 y_2''$

$$u_1 y_1'' + u_2 y_2''$$
$$= u_1(-P(x) y_1' - Q(x) y_1) + u_2(-P(x) y_2' - Q(x) y_2)$$
$$= -P(u_1 y_1' + u_2 y_2') - Q(u_1 y_1 + u_2 y_2) \tag{3.152}$$
$$= -P y_P' - Q y_P$$

Finally, plugging back (3.152) into (3.150)

$$y_P'' = -P y_P' - Q y_P + u_1' y_1' + u_2' y_2'$$
$$y_P'' + P y_P' + Q y_P = u_1' y_1' + u_2' y_2' \tag{3.153}$$

But, according to (3.145), we have

$$y_P'' + Py_P' + Qy_P = u_1'y_1' + u_2'y_2' \qquad (3.154)$$

$$y_P'' + Py_P' + Qy_P = f(x) \qquad (3.155)$$

Thus,

$$u_1'y_1' + u_2'y_2' = f(x) \qquad (3.156)$$

Writing (3.148) and (3.156) in a matrix form, we get

$$\begin{pmatrix} y_1 & y_2 \\ y_1' & y_2' \end{pmatrix} \begin{pmatrix} u_1' \\ u_2' \end{pmatrix} = \begin{pmatrix} 0 \\ f(x) \end{pmatrix} \qquad (3.157)$$

One may easily generalize the above derivation for the higher-order DEs.

$$y^{(n)} + P_1 y^{(n-1)} + \cdots + P_n y = f(x)$$

$$y_c = c_1 y_1 + \cdots + c_n y_n$$

$$y_P = u_1 y_1 + \cdots + u_n y_n$$

$$\begin{pmatrix} y_1 & y_2 & \cdots & y_n \\ y_1' & y_2' & \cdots & y_n' \\ \vdots & \vdots & \ddots & \vdots \\ y_1^{(n-1)} & y_2^{(n-1)} & \cdots & y_n^{(n-1)} \end{pmatrix} \begin{pmatrix} u_1' \\ u_2' \\ \vdots \\ u_n' \end{pmatrix} \qquad (3.158)$$

$$= \begin{pmatrix} 0 \\ \vdots \\ 0 \\ f(x) \end{pmatrix}$$

Next, let's derive a formula for u_1' and u_2' and, ultimately, for y_P. From (3.148), we have

$$u_1' = -\frac{y_2}{y_1} u_2{}' \qquad (3.159)$$

and plugging (3.159) into (3.156) gives

$$f(x) = -y_1' \left(\frac{y_2}{y_1} u_2' \right) + y_2' u_2'$$

$$= \frac{u_2'}{y_1} \begin{vmatrix} y_1 & y_2 \\ y_1' & y_2' \end{vmatrix}$$

$$= \frac{u_2'}{y_1} W(y_1, y_2) \tag{3.160}$$

where $W(y_1, y_2)$ is the Wronskian. Thus, we have

$$u_2'(x) = \frac{y_1(x) f(x)}{W(y_1, y_2)}$$

$$u_2(x) = \int \frac{y_1(t) f(t)}{W(y_1(t), y_2(t))} dt \tag{3.161}$$

The above indefinite integral used the integrating variable t to reduce confusion and it can be considered as a definite integral integrating from some arbitrary point to x. This arbitrary point adds a constant to $u_2(x)$ but it does not impact the PS. The same scheme can be used for both $u_1(x)$ and $u_2(x)$.

Plugging u_2' back to (3.159), we have

$$u_1' = -\frac{y_2(x) f(x)}{W(y_1, y_2)} \tag{3.162}$$

and

$$u_1(x) = -\int \frac{y_2(t) f(t)}{W(y_1(t), y_2(t))} dt \tag{3.163}$$

To summarize, we have

$$\begin{cases} u_1(x) = -\int \dfrac{y_2(t)f(t)}{W\big(y_1(t), y_2(t)\big)}\,dt \\[4mm] u_2(x) = \int \dfrac{y_1(t)f(t)}{W\big(y_1(t), y_2(t)\big)}\,dt \end{cases} \qquad (3.164)$$

Plugging (3.161) and (3.163) into (3.147), we find the PS

$$\begin{aligned} y_P(x) &= u_1(x)y_1(x) + u_2(x)y_2(x) \\ &= -y_1(x)\int \frac{y_2(t)f(t)}{W\big(y_1(t), y_2(t)\big)}\,dt + y_2(x)\int \frac{y_1(t)f(t)}{W\big(y_1(t), y_2(t)\big)}\,dt \\ &= \int \frac{y_2(x)y_1(t) - y_1(x)y_2(t)}{W\big(y_1(t), y_2(t)\big)}\,f(t)\,dt \end{aligned}$$

$$y_P(x) = \int K(x,t)f(t)\,dt \qquad (3.165)$$

where

$$K(x,t) = \frac{y_2(x)y_1(t) - y_1(x)y_2(t)}{W\big(y_1(t), y_2(t)\big)} \qquad (3.166)$$

is called the *kernel* in mathematics. The kernel is also called *Green's function* in physics. This function, depending only on the Homo portion of the DE, is the impulse response of an InHomo DE with specific ICs or BCs. The basic concept was introduced by G. Green (1793-1841) in the 1830s and it has been extended to other fields where it is given other names: *correlation function* in many-body theory and *propagator* in quantum field theory.

Summary

The methods of finding the PS $y_P(x)$ of an InHomo DE depend highly on the nature of the source term of the original DE. This is obvious for the MUC although VOP is much more general; one may find the PS by VOP as well as the MUC. As illustrated by the Venn diagram

(Figure 3.1), the PS that can be found by the MUC can also be found by the VOP although the latter may be much more cumbersome algebraically. However, PS that can be found by VOP is not guaranteed to be found by the MUC. Therefore, when finding the PS to an InHomo DE, the first thing one should do is to analyze the source term for selecting an effective method.

Consider a 2nd.O c-coeff DE

$$y'' + ay' + by = f(x) \tag{3.167}$$

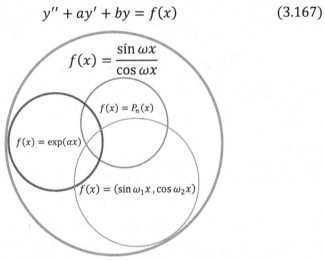

Figure 3.1 This Venn diagram illustrates the logical relationships of the sets for source term $f(x)$ and the applicable methods.

Step 1: Find the C-Eq for the Homo portion

$$Ly = 0 \tag{3.168}$$

That is

$$r^2 + ar + b = 0 \tag{3.169}$$

Step 2: Find the two solutions y_1 and y_2 and compose the complementary solutions y_c of the Homo DE.

$$y_c(x) = c_1 y_1(x) + c_2 y_2(x) \tag{3.170}$$

Step 3: Compose the TS to the InHomo DE $Ly = f(x)$

$$y_P(x) = u_1(x)y_1(x) + u_2(x)y_2(x) \qquad (3.171)$$

where $u_1(x) \neq u_2(x)$ and are functions of x.

Step 4: Solve the equation to find u_1 and u_2

$$\begin{aligned} u_1'y_1 + u_2'y_2 &= 0 \\ u_1'y_1' + u_2'y_2' &= f(x) \end{aligned} \qquad (3.172)$$

OR: Solve the matrix below to find u_1 and u_2

$$\begin{pmatrix} y_1 & y_2 \\ y_1' & y_2' \end{pmatrix} \begin{pmatrix} u_1' \\ u_2' \end{pmatrix} = \begin{pmatrix} 0 \\ f(x) \end{pmatrix} \qquad (3.173)$$

OR: We can find the values of u_1 and u_2 from the formulas given below

$$\begin{aligned} u_1(x) &= -\int^x \frac{y_2(t)f(t)}{W(y_1(t), y_2(t))} dt \\ u_2(x) &= \int^x \frac{y_1(t)f(t)}{W(y_1(t), y_2(t))} dt \end{aligned} \qquad (3.174)$$

where $W(y_1, y_2)$ is the Wronskian of y_1 and y_2.

Step 5: Find y_P by substituting the values of u_1 and u_2 in the equation below

$$y_P = u_1 y_1 + u_2 y_2 \qquad (3.175)$$

Or: Find y_P by using the equation below

$$y_P(x) = \int^x \frac{y_2(x)y_1(t) - y_1(x)y_2(t)}{W(y_1(t), y_2(t))} f(t)dt \qquad (3.176)$$

Now find the GS

$$y(x) = c_1 y_1 + c_2 y_2 + u_1 y_1 + u_2 y_2 \qquad (3.177)$$

Example 2
Find the GS to the following DE

$$y'' - 3y' - 4y = 15\exp(4x) \qquad (3.178)$$

by VOP.

Solution

Step 1: The C-Eq is
$$r^2 - 3r - 4 = 0$$
$$r_1 = 4, \qquad r_2 = -1$$
The solution to the Homo DE is
$$y_1 = \exp(4x), \qquad y_2 = \exp(-x)$$
Step 2: The complementary solution is
$$y_C = c_1\exp(4x) + c_2\exp(-x)$$
Step 3: Composing the TS
$$y_P = u_1\exp(4x) + u_2\exp(-x)$$
Step 4: Find u_1 and u_2
$$\exp(4x)\,u_1' + \exp(-x)\,u_2' = 0 \qquad (3.179)$$
$$4\exp(4x)\,u_1' - \exp(-x)\,u_2' = 15\exp(4x) \qquad (3.180)$$
$$5\exp(4x)\,u_1' = 15\exp(4x) \qquad (3.179)+(3.180)$$
$$u_1' = 3$$
$$u_1(x) = \int 3dx = 3x$$
$$-5u_2'\exp(-x) = 15\exp(4x) \qquad (3.180)-4(3.179)$$
$$u_2' = -3\exp(5x)$$
$$u_2(x) = \int(-3\exp(5x))dx = -\frac{3}{5}\exp(5x)$$
Step 5: We now find the PS
$$y_P(x) = 3x\exp(4x) - \frac{3}{5}\exp(4x)$$
We can find $y_P(x)$ directly using the formula
$$y_P(x) = \int^x \frac{y_2(x)y_1(t) - y_1(x)y_2(t)}{W(y_1(t), y_2(t))} f(t)dt$$

$$y_1(x) = \exp(4x)$$
$$y_2(x) = \exp(-x)$$

$$W[t] = \begin{vmatrix} y_1(t) & y_2(t) \\ y_1'(t) & y_2'(t) \end{vmatrix} = \begin{vmatrix} \exp(4t) & \exp(-t) \\ 4\exp(4t) & -\exp(-t) \end{vmatrix} = -5\exp(3t)$$

$$y_P(x) = \int^x \frac{y_2(x)y_1(t) - y_1(x)y_2(t)}{W[t]} f(t)dt$$
$$= \int^x (y_2(x)y_1(t) - y_1(x)y_2(t))\frac{f(t)}{W[t]}dt$$

$$= \int^x (\exp(-x)\exp(4t) - \exp(4x)\exp(-t)) \frac{15\exp(4t)}{(-5\exp(3t))} dt$$

$$= -3 \int^x (\exp(-x)\exp(5t) - \exp(4x)) dt$$

$$= -3 \exp(-x) \left(\frac{1}{5}\exp(5x) \right) + 3x\exp(4x)$$

$$= -\frac{3}{5}\exp(4x) + 3x\exp(4x)$$

which is the same as the y_p found earlier. Note: we neglected the constant.

Step 6: We now find the GS to the problem

$$y(x) = y_C(x) + y_P(x)$$

$$= C_1\exp(4x) + C_2\exp(-x) + 3x\exp(4x) - \frac{3}{5}\exp(4x)$$

$$= C_3\exp(4x) + C_2\exp(-x) + 3x\exp(4x)$$

where

$$C_3 = C_1 - \frac{3}{5}$$

This GS is the same as we obtained using the MUC.

Example 3

Find the GS to the following DE

$$y'' + y = \tan x$$

Solution

Method 1: VOP (detailed steps).

Here we have $f(x) = \tan x$. The C-Eq to the Homo DE is

$$r^2 + 1 = 0$$

$$r_{1,2} = \pm i$$

$$y_C(x) = C_1 y_1(x) + C_2 y_2(x)$$

$$= C_1 \cos x + C_2 \sin x$$

Thus,

$$y_1'(x) = -\sin x, \qquad y_2'(x) = \cos x$$

Plugging these into

$$u_1' y_1 + u_2' y_2 = 0$$

$$u_1' y_1' + u_2' y_2' = f(x)$$

Solving for $u_1'(x)$ and $u_2'(x)$, we have

$$u_1' = -\sin x \tan x$$

$$= \cos x - \sec x$$

$$u_2' = \cos x \tan x$$

$$= \sin x$$

Therefore

$$u_1(x) = \int^x (\cos t - \sec t)dt$$

$$= \sin x - \ln|\sec x + \tan x|$$

$$u_2(x) = \int^x \sin t \, dt$$

$$= -\cos x$$

$$y_P(x) = u_1 y_1 + u_2 y_2$$

$$= (\sin x - \ln|\sec x + \tan x|)\cos x - \cos x \sin x$$

$$= -\cos x \ln|\sec x + \tan x|$$

Method 2: VOP (using formula).

$$y_1(x) = \cos x$$
$$y_2(x) = \sin x$$
$$K(x,t) = \frac{y_2(x)y_1(t) - y_1(x)y_2(t)}{W[t]}$$
$$= \frac{\sin x \cos t - \cos x \sin t}{1}$$
$$= \sin(x - t)$$

$$y_P(x) = \int^x K(x,t)f(t)dt$$

$$= \int^x \sin(x - t)\tan t \, dt$$

$$= -\cos x \ln|\sec x + \tan x|$$

Finally, we have the GS

$$y(x) = y_C(x) + y_P(x)$$
$$= (C_1 - \ln|\sec x + \tan x|)\cos x + C_2 \sin x$$

Remark: While computing $\int \sec x \, dx$, we usually introduce substitution $u = \sec x + \tan x$ with which we have

$$du = \sec x \,(\sec x + \tan x)dx$$

or

$$\frac{du}{u} = \sec x \, dx$$

Thus,

$$\int \sec x \, dx = \int \frac{du}{u} = \ln|u| + c = \ln|\sec x + \tan x| + c$$

Problems

Problem 3.4.1 Find the G.S of the following DE
$$y'' - 2y' - 8y = \exp(4x)$$

Problem 3.4.2 Find the GS to the following DE
$$y'''' - y = 1$$

Problem 3.4.3 Find the GS to the following DE
$$y'''' - y = 4\exp(x)$$

Problem 3.4.4 Find the GS to the following DE
$$y^{(4)} - y''' - y'' - y' - 2y = 18x^5$$

Problem 3.4.5 Find the PS to the following DE
$$y''' + y'' + y' + y = x^3 + x^2$$

Problem 3.4.6 Find the PS to the following DE
$$y''' + y'' + y' + y = 1 + \exp(x) + \exp(2x) + \exp(3x)$$

Problem 3.4.7 Find the GS to the following DE
$$y^{(4)} + \omega^2 y'' = (x^2 + 1)\sin(\omega x)$$

Problem 3.4.8 Find the GS to the following DE for conditions $\omega \neq \omega_0$ and $\omega = \omega_0$
$$y'' + \omega^2 y = \exp(i\omega_0 x)$$

Problem 3.4.9 Find the GS to the following DE
$$y'' + 9y = \cos 3x$$

Problem 3.4.10 Find the GS to the following DE
$$y'' + 2y' + y = 5(\sin x + \cos x)$$

Problem 3.4.11 Find the GS to the following DE
$$y''' + y'' + y' + y = 1 + \cos x + \sin 2x + \exp(-x)$$

Problem 3.4.12 Find the GS to the following DE
$$x^2 y'' - 4xy' + 6y = x^3$$

Problem 3.4.13 Find the GS to the following DE
$$x^6 y'' + 2x^5 y' - 12x^4 y = 1$$

Problem 3.4.14 Find the GS to the following DE
$$x^2 y'' + 2xy' - 6y = 72x^5$$

Problem 3.4.15 Find the GS to the following DE
$$4x^2 y'' - 4xy' + 3y = 8x^{\frac{4}{3}}$$

Problem 3.4.16 Find the GS to the following DE
$$x^2 y'' + xy' + y = \ln x$$

Problem 3.4.17 Find the GS to the following DE
$$x^3 y''' + x^2 y'' + xy' - y = 1 + x + x^2$$

Problem 3.4.18 Find the GS to the following DE
$$x^2 y'' + xy' - 4y = x^2 + x^{-2}$$

Problem 3.4.19 Given the three LI solutions of the Homo portion of $y^{(3)} + \alpha y^{(2)} + \beta y^{(1)} + \gamma y = f(x)$ as $y_1(x)$, $y_2(x)$, $y_3(x)$, find the GS for the InHomo DE in terms of α, β, γ, and the given functions.

Problem 3.4.20 Use any method of your choice to find a PS $x_P(t)$ for the following DE
$$\frac{d^2 x}{dt^2} + \omega^2 x(t) = f(t)$$
where ω is a constant and
(1) $f(t) = \exp(i\omega t);$

(2) $f(t) = \exp(\omega t)$;

(3) $f(t)$ is a general function.

You may express the PS $x_P(t)$ in terms of the given $f(t)$ in some integral form if you do not have enough information to integrate it.

Problem 3.4.21 For a 2nd.O linear c-coeff InHomo DE
$$y'' - (r_1 + r_2)y' + r_1 r_2 y = f(x)$$
where r_1 and r_2 are two different real constants while $f(x)$ is a real function, find

(1) the two LI solutions $y_1(x)$ and $y_2(x)$ for the Homo portion of the DE in terms of r_1 and r_2;

(2) the PS $y_P(x) = u_1(x)y_1(x) + u_2(x)y_2(x)$ by VOP;

(3) the GS for the original DE.

Problem 3.4.22 For a given Homo DE
$$\frac{d^2 y}{dx^2} + P(x)\frac{dy}{dx} + Q(x)y = 0$$
one solution is given as $y_1(x) \neq 0$. Find the GS to this DE by finding another LI solution $y_2(x) = Z(x)y_1(x)$. Note that $y_1(x), P(x)$ and $Q(x)$ are known functions while $Z(x)$ is not.

Problem 3.4.23 Find the PS to the following DE
$$y'' + y = \cot x$$

Problem 3.4.24 Find the PS to the following DE
$$y''' + y'' + y' + y = f(x)$$
where $f(x)$ is a well-defined function.

Problem 3.4.25 Find a PS to the following DE
$$y'' + 9y = \sin x \tan x$$

Problem 3.4.26 Find a PS to the following DE
$$y'' + \omega^2 y = \sin \omega x \tan \omega x$$

Problem 3.4.27 Find a PS (with constants a, b) of the following DE
$$y'' + a^2 y = \tan bx$$

Problem 3.4.28 For InHomo DE
$$y'' + y = 2\sin x$$
Find
   (1)   the two LI solutions to the Homo portion of the DE;
   (2)   the PS $y_P(x)$ by VOP;
   (3)   the GS for the original DE.

Problem 3.4.29 You can verify by substitution that $y_C = C_1 x + C_2 x^{-1}$ is a complementary function for the 2nd.O InHomo DE
$$x^2 y'' + xy' - y = 72x^5$$
Before applying VOP, you must divide this DE by its leading coefficient x^2 to rewrite it in the standard form
$$y'' + \frac{1}{x}y' - \frac{1}{x^2}y = 72x^3$$
resulting in $f(x) = 72x^3$.
$$L(y) = y'' + P(x)y' + Q(x)y = f(x)$$
Now, find u_1 and u_2 by solving
$$\begin{cases} u_1' y_1 + u_2' y_2 = 0 \\ u_1' y_1' + u_2' y_2' = f(x) \end{cases}$$
and thereby derive the PS
$$y_P = 3x^5$$

Problem 3.4.30 Find the GS to the following DE using two methods,
$$x'' - 3x' - 4x = 15\exp(4t) + 5\exp(-t)$$
(1) The MUC;
(2) VOP.

Problem 3.4.31 Find a PS to the following DE
$$x''' + P_1(t)x'' + P_2(t)x' + P_3(t)x = f(t),$$
whose complementary solution is given as
$$x_C(t) = c_1 x_1(t) + c_2 x_2(t) + c_3 x_3(t).$$
Your solution should be expressed in terms of the given functions.

Chapter 4
Systems of Linear DEs

4.1 Basics of System of DEs

Given below is the general form of a 1st.O DE

$$f(t, x, x') = 0 \qquad (4.1)$$

where t is the IV and x is the DV.

The general system of two 1st.O DEs with two DVs x_1 and x_2 is

$$\begin{cases} f(t, x_1, x_2, x_1', x_2') = 0 \\ g(t, x_1, x_2, x_1', x_2') = 0 \end{cases} \qquad (4.2)$$

where $t =$ Time, the IV, $x_1 =$ Function of $t =$ 1st DV, and $x_2 =$ Function of $t =$ 2nd DV.

Example 1

$$\begin{cases} x_1 + 2x_2 + 3x_1' + 4x_2' = f(t) \\ x_1 + x_2 + x_1' + x_2' = g(t) \end{cases} \tag{4.3}$$

is a coupled system of ODEs.

Example 2

The following two DEs

$$\begin{cases} x_1 + x_2' + x_1'' = f(t) \\ x_2 + x_1' + x_2'' = g(t) \end{cases} \tag{4.4}$$

are coupled because x_1 and x_2 are mixed in the second DE although there is no x_2 in the first DE.

Example 3

While DEs

$$\begin{cases} x_1 + 3x_1' + 5x_1'' = f(t) \\ x_2 + 2x_2' + 4x_2'' = g(t) \end{cases} \tag{4.5}$$

are two independent (uncoupled) DEs because x_1 and x_2 are in no way related by the above two DEs. Solving the two DEs, independently, should yield solutions to the two DEs.

Generalization

A generalization of (4.2) to n^{th} order, m DEs and m DVs can be written as

$$f_1(t; x_1, x_1', x_1'', \ldots, x_1^{(n)}; x_2, x_2', x_2'', \ldots, x_2^{(n)};$$

$$\ldots; x_m, x_m', x_m'', \ldots, x_m^{(n)}) = 0$$

$$f_2(t; x_1, x_1', x_1'', \ldots, x_1^{(n)}; x_2, x_2', x_2'', \ldots, x_2^{(n)};$$

$$\ldots; x_m, x_m', x_m'', \ldots, x_m^{(n)}) = 0$$

$$f_n(t; x_1, x_1', x_1'', \ldots, x_1^{(n)}; x_2, x_2', x_2'', \ldots, x_2^{(n)};$$

$$\ldots; x_m, x_m', x_m'', \ldots, x_m^{(n)}) = 0$$

Matrix format

A system of DEs (DE.Syst) is more compact and, sometimes, more convenient to solve, when expressed in matrix format.

Example 4

A DE.Syst

$$\begin{cases} x_1' = a_{11}x_1 + a_{12}x_2 + b_1(t) \\ x_2' = a_{21}x_1 + a_{22}x_2 + b_1(t) \end{cases} \tag{4.6}$$

can be expressed in the following matrix form

$$\begin{pmatrix} x_1' \\ x_2' \end{pmatrix} = \begin{pmatrix} a_{11} & a_{12} \\ a_{21} & a_{22} \end{pmatrix} \begin{pmatrix} x_1 \\ x_2 \end{pmatrix} + \begin{pmatrix} b_1(t) \\ b_2(t) \end{pmatrix} \tag{4.7}$$

If we define the following

$$X(t) = \big(x_1(t), x_2(t)\big)^T \text{ and } B(t) = \big(b_1(t), b_2(t)\big)^T$$

the DE.Syst can be converted into matrix format

$$X'(t) = AX(t) + B(t) \tag{4.8}$$

where $X(t)$ is the unknown vector, $B(t)$ is the source vector (and is called by at least five other names, *e.g.*, the external force vector, the right-hand side, etc), and A is the matrix.

Problems

Problem 4.1.1 Convert the following DE.Syst into matrix format
$$\begin{cases} x_1' = 4x_1 - 7x_2 \\ x_2' = 2x_1 - 5x_2 \end{cases}$$

Problem 4.1.2 Convert the following DE.Syst into matrix format
$$\begin{cases} x_1' = -x_1 + 5x_2 + 3x_3 + \sin t \\ x_2' = x_2 + x_3 \\ x_3' = -2x_2 - 2x_3 - t \end{cases}$$

4.2   First-Order Systems and Applications

4.2.1   One Block and One Spring

Example 1

One spring and one block without external force:

Consider a system of a block of mass m and a spring with constant k that expands and stretches from the natural length of the spring. The displacement of the center of mass of the block from its naturally resting spot is assumed to be x.

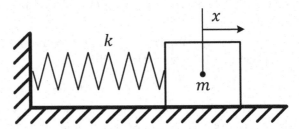

Figure 4.1 A block-spring system with mass m and spring constant k.

Solution

Given the force on the block by the spring is $F = -kx$, we can compose the block's equation of motion

$$mx'' = -kx$$

according to Newton's 2nd law $F = ma = mx''$.

Thus,

$$x'' = -\frac{k}{m}x$$

Defining $\omega^2 = \frac{k}{m}$ as a positive constant, we have

$$x'' = -\omega^2 x$$
$$x'' + \omega^2 x = 0$$

So, the C-Eq is

$$r^2 + \omega^2 = 0$$

Thus,

$$r_{1,2} = \pm i\omega$$

The GS is

$$x(t) = C_1 \cos(\omega t) + C_2 \sin(\omega t)$$

Applying the ICs

$$\begin{cases} x(0) = x_0 \\ x'(0) = v_0 \end{cases}$$

we get $C_1 = x_0$ and $C_2 = \frac{v_0}{\omega}$. Thus, the PS is

$$x(t) = x_0 \cos \omega t + \frac{v_0}{\omega} \sin \omega t$$

Example 2

In this example, we study the motion of the block under the influence of an external force $f(t)$ applied on the block.

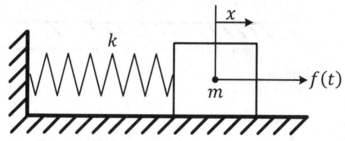

Figure 4.2 A block-spring system with an external force $f(t)$.

The EoM is

$$mx'' = -kx + f(t)$$
$$x'' = -\frac{k}{m}x + \frac{f(t)}{m} \tag{4.9}$$

Let $\omega^2 = \frac{k}{m}$ and $A(t) = \frac{f(t)}{m}$. Then, we have

$$x'' + \omega^2 x = A(t) \tag{4.10}$$

which can be solved using the VOP learned previously

$$x(t) = x_c(t) + x_p(t) \tag{4.11}$$

Obviously, one can compose a much more sophisticated system by tangling it with more springs or more blocks or both. Let's consider a setup with two vibrating blocks.

4.2.2   Two Blocks and Two Springs

Example 1

We move Block 1 by x_1 and, thus, Spring 1 is extended by x_1.

We move Block 2 by x_2 and, thus, Spring 2 is extended by $(x_2 - x_1)$.

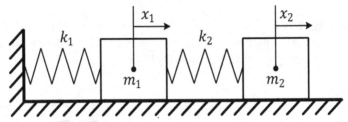

Figure 4.3 A system of two blocks connected by two springs.

Solution

Force on Block 2 is

$$-k_2(x_2 - x_1)$$

Force on Block 1 is

$$-k_1 x_1 + k_2(x_2 - x_1)$$

The H.Syst is

$$\begin{cases} m_1 x_1'' = -k_1 x_1 + k_2(x_2 - x_1) \\ m_2 x_2'' = -k_2(x_2 - x_1) \end{cases}$$

Simplifying the first DE yields

$$x_1'' = -\frac{k_1}{m_1} x_1 + \frac{k_2}{m_1}(x_2 - x_1)$$

Simplifying the second DE yields

$$x_2'' = -\frac{k_2}{m_2}(x_2 - x_1)$$

Let $\omega_1^2 = \frac{k_1}{m_1}$, $\omega_2^2 = \frac{k_2}{m_2}$ and $\omega_{12}^2 = \frac{k_2}{m_1}$ (coupled frequency), one can simplify the Homo system as

$$\begin{cases} x_1'' = -\omega_1^2 x_1 + \omega_{12}^2(x_2 - x_1) \\ x_2'' = -\omega_2^2(x_2 - x_1) \end{cases}$$

Example 2

We study the motion of the two blocks under the influence of a force $f(t)$.

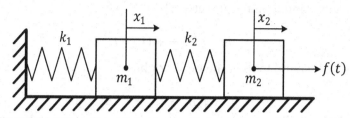

Figure 4.4 A system of two blocks connected by two springs and an external force on block-2.

Solution

The force on m_2 is $-k_2(x_2 - x_1) + f(t)$ while the force on m_1 is still
$$-k_1 x_1 + k_2(x_2 - x_1)$$
The InHomo system now becomes
$$\begin{cases} m_1 x_2'' = -k_2(x_2 - x_1) + f(t) \\ m_2 x_1'' = -k_1 x_1 + k_2(x_2 - x_1) \end{cases}$$
This gives
$$x_1'' = -\frac{k_1}{m_1} x_1 + \frac{k_2}{m_1}(x_2 - x_1)$$
$$x_2'' = -\frac{k_2}{m_2}(x_2 - x_1) + \frac{f(t)}{m_2}$$
$$= -\frac{k_2}{m_2}(x_2 - x_1) + B(t)$$
where $B(t) = \frac{f(t)}{m_2}$
Let $\omega_1^2 = \frac{k_1}{m_1}$, $\omega_2^2 = \frac{k_2}{m_2}$ and $\omega_{12}^2 = \frac{k_2}{m_1}$ (coupled frequency), we have
$$\begin{cases} x_1'' = -\omega_1^2 x_1 + \omega_{12}^2(x_2 - x_1) \\ x_2'' = -\omega_2^2(x_2 - x_1) + B(t) \end{cases}$$

4.2.3   Kirchhoff Circuit Laws

Kirchhoff's circuit laws relate the current and voltage in the lumped element of electric circuits. German physicist G. Kirchhoff (1824-1887) first introduced the laws in 1845 after extending the works of German physicist G. Ohm (1789-1854) and of Scottish physicist J. C. Maxwell (1831-1879). The laws are simply called Kirchhoff's laws or Kirchhoff's rules.

The left-hand loop of the network has

$$E_0 = L\frac{dI_1}{dt} + (I_1 - I_2)R_2 \qquad (4.12)$$

The right-hand loop of the network has

$$\frac{1}{C}Q + I_2 R_1 - (I_1 - I_2)R_2 = 0 \qquad (4.13)$$

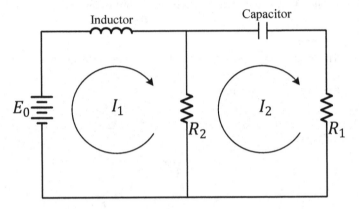

Figure 4.5 An example of Kirchhoff circuit.

Circuit Element	Voltage Drop
Inductor	$L\dfrac{dI}{dt}$
Resistor	RI
Capacitor	$\dfrac{1}{C}Q$

Table: Voltage drops across common circuit element.

Since $\frac{dQ}{dt} = I_2$, differentiation of (4.12) and substitution of $\frac{dQ}{dt}$ yields

$$\frac{1}{C}\frac{dQ}{dt} + (R_1 + R_2)\frac{dI_2}{dt} - R_2\frac{dI_1}{dt} = 0 \qquad (4.14)$$

This gives

$$\frac{dI_2}{dt} = \frac{R_2}{R_1 + R_2}\frac{dI_1}{dt} - \frac{1}{C(R_1 + R_2)}I_2 \qquad (4.15)$$

And from (4.13) we have

$$\frac{dI_1}{dt} = -\frac{R_2}{L}I_1 + \frac{R_2}{L}I_2 + \frac{E_0}{L} \tag{4.16}$$

Substituting $\frac{dI_1}{dt}$ into (4.15), we get

$$\frac{dI_2}{dt} = \frac{R_2}{R_1 + R_2}\left(-\frac{R_2}{L}I_1 + \frac{R_2}{L}I_2 + \frac{E_0}{L}\right) - \frac{1}{C(R_1 + R_2)}I_2$$

$$\frac{dI_2}{dt} = -\frac{R_2^2}{L(R_1 + R_2)}I_1 + \frac{1}{R_1 + R_2}\left(\frac{R_2^2}{L} - \frac{1}{C}\right)I_2 + \frac{R_2}{R_1 + R_2}\frac{E_0}{L} \tag{4.17}$$

From (4.16) and (4.17), we have

$$\frac{d}{dt}\binom{I_1}{I_2} = \begin{pmatrix} -\dfrac{R_2}{L} & \dfrac{R_2}{L} \\ -\dfrac{R_2^2}{L(R_1 + R_2)} & \dfrac{1}{R_1 + R_2}\left(\dfrac{R_2^2}{L} - \dfrac{1}{C}\right) \end{pmatrix}\binom{I_1}{I_2} \tag{4.18}$$

$$+ \begin{pmatrix} \dfrac{E_0}{L} \\ \dfrac{R_2}{R_1 + R_2}\dfrac{E_0}{L} \end{pmatrix}$$

Since, L, C, E_0, R_1 and R_2 are all constants, the above DEs can be condensed as

$$\frac{dI}{dt} = AI + E \tag{4.19}$$

where vector $I = (I_1, I_2)^T$ and A and E are matrices with constant elements.

Problems

Problem 4.2.1 For the setting with typical assumptions that the springs are massless, and the surface the two massive blocks are placed is frictionless, please

(1) construct the DE that governs the motion of the two blocks;

(2) find the GS for the blocks if we break the middle spring;

(3) find the GS for the blocks if $k_1 = k_2 = k_3$ for all cases, you may assume arbitrary IC.

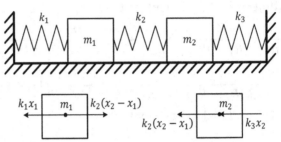

Figure 4.6 The block-spring system for Problem 4.2.1.

Problem 4.2.2 Two particles each of mass m move in a plane with co-ordinates $(x(t), y(t))$ under the influence of a force that is directed toward the origin and has an inverse-square central force field of

$$\frac{k}{x^2 + y^2}.$$

Show that

$$\begin{cases} mx'' = -\dfrac{kx}{r^3} \\ my'' = -\dfrac{ky}{r^3} \end{cases}$$

where $r = \sqrt{x^2 + y^2}$.

Problem 4.2.3 Suppose that a projectile of mass m moves in a vertical plane in the atmosphere near the surface of the Earth under the influence of two forces: a downward gravitational force of magnitude mg, and a resistive force F_R that is directed opposite to the velocity vector v and has magnitude kv^2 (where $v = |v|$ is the speed of the projectile). Show that the EoMs of the projectile are

$$\begin{cases} mx'' = -kvx' \\ my'' = -kvy' - mg \end{cases}$$

Problem 4.2.4 Two massless springs with spring constants k_1 and k_2 are hanging from a ceiling. The top end of the spring-1 is fixed at the ceiling, while the other end is connected to a block-1. The top end of the spring-2 is connected to block-1, while the other end is connected to block-2. The mass for block-1 is m_1, and that for block-2 is m_2. Initially, we hold the blocks so that the springs will stay at their natural lengths. At time $t = 0$, we will release both blocks to let them vibrate under the influence of the spring and gravity with gravitational constant g. Find the motion of the blocks.

Problem 4.2.5 Three massless springs with spring constants k_1, k_2 and k_3 are hanging on a ceiling. The top end of spring-1 is fixed at the ceiling while the other end is connected to block-1. The end of spring-2 is connected to block-1 while the other end is connected to block-2. The top end of spring-3 is connected to block-2 and the other end is connected to block-3. The mass for blocks are m_1, m_2 and m_3. Initially, we hold the blocks so that the springs will stay at their natural lengths. At time $t=0$, we release the blocks to let them vibrate under the influence of the spring and gravity with gravitational constant g. Find the motion of the blocks.

Problem 4.2.6 One end of a massless spring with spring constant k is attached to a ceiling and the other end to a small ball of mass m. Normally, the spring stretches down a little, due to gravity, but stays perfectly vertical. Now, someone moves the ball up a little and the spring recoils to its natural length of L_0. If the ball is moved a little from its natural spot, without stretching or squeezing the spring, the spring stays on a line about θ_0 degrees from the vertical line. After being released, the ball starts some magic motion. Compute its motion with a viable approximation that $\sin \theta \approx \theta$ for a small angle θ.

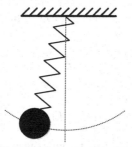

Figure 4.7 The pendulum system for Problem 4.2.6.

4.3   The Substitution Method

Example 1

Find the GS to the following two coupled DEs with ICs, *i.e.*, Homo system

$$\begin{cases} x_1' = x_1 - 8x_2 \\ x_2' = -x_1 + 3x_2 \\ x_1(0) = x_2(0) = 2 \end{cases} \tag{4.20}$$

Solution

Step 1: Solve for one of the two DVs x_1 or x_2 from the given DEs.
The second DE can be written as

$$x_1 = 3x_2 - x_2' \tag{4.21}$$

Differentiating it gets

$$x_1' = 3x_2' - x_2''$$

Plugging x_1 and x_1' into the first DE, we get

$$(3x_2 - x_2')' = (3x_2 - x_2') - 8x_2$$

Therefore

$$3x_2' - x_2'' = 3x_2 - x_2' - 8x_2$$

Thus, we get a 2nd.O DE for x_2

$$x_2'' - 4x_2' - 5x_2 = 0 \tag{4.22}$$

Step 2: Solving the resulting single-variable, decoupled, DE by the C-Eq method, we have

$$r^2 - 4r - 5 = (r - 5)(r + 1) = 0$$

whose roots are $r = 5$ and $r = -1$. Thus, the solution for x_2 is

$$x_2 = C_1 \exp(5t) + C_2 \exp(-t)$$

Step 3: Back substitution to solve for the other DVs. From (4.21), we get

$$\begin{aligned} x_1 &= 3x_2 - x_2' \\ &= 3(C_1 \exp(5t) + C_2 \exp(-t)) - (C_1 \exp(5t) + C_2 \exp(-t))' \\ &= 3C_1 \exp(5t) + 3C_2 \exp(-t) - (5C_1 \exp(5t) - C_2 \exp(-t)) \\ &= -2C_1 \exp(5t) + 4C_2 \exp(-t) \end{aligned}$$

Thus, we have the GS

$$\begin{cases} x_1 = -2C_1 \exp(5t) + 4C_2 \exp(-t) \\ x_2 = C_1 \exp(5t) + C_2 \exp(-t) \end{cases}$$

Step 4: Use the IC's to determine the integration constants.

$$\begin{cases} x_1(0) = 2 = -2C_1 + 4C_2 \\ x_2(0) = 2 = C_1 + C_2 \end{cases}$$

This gives $C_1 = 1$ and $C_2 = 1$. Thus, the PS is
$$\begin{cases} x_1 = -2\exp(5t) + 4\exp(-t) \\ x_2 = \exp(5t) + \exp(-t) \end{cases}$$

Solution Steps:

Step 1 Express one DV by the other DV and its derivatives.

Step 2 Decouple the two DEs by substitution, *i.e.,* obtaining a single-DV DE.

Step 3 Solve the resulting single-DV DE.

Step 4 Back substitution to solve for the other DV.

Example 2
Find the GS to the following Homo system
$$\left\{\begin{array}{r} x' = 4x - 3y \\ y' = 6x - 7y \\ x(0) = 2 \\ y(0) = -1 \end{array}\right. \tag{4.23}$$

Solution
Step 1: Solve the first DE for y in terms of x and x', or solve the second DE for x in terms of y and y'.

There is no written rule on which variable is better to eliminate first. This depends highly on experience and mathematical perception. Fortunately, very often, it does not matter which order to take.

From the first DE $x' = 4x - 3y$, we get
$$y = \frac{4}{3}x - \frac{1}{3}x'$$
and
$$y' = \frac{4}{3}x' - \frac{1}{3}x''$$

Step 2: Plugging y and y' into the second DE, we get
$$\frac{4}{3}x' - \frac{1}{3}x'' = 6x - 7\left(\frac{4}{3}x - \frac{1}{3}x'\right)$$
or
$$x'' + 3x' - 10x = 0$$

Step 3: Solving the resulting single-dependent-variable DE by the C-Eq method, we have

$$r^2 + 3r - 10 = 0$$

Solving this C-Eq gives

$$r_{1,2} = -5, 2$$

The GS for $x(t)$

$$x(t) = C_1 \exp(-5t) + C_2 \exp(2t)$$

Step 4: Back substituting to solve for $y(t)$,

$$y = \frac{4}{3}x - \frac{1}{3}x'$$

$$= \frac{4}{3}(C_1 \exp(-5t) + C_2 \exp(2t)) - \frac{1}{3}(C_1 \exp(-5t) + C_2 \exp(2t))'$$

$$= \frac{4}{3}(C_1 \exp(-5t) + C_2 \exp(2t)) - \frac{1}{3}(-5C_1 \exp(-5t) + 2C_2 \exp(2t))$$

$$= 3C_1 \exp(-5t) + \frac{2}{3}C_2 \exp(2t)$$

The GS is

$$\begin{cases} x(t) = C_1 \exp(-5t) + C_2 \exp(2t) \\ y(t) = 3C_1 \exp(-5t) + \frac{2}{3}C_2 \exp(2t) \end{cases}$$

Step 5: Applying the ICs

$$\begin{cases} x(0) = 2 = C_1 + C_2 \\ y(0) = -1 = 3C_1 + \frac{2}{3}C_2 \end{cases}$$

to generate $C_1 = -1$ and $C_2 = 3$. Thus, the PS is

$$\begin{cases} x = -\exp(-5t) + 3\exp(2t) \\ y = -3\exp(-5t) + 2\exp(2t) \end{cases}$$

Example 3

Find the GS to the following InHomo system

$$\begin{cases} x + 2y' = 3 \\ 2x' - y = 1 \end{cases} \tag{4.24}$$

Solution

Step 1: From the first DE, we get

$$x = 3 - 2y'$$

$$x' = (3 - 2y')' = -2y''$$

Step 2: Substituting the above to the second DE, we get

$$4y'' + y = -1$$

Step 3: Solving the above single-DV DE. The Homo portion's C-Eq roots are $r_{1,2} = \pm\frac{1}{2}i$ and its GS is

$$y_C(t) = C_1 \cos\frac{t}{2} + C_2 \sin\frac{t}{2}$$

Its PS can be obtained by the MUC via the TS

$$y_P(t) = A$$

and, easily, we found $A = -1$. Thus, the GS is

$$y(t) = C_1 \cos\frac{t}{2} + C_2 \sin\frac{t}{2} - 1$$

Step 4: Back substituting into $x = 3 - 2y'$, we get

$$x(t) = -C_2 \cos\frac{t}{2} + C_1 \sin\frac{t}{2} + 3$$

Finally, the GS for the InHomo system is

$$\begin{cases} y(t) = C_1 \cos\frac{t}{2} + C_2 \sin\frac{t}{2} - 1 \\ x(t) = -C_2 \cos\frac{t}{2} + C_1 \sin\frac{t}{2} + 3 \end{cases}$$

Problems

Problem 4.3.1 Solve the following DE system

$$\begin{cases} x'' - y'' + x = 2\exp(-t) \\ x'' + y'' - x = 0 \end{cases}$$

Problem 4.3.2 Solve the following DE system

$$\begin{cases} x' = -y \\ y' = 10x - 7y \\ x(0) = 2 \\ y(0) = -7 \end{cases}$$

Problem 4.3.3 Solve the following DE system

$$\begin{cases} x' = x + 2y \\ y' = 2x - 2y \\ x(0) = 1 \\ y(0) = 2 \end{cases}$$

Problem 4.3.4 Find the GS to the following DE.Syst with the given masses and spring constants.

$$\begin{cases} m_1 x_1'' = -(k_1 + k_2)x_1 + k_2 x_2 \\ m_2 x_2'' = k_2 x_1 - (k_2 + k_3)x_2 \end{cases}$$

where $m_1 = m_2 = 1, k_1 = 1, k_2 = 4$ and $k_3 = 1$

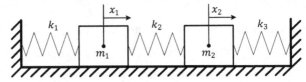

Figure 4.8 The block-spring system for Problem 4.3.4.

Problem 4.3.5 Find the GS to the following InHomo system

$$\begin{cases} x_1' = x_1 + 2x_2 + t \\ x_2' = 2x_1 + x_2 + t^2 \end{cases}$$

Problem 4.3.6 Find the GS to the following Homo system

$$\begin{cases} x' = y \\ y' = -9x + 6y \end{cases}$$

Problem 4.3.7 Solve the following Homo system

$$\begin{cases} x' = x - 2y \\ y' = x - y \\ x(0) = 1 \\ y(0) = 2 \end{cases}$$

Problem 4.3.8 Solve the following Homo system

$$\begin{cases} x' = 3x + 4y \\ y' = 3x + 2y \\ x(0) = 1 \\ y(0) = 1 \end{cases}$$

Problem 4.3.9 Solve the following Homo system

$$\begin{cases} x' = 2x \\ y' = 3x + 3y \\ z' = 4x + 4y + 4z \end{cases}$$

4.4   The Operator Method

In this operator method (O-method) we introduce a new operator

$$D \equiv \frac{d}{dt} \qquad (4.25)$$

Linear Properties of the Operators

$$(1)\ D(x + y) = Dx + Dy \qquad (4.26)$$
$$(2)\ D(\alpha x) = \alpha Dx \qquad (4.27)$$

Combining the above two properties, one can construct the following

$$D(\alpha x + \beta y) = \alpha Dx + \beta Dy \qquad (4.28)$$

where α and β are constants.

Example 1

Solve the following Homo system by the O-method. This problem was solved earlier by the S-method.

$$\begin{cases} x_1' = x_1 - 8x_2 \\ x_2' = -x_1 + 3x_2 \\ x_1(0) = x_2(0) = 2 \end{cases} \qquad (4.29)$$

Solution

Let's rewrite the DEs by using the operator

$$\begin{cases} Dx_1 = x_1 - 8x_2 \\ Dx_2 = -x_1 + 3x_2 \end{cases}$$

That is

$$\begin{cases} (D - 1)x_1 + 8x_2 = 0 & (4.30) \\ x_1 + (D - 3)x_2 = 0 & (4.31) \end{cases}$$

Applying operator $(D - 1)$ to (4.31) and subtracting it by (4.30), we get

$$\big((D - 3)(D - 1) - 8\big)x_2 = 0$$
$$(D^2 - 4D - 5)x_2 = 0$$

whose C-Eq is

$$r^2 - 4r - 5 = 0$$
$$r_1 = 5, r_2 = -1$$

Thus, the GS for x_2 is

$$x_2 = C_1 \exp(5t) + C_2 \exp(-t)$$

Substituting x_2 into (4.31) to solve for x_1, we get

$$x_1 = -2C_1 \exp(5t) + 4C_2 \exp(-t)$$

Applying the ICs, we get $C_1 = 1$ and $C_2 = 1$.

Finally, the PS for the Homo system is

$$\begin{cases} x_1 = -2\exp(5t) + 4\exp(-t) \\ x_2 = \exp(5t) + \exp(-t) \end{cases}$$

Example 2

Solve the following Homo system by the Operator Method

$$\begin{cases} x' = 4x - 3y \\ y' = 6x - 7y \\ x(0) = 2 \\ y(0) = -1 \end{cases} \tag{4.32}$$

Solution

We can rewrite the Homo system by using the time-differential operator

$$\begin{cases} Dx = 4x - 3y \\ Dy = 6x - 7y \end{cases}$$

Thus, two DEs can be written as

$$\begin{cases} (D - 4)x + 3y = 0 & (4.33) \\ 6x - (D + 7)y = 0 & (4.34) \end{cases}$$

Operating on the two DE by $3 \times (4.33) + (D + 7) \times (4.34)$, we get

$$\big((D + 7)(D - 4) + 18\big)x = (D^2 + 3D - 10)x = 0$$

whose C-Eq is

$$r^2 + 3r - 10 = 0$$
$$r_1 = 2$$
$$r_2 = -5$$

This gives

$$x = C_1 \exp(2t) + C_2 \exp(-5t)$$

Back substituting into (4.33) to solve for y.

$$y = \frac{2}{3}C_1 \exp(2t) + 3C_2 \exp(-5t)$$

The GS is

$$\begin{cases} x = C_1 \exp(2t) + C_2 \exp(-5t) \\ y = \frac{2}{3}C_1 \exp(2t) + 3C_2 \exp(-5t) \end{cases}$$

Plugging the IC, we have

$$\begin{cases} x(0) = C_1 + C_2 = 2 \\ y(0) = \dfrac{2}{3}C_1 + 3C_2 = -1 \end{cases}$$

This gives

$$\begin{cases} C_1 = 3 \\ C_2 = -1 \end{cases}$$

The PS is

$$\begin{cases} x = 3\exp(2t) - \exp(-5t) \\ y = 2\exp(2t) - 3\exp(-5t) \end{cases}$$

One application of solving DE.Syst is that one can consider solving DEs of higher order by converting them into a system of 1st.O DEs.

Given a function $f(y''', y'', y', y, x) = 0$

Let

$$\begin{aligned} y_1 &= y \\ y_2 &= y' \\ y_3 &= y'' \\ y_4 &= y''' \end{aligned} \qquad (4.35)$$

Thus, the DE becomes

$$f(y_4, y_3, y_2, y_1, x) = 0 \qquad (4.36)$$

Coupling with the DVs introduced earlier, we can form a DE.Syst,

$$\begin{cases} f(y_4, y_3, y_2, y_1, x) = 0 \\ \qquad\quad y_1' = y_2 \\ \qquad\quad y_2' = y_3 \\ \qquad\quad y_3' = y_4 \end{cases} \qquad (4.37)$$

Solving this DE.Syst can help solve the original DE of higher order.

Problems

Problem 4.4.1 Define two operators
$$L_1 \equiv D^2 + D + 1$$
$$L_2 \equiv D^2 + 2D$$
Applying these two operators to a function $x(t)$, we may get $L_1 L_2 x(t)$ and $L_2 L_1 x(t)$. Check if $L_1 L_2 x(t) = L_2 L_1 x(t)$. What if we define the two operators as follows, check it again?
$$L_1 x(t) \equiv Dx(t) + tx(t)$$
$$L_2 x(t) \equiv tDx(t) + x(t)$$

Problem 4.4.2 Find the GS to the following Homo system
$$\begin{cases} x' = x + 2y + z \\ y' = 6x - y \\ z' = -x - 2y - z \end{cases}$$

Problem 4.4.3 Find the GS to the following InHomo system
$$\begin{cases} (D^2 + D)x + D^2 y = 2 \exp(-t) \\ (D^2 - 1)x + (D^2 - D)y = 0 \end{cases}$$

Problem 4.4.4 Solve the following Homo system
$$\begin{cases} x' = 3x + 9y \\ y' = 2x + 2y \\ x(0) = y(0) = 2 \end{cases}$$

Problem 4.4.5 Find the GS to the following InHomo system
$$\begin{cases} x_1' = x_1 + 2x_2 + t \\ x_2' = 2x_1 + x_2 + t^2 \end{cases}$$

Problem 4.4.6 Find the GS to the following InHomo system
$$\begin{cases} x' = y + z + \exp(-t) \\ y' = x + z \\ z' = x + y \end{cases}$$

Problem 4.4.7 Solve the following Homo system by (1) S-method and (2) the operator method

$$\begin{cases} x' = x + y \\ y' = 6x - y \end{cases}$$

Problem 4.4.8 Find the GS to the following InHomo system

$$\begin{cases} (D + 2)x + (D - 3)y = \exp(-2t) \\ (D - 2)x + (D + 3)y = \exp(3t) \end{cases}$$

Problem 4.4.9 Find the GS to the following InHomo system

$$\begin{cases} x'' + 13y' - 4x = 6\sin t \\ 2x' - y'' + 9y = 0 \end{cases}$$

Problem 4.4.10 Solve the following Homo system by (1) the S-method and (2) the operator method

$$\begin{cases} x' = 3x - 2y \\ y' = 2x + y \end{cases}$$

4.5   The Eigen-Analysis Method

All systems of linear DEs can be expressed in terms of matrices and vectors, which are widely used in linear algebra and multivariable calculus.

For example, we can express the system of three 1st.O DEs below in a matrix format:

$$\begin{pmatrix} x_1 \\ x_2 \\ x_3 \end{pmatrix}' = \begin{pmatrix} a_{11} & a_{12} & a_{13} \\ a_{21} & a_{22} & a_{23} \\ a_{31} & a_{32} & a_{33} \end{pmatrix} \begin{pmatrix} x_1 \\ x_2 \\ x_3 \end{pmatrix} + \begin{pmatrix} f_1(t) \\ f_2(t) \\ f_3(t) \end{pmatrix} \tag{4.38}$$

by a general (dimensionally implicit) notation

$$X'(t) = AX(t) + F(t) \tag{4.39}$$

where

$$X(t) = \begin{pmatrix} x_1 \\ x_2 \\ x_3 \end{pmatrix}, A = \begin{pmatrix} a_{11} & a_{12} & a_{13} \\ a_{21} & a_{22} & a_{23} \\ a_{31} & a_{32} & a_{33} \end{pmatrix}, F(t) = \begin{pmatrix} f_1(t) \\ f_2(t) \\ f_3(t) \end{pmatrix} \tag{4.40}$$

where X is the vector of DVs, A is the coefficient matrix, and $F(t)$ is the InHomo term or the source term or the RHS.

As in the discussions of single DEs, if $F(t) = (0,0,0)^T$, it is a Homo system. O.W., it is an InHomo system. Similar to solving single DEs, a Homo system is much easier to solve than InHomo system. We discuss a method that requires analysis of the eigen properties of the coefficient matrix and we call this method, for lack of widely accepted name, eigen-analysis method (E-method).

For a Homo system:

$$X' = AX \tag{4.41}$$

we select a TS

$$X = V \exp(\lambda t) \tag{4.42}$$

where V is a vector and λ is a scalar and both are determined by the matrix A

$$X' = V\lambda \exp(\lambda t) \tag{4.43}$$

Plugging it into the above Homo system, we get

$$V\lambda \exp(\lambda t) = AV \exp(\lambda t) \tag{4.44}$$

or

$$(A - \lambda I)V \exp(\lambda t) = 0 \tag{4.45}$$

Since $\exp(\lambda t) \neq 0$, we must have $(A - \lambda I)V = 0$. Considering $V = 0$ gives only trivial solution, we are only interested in the case where

$$\det(A - \lambda I) = 0 \tag{4.46}$$

That is, λ is the eigenvalue (e-value) and V is the associated eigenvector (e-vector) of the matrix A. Thus, the GS to a Homo system can be written as

$$X = \sum c_i V_i \exp(\lambda_i t) \tag{4.47}$$

Now, for three different cases of the e-values, we have three different solution methods.

Case 1: Non-repeated real e-values. The above linear combination can be used to compose the GS. See Example 1 below.

Case 2: Repeated real e-values. Composing the second or third or more solutions is a much trickier process which is demonstrated by Example 3.

Case 3: Non-repeated complex e-values. The above linear combination for composing the GS holds except when the term involving $\exp(\lambda_i t)$ or the e-vector V_i or both is complex. See Examples 4 and 5 below.

For an InHomo system, we need to find a PS and, then, add it to the solution to the Homo portion of the system. To find the PS, we can use the MUC, as well as VOP. A few examples will demonstrate the procedure.

Example 1

Find the GS to the following Homo system

$$\begin{pmatrix} x \\ y \end{pmatrix}' = \begin{pmatrix} 4 & -3 \\ 6 & -7 \end{pmatrix} \begin{pmatrix} x \\ y \end{pmatrix} \tag{4.48}$$

Solution

The e-values of coefficient matrix

$$A = \begin{pmatrix} 4 & -3 \\ 6 & -7 \end{pmatrix}$$

can be found by solving

$$\begin{aligned} \det(A - \lambda I) = \det \begin{pmatrix} 4-\lambda & -3 \\ 6 & -7-\lambda \end{pmatrix} \\ = (4-\lambda)(-7-\lambda) + 18 \\ = \lambda^2 + 3\lambda - 10 \\ = (\lambda - 2)(\lambda + 5) \\ = 0 \end{aligned}$$

which are $\lambda_1 = 2$ and $\lambda_2 = -5$.

For $\lambda_1 = 2$, we can find the e-vector V_1 via

$$AV_1 = \lambda_1 V_1$$

$$\begin{pmatrix} 4 & -3 \\ 6 & -7 \end{pmatrix} V_1 = 2V_1$$

as $V_1 = (3,2)^T$. Similarly, for $\lambda_2 = -5$, the e-vector is $V_2 = (1,3)^T$.

Thus, the GS to the Homo system is

$$\begin{pmatrix} x \\ y \end{pmatrix} = c_1 \begin{pmatrix} 3 \\ 2 \end{pmatrix} \exp(2t) + c_2 \begin{pmatrix} 1 \\ 3 \end{pmatrix} \exp(-5t)$$

Example 2

Find the GS to the following Homo system

$$\begin{cases} x' = 3x + 4y \\ y' = 3x + 2y \end{cases} \tag{4.49}$$

Solution

The original Homo system can be written as $X' = AX$ where

$$A = \begin{pmatrix} 3 & 4 \\ 3 & 2 \end{pmatrix}$$

We can easily find its e-values by

$$\det(A - \lambda I) = \det \begin{vmatrix} 3 - \lambda & 4 \\ 3 & 2 - \lambda \end{vmatrix} = 0$$

with $\lambda_1 = -1$, $\lambda_2 = 6$. Next, we find the associated e-vectors.

For $\lambda_1 = -1$, we get

$$\begin{pmatrix} 3 - (-1) & 4 \\ 3 & 2 - (-1) \end{pmatrix} V_1 = \begin{pmatrix} 0 \\ 0 \end{pmatrix}$$

Thus, the e-vector is

$$V_1 = \begin{pmatrix} 1 \\ -1 \end{pmatrix}$$

So, we get

$$X_1 = \begin{pmatrix} 1 \\ -1 \end{pmatrix} \exp(-t)$$

For $\lambda_2 = 6$, we get

$$\begin{pmatrix} 3 - 6 & 4 \\ 3 & 2 - 6 \end{pmatrix} V_2 = \begin{pmatrix} 0 \\ 0 \end{pmatrix}$$

Thus, the e-vector is

$$V_2 = \begin{pmatrix} 4 \\ 3 \end{pmatrix}$$

So, we get

$$X_2 = \begin{pmatrix} 4 \\ 3 \end{pmatrix} \exp(6t)$$

Finally, the GS is

$$X = C_1 \begin{pmatrix} 1 \\ -1 \end{pmatrix} \exp(-t) + C_2 \begin{pmatrix} 4 \\ 3 \end{pmatrix} \exp(6t)$$

Example 3

Find the GS to the following Homo system.

$$\begin{pmatrix} x \\ y \end{pmatrix}' = \begin{pmatrix} 1 & -3 \\ 3 & 7 \end{pmatrix} \begin{pmatrix} x \\ y \end{pmatrix} \tag{4.50}$$

Solution

The A matrix has two identical e-values

$$\lambda_1 = 4, \qquad \lambda_2 = 4$$

The associated e-vector V_1 for the e-value λ_1 is

$$V_1 = \begin{pmatrix} 1 \\ -1 \end{pmatrix}$$

and the solution is

$$X_1(t) = V_1 \exp(\lambda_1 t) = \begin{pmatrix} 1 \\ -1 \end{pmatrix} \exp(4t)$$

To find the GS for the Homo system, we need to find another solution $X_2(t)$ that is LI of $X_1(t)$. We may propose a TS by imitating our solution method for a single DE

$$X_2(t) = t\, V_2 \exp(\lambda t)$$

which fails instantly because one cannot find such a vector V_2.

A more suitable TS can be

$$X_2(t) = (t\, V_1 + V_2)\exp(\lambda t)$$
$$= tV_1 \exp(\lambda t) + V_2 \exp(\lambda t)$$

and its derivative is

$$X_2' = (V_1)\exp(\lambda t) + (V_1 t + V_2)\lambda \exp(\lambda t)$$

Plugging them into the original Homo system, we get

$$X_2' = (V_1)\exp(\lambda t) + (V_1 t + V_2)\lambda \exp(\lambda t) = A(V_1 t + V_2)\exp(\lambda t)$$
$$(A - \lambda I)V_2 \exp(\lambda t) = V_1 \exp(\lambda t)$$
$$(A - \lambda I)V_2 = V_1$$

The vector V_2 satisfying the above system can be used to compose $X_2(t)$. Additionally, since $(A - \lambda I)V_1 = 0$, we have

$$(A - \lambda I)(A - \lambda I)V_2 = (A - \lambda I)V_1 = 0$$
$$(A - \lambda I)^2 V_2 = 0$$

The vector V_2 satisfying the above system can also be used to compose $X_2(t)$.

Now, with two LI solutions $X_1(t)$ and $X_2(t)$, we can compose the GS for the original Homo system as

$$X(t) = C_1 X_1(t) + C_2 X_2(t)$$

where C_1 and C_2 are usual constants.

We may find V_2 using two approaches. The first approach is to solve the following

$$(A - \lambda I)V_2 = V_1$$

or

$$\begin{pmatrix} -3 & -3 \\ 3 & 3 \end{pmatrix} V_2 = \begin{pmatrix} 1 \\ -1 \end{pmatrix}$$

which has infinitely many solutions for V_2 and we can, for example, set it as

$$V_2 = \begin{pmatrix} 0 \\ -1/3 \end{pmatrix}$$

The second approach is to solve the following

$$(A - \lambda I)^2 V_2 = 0$$

We notice

$$(A - \lambda I)^2 V_2 = \begin{pmatrix} -3 & -3 \\ 3 & 3 \end{pmatrix}\begin{pmatrix} -3 & -3 \\ 3 & 3 \end{pmatrix} V_2 = \begin{pmatrix} 0 & 0 \\ 0 & 0 \end{pmatrix} V_2 = 0$$

which means any solution V_2 will satisfy the above linear system. For convenience, we may select the following possibly the simplest solution

$$V_2 = \begin{pmatrix} 0 \\ 1 \end{pmatrix}$$

Finally, we have the two LI solutions

$$X_1(t) = \begin{pmatrix} 1 \\ -1 \end{pmatrix} \exp(4t)$$

$$X_2(t) = \left(\begin{pmatrix} 1 \\ -1 \end{pmatrix} t + \begin{pmatrix} 0 \\ 1 \end{pmatrix} \right) \exp(4t) = \begin{pmatrix} t \\ -t+1 \end{pmatrix} \exp(4t)$$

So, finally, the GS for the original Homo system is

$$X = C_1 X_1 + C_2 X_2 = C_1 \begin{pmatrix} 1 \\ -1 \end{pmatrix} \exp(4t) + C_2 \begin{pmatrix} t \\ -t+1 \end{pmatrix} \exp(4t)$$

Example 4

Find the GS to the following Homo system.

$$\begin{pmatrix} x \\ y \end{pmatrix}' = \begin{pmatrix} 1 & -2 \\ 1 & -1 \end{pmatrix} \begin{pmatrix} x \\ y \end{pmatrix} \tag{4.51}$$

Solution

We can find the e-values by

$$\det(A - \lambda I) = \det\begin{pmatrix} 1-\lambda & -2 \\ 1 & -1-\lambda \end{pmatrix} = \lambda^2 + 1 = 0$$

whose roots are complex numbers $\lambda_{1,2} = \mp i$.

Next, we find the associated e-vectors.

For $\lambda_1 = -i$, we get

$$\begin{pmatrix} 1-(-i) & -2 \\ 1 & -1-(-i) \end{pmatrix} \begin{pmatrix} v_{1a} \\ v_{1b} \end{pmatrix} = \begin{pmatrix} 0 \\ 0 \end{pmatrix}$$

Thus, $(1+i)v_{1a} - 2v_{1b} = 0$, we may select

$$V_1 = \begin{pmatrix} 1 \\ (1+i)/2 \end{pmatrix}$$

So, we get

$$X_1(t) = \begin{pmatrix} 1 \\ (1+i)/2 \end{pmatrix} \exp(-it)$$

For $\lambda_2 = i$, we get

$$\begin{pmatrix} 1-i & -2 \\ 1 & -1-i \end{pmatrix} \begin{pmatrix} v_{2a} \\ v_{2b} \end{pmatrix} = \begin{pmatrix} 0 \\ 0 \end{pmatrix}$$

Thus, $(1-i)v_{2a} - 2v_{2b} = 0$, we may select

$$V_2 = \begin{pmatrix} 1 \\ (1-i)/2 \end{pmatrix}$$

So, we get

$$X_2(t) = \begin{pmatrix} 1 \\ (1-i)/2 \end{pmatrix} \exp(it)$$

Finally, the GS is

$$X(t) = C_1 \begin{pmatrix} 1 \\ (1+i)/2 \end{pmatrix} \exp(-it) + C_2 \begin{pmatrix} 1 \\ (1-i)/2 \end{pmatrix} \exp(it)$$

which may be converted into a trigonometric format. If one wishes to do so, the Euler's formula, $\exp(it) = \cos x + i \sin x$, is needed.

Example 5
Find the GS to the following Homo system.

$$\begin{pmatrix} x \\ y \end{pmatrix}' = \begin{pmatrix} 1 & -2 \\ 2 & 1 \end{pmatrix}\begin{pmatrix} x \\ y \end{pmatrix} \tag{4.52}$$

Solution
We can find the e-values by

$$\det(A - \lambda I) = \det\begin{pmatrix} 1-\lambda & -2 \\ 2 & 1-\lambda \end{pmatrix} = (\lambda - 1)^2 + 4 = 0$$

whose roots are complex numbers

$$\lambda_1 = 1 - 2i, \qquad \lambda_2 = 1 + 2i$$

Next, we find the associated e-vectors.
For $\lambda_1 = 1 - 2i$, we get

$$\begin{pmatrix} 1-(1-2i) & -2 \\ 2 & 1-(1-2i) \end{pmatrix}\begin{pmatrix} v_{1a} \\ v_{1b} \end{pmatrix} = \begin{pmatrix} 0 \\ 0 \end{pmatrix}$$

Thus, $(i)v_{1a} - v_{1b} = 0$, we may select

$$V_1 = \begin{pmatrix} 1 \\ i \end{pmatrix}$$

So, we get

$$X_1(t) = \begin{pmatrix} 1 \\ i \end{pmatrix}\exp((1-2i)t) = \begin{pmatrix} 1 \\ i \end{pmatrix}\exp(t)\exp(-2it)$$

For $\lambda_2 = 1 + 2i$, we get

$$\begin{pmatrix} 1-(1+2i) & -2 \\ 2 & 1-(1+2i) \end{pmatrix}\begin{pmatrix} v_{2a} \\ v_{2b} \end{pmatrix} = \begin{pmatrix} 0 \\ 0 \end{pmatrix}$$

Thus, $(i)v_{2a} + v_{2b} = 0$, we may select

$$V_1 = \begin{pmatrix} 1 \\ -i \end{pmatrix}$$

So, we get

$$X_2(t) = \begin{pmatrix} 1 \\ -i \end{pmatrix}\exp((1+2i)t) = \begin{pmatrix} 1 \\ -i \end{pmatrix}\exp(t)\exp(2it)$$

Finally, the GS is

$$X(t) = C_1\begin{pmatrix} 1 \\ i \end{pmatrix}\exp(t)\exp(-2it) + C_2\begin{pmatrix} 1 \\ -i \end{pmatrix}\exp(t)\exp(2it)$$

$$= \exp(t)\left(C_1\begin{pmatrix} 1 \\ i \end{pmatrix}\exp(-2it) + C_2\begin{pmatrix} 1 \\ -i \end{pmatrix}\exp(2it)\right)$$

Again, one may convert to express it in terms of trigonometric functions but it is highly unnecessary.

Example 6

Find the GS to the following InHomo system.

$$\begin{pmatrix} x_1 \\ x_2 \end{pmatrix}' = \begin{pmatrix} 1 & 2 \\ 2 & 1 \end{pmatrix} \begin{pmatrix} x_1 \\ x_2 \end{pmatrix} + \begin{pmatrix} t \\ t^2 \end{pmatrix} \tag{4.53}$$

Solution

First, we solve the Homo system

$$\begin{pmatrix} x_1 \\ x_2 \end{pmatrix}' = \begin{pmatrix} 1 & 2 \\ 2 & 1 \end{pmatrix} \begin{pmatrix} x_1 \\ x_2 \end{pmatrix}$$

It can be written as

$$X' = AX$$

where

$$A = \begin{pmatrix} 1 & 2 \\ 2 & 1 \end{pmatrix}$$

This Homo system can be solved by any of the three methods discussed earlier while the E-method is demonstrated here.

$$\det(A - \lambda I) = \det \begin{pmatrix} 1-\lambda & 2 \\ 2 & 1-\lambda \end{pmatrix} = (1-\lambda)^2 - 4 = 0$$

whose roots, or A's e-values, are

$$\lambda_1 = 3, \qquad \lambda_2 = -1$$

For $\lambda_1 = 3$, we get the e-vector by solving

$$\begin{pmatrix} -2 & 2 \\ 2 & -2 \end{pmatrix} V_1 = \begin{pmatrix} 0 \\ 0 \end{pmatrix}$$

and the e-vector is

$$V_1 = \begin{pmatrix} 1 \\ 1 \end{pmatrix}$$

For $\lambda_1 = -1$, we get the e-vector by solving

$$\begin{pmatrix} 2 & 2 \\ 2 & 2 \end{pmatrix} V_2 = \begin{pmatrix} 0 \\ 0 \end{pmatrix}$$

and the e-vector is

$$V_2 = \begin{pmatrix} -1 \\ 1 \end{pmatrix}$$

So, the GS to the Homo system is

$$X = \begin{pmatrix} x_1 \\ x_2 \end{pmatrix} = C_1 \begin{pmatrix} 1 \\ 1 \end{pmatrix} \exp(3t) + C_2 \begin{pmatrix} -1 \\ 1 \end{pmatrix} \exp(-t)$$

The InHomo term

$$F(t) = \begin{pmatrix} t \\ t^2 \end{pmatrix} = \begin{pmatrix} 0 \\ 1 \end{pmatrix} t^2 + \begin{pmatrix} 1 \\ 0 \end{pmatrix} t$$

The TS is

$$X_P(t) = E_2 t^2 + E_1 t + E_0$$

where E_2, E_1, E_0 are vectors. The derivative of the above is

$$X_P'(t) = 2E_2 t + E_1$$

Plugging them into the original system, we get

$$2E_2 t + E_1 = A(E_2 t^2 + E_1 t + E_0) + \begin{pmatrix} 0 \\ 1 \end{pmatrix} t^2 + \begin{pmatrix} 1 \\ 0 \end{pmatrix} t$$

Matching the coefficients of the polynomial of t, we have

$$\begin{cases} AE_2 + \begin{pmatrix} 0 \\ 1 \end{pmatrix} = \begin{pmatrix} 0 \\ 0 \end{pmatrix} \\ 2E_2 = AE_1 + \begin{pmatrix} 1 \\ 0 \end{pmatrix} \\ E_1 = AE_0 \end{cases}$$

Solving the above, we get

$$E_0 = \frac{1}{27} \begin{pmatrix} -43 \\ 38 \end{pmatrix}$$

$$E_1 = \frac{1}{9} \begin{pmatrix} 11 \\ -16 \end{pmatrix}$$

$$E_2 = \frac{1}{3} \begin{pmatrix} -2 \\ 1 \end{pmatrix}$$

Then, the GS to the original InHomo system is

$$X(t) = \begin{pmatrix} x_1 \\ x_2 \end{pmatrix}$$

$$= C_1 \begin{pmatrix} 1 \\ 1 \end{pmatrix} \exp(3t) + C_2 \begin{pmatrix} -1 \\ 1 \end{pmatrix} \exp(-t)$$

$$+ \frac{1}{3} \begin{pmatrix} -2 \\ 1 \end{pmatrix} t^2 + \frac{1}{9} \begin{pmatrix} 11 \\ -16 \end{pmatrix} t + \frac{1}{27} \begin{pmatrix} -43 \\ 38 \end{pmatrix}$$

Example 7

Find the GS to the following InHomo system

$$X' = \begin{pmatrix} 1 & 3 \\ 1 & -1 \end{pmatrix} X + \begin{pmatrix} 1 \\ 0 \end{pmatrix} \exp(-t) \tag{4.54}$$

Solution

The coefficient matrix of the system is

$$A = \begin{pmatrix} 1 & 3 \\ 1 & -1 \end{pmatrix}$$

whose e-value equation is

$$\det(A - \lambda I) = \det \begin{pmatrix} 1 - \lambda & 3 \\ 1 & -1 - \lambda \end{pmatrix} = \lambda^2 - 4 = 0$$

whose roots, or A's e-values, are

$$\lambda_1 = 2, \qquad \lambda_2 = -2$$

For $\lambda_1 = 2$, we get the e-vector by solving

$$\begin{pmatrix} -1 & 3 \\ 1 & -3 \end{pmatrix} V_1 = \begin{pmatrix} 0 \\ 0 \end{pmatrix}$$

and the e-vector is

$$V_1 = \begin{pmatrix} 3 \\ 1 \end{pmatrix}$$

For $\lambda_1 = -2$, we get the e-vector by solving

$$\begin{pmatrix} 3 & 3 \\ 1 & 1 \end{pmatrix} V_2 = \begin{pmatrix} 0 \\ 0 \end{pmatrix}$$

and the e-vector is

$$V_2 = \begin{pmatrix} -1 \\ 1 \end{pmatrix}$$

So, the GS to the Homo system is

$$X = \begin{pmatrix} x_1 \\ x_2 \end{pmatrix} = C_1 \begin{pmatrix} 3 \\ 1 \end{pmatrix} \exp(2t) + C_2 \begin{pmatrix} -1 \\ 1 \end{pmatrix} \exp(-2t)$$

The InHomo term is

$$F(t) = \begin{pmatrix} 1 \\ 0 \end{pmatrix} \exp(-t)$$

The TS is

$$X_P(t) = E \exp(-t)$$

where E is a vector. The derivative of the above is

$$X_P'(t) = -E \exp(-t)$$

Plugging them into the original system, we get

$$-E \exp(-t) = A(E \exp(-t)) + \begin{pmatrix} 1 \\ 0 \end{pmatrix} \exp(-t)$$

Matching the coefficients of the terms of $\exp(-t)$, we have

$$-E = AE + \begin{pmatrix} 1 \\ 0 \end{pmatrix}$$

Or

$$(A + I)E = \begin{pmatrix} -1 \\ 0 \end{pmatrix}$$

recalling I is the identity matrix

$$I = \begin{pmatrix} 1 & 0 \\ 0 & 1 \end{pmatrix}$$

Or

$$\begin{pmatrix} 1+1 & 3 \\ 1 & -1+1 \end{pmatrix} E = \begin{pmatrix} -1 \\ 0 \end{pmatrix}$$

Solving the above, we get

$$E = -\frac{1}{3} \begin{pmatrix} 0 \\ 1 \end{pmatrix}$$

A PS to the original InHomo system is

$$X_P(t) = -\frac{1}{3} \begin{pmatrix} 0 \\ 1 \end{pmatrix} \exp(-t)$$

Then, the GS to the original InHomo system is

$$X(t) = C_1 \begin{pmatrix} 3 \\ 1 \end{pmatrix} \exp(2t) + C_2 \begin{pmatrix} -1 \\ 1 \end{pmatrix} \exp(-2t) - \frac{1}{3} \begin{pmatrix} 0 \\ 1 \end{pmatrix} \exp(-t)$$

In general, for the following InHomo system

$$X' = AX + B \exp(\alpha t)$$

we may select a TS

$$X_P(t) = E \exp(\alpha t)$$

such that

$$X'_P(t) = \alpha\, E \exp(\alpha t)$$

Plugging into the original InHomo system, we have

$$\alpha\, E \exp(\alpha t) = AE \exp(\alpha t) + B \exp(\alpha t)$$

Matching coefficients, we have

$$\alpha\, E = AE + B$$

or

$$(A - \alpha I)E = -B$$

As long as $(A - \alpha I)$ is not singular, *i.e.*, α is not an e-value of A, we have

$$E = -(A - \alpha I)^{-1}B$$

The PS is

$$X_P(t) = (-(A - \alpha I)^{-1}B) \exp(\alpha t)$$

However, if α is an e-value of A, the problem is much subtler.

Example 8

Find the GS to the following InHomo system.

$$X' = \begin{pmatrix} 1 & 3 \\ 1 & -1 \end{pmatrix} X + \begin{pmatrix} 1 \\ 0 \end{pmatrix} \exp(-2t) \tag{4.55}$$

Solution

Method 1: The S-method.

We may construct the following

$$\begin{cases} x'_1 = x_1 + 3x_2 + \exp(-2t) \\ x'_2 = x_1 - x_2 \end{cases}$$

From the second DE, we get

$$x_1 = x'_2 + x_2$$
$$x'_1 = x''_2 + x'_2$$

Inserting them to the first DE, we get

$$x''_2 - 4x_2 = \exp(-2t)$$

whose GS can be obtained using the C-Eq method and the MUC as

$$x_2(t) = C_1 \exp(2t) + C_2 \exp(-2t) - \frac{1}{4}t \exp(-2t)$$

Plugging it into the second DE, we get

$$x_1(t) = 3C_1 \exp(2t) - C_2 \exp(-2t) - \frac{1}{4}\exp(-2t) + \frac{1}{4}t \exp(-2t)$$

The GS's can be expressed in the following vector form

$$\begin{pmatrix} x_1 \\ x_2 \end{pmatrix} = C_1 \begin{pmatrix} 3 \\ 1 \end{pmatrix} \exp(2t) + C_2 \begin{pmatrix} -1 \\ 1 \end{pmatrix} \exp(-2t) - \frac{1}{4}\begin{pmatrix} 1 \\ 0 \end{pmatrix} \exp(-2t)$$
$$+ \frac{1}{4}\begin{pmatrix} 1 \\ -1 \end{pmatrix} t \exp(-2t)$$

Method 2: The E-method and the MUC.

Using the E-method, we find the GS to the Homo system

$$X = \begin{pmatrix} x_1 \\ x_2 \end{pmatrix} = C_1 \begin{pmatrix} 3 \\ 1 \end{pmatrix} \exp(2t) + C_2 \begin{pmatrix} -1 \\ 1 \end{pmatrix} \exp(-2t)$$

The InHomo term is

$$F(t) = \begin{pmatrix} 1 \\ 0 \end{pmatrix} \exp(-2t)$$

and $\exp(-2t)$ is a part of the solution to the Homo system.
The TS is

$$X_P(t) = E \exp(-2t) + Gt \exp(-2t)$$

where E and G are vectors. The derivative of the above is

$$X_P'(t) = -2E \exp(-2t) + G \exp(-2t) - 2Gt \exp(-2t)$$

Plugging them into the original system, we get

$$-2E \exp(-2t) + G \exp(-2t) - 2Gt \exp(-2t)$$
$$= A(E \exp(-2t) + Gt \exp(-2t)) + \begin{pmatrix} 1 \\ 0 \end{pmatrix} \exp(-2t)$$

Matching the coefficients of the terms of $\exp(-2t)$ and those of $t \exp(-2t)$, we have

$$\begin{cases} -2E + G = AE + \begin{pmatrix} 1 \\ 0 \end{pmatrix} \\ -2G = AG \end{cases}$$

Solving the above, we get

$$E = -\frac{1}{4} \begin{pmatrix} 1 \\ 0 \end{pmatrix}$$

$$G = \frac{1}{4} \begin{pmatrix} 1 \\ -1 \end{pmatrix}$$

One PS to the original InHomo system is

$$X_P(t) = -\frac{1}{4} \begin{pmatrix} 1 \\ 0 \end{pmatrix} \exp(-2t) + \frac{1}{4} \begin{pmatrix} 1 \\ -1 \end{pmatrix} t \exp(-2t)$$

Then, the GS to the original InHomo system is

$$X(t) = C_1 \begin{pmatrix} 3 \\ 1 \end{pmatrix} \exp(2t) + C_2 \begin{pmatrix} -1 \\ 1 \end{pmatrix} \exp(-2t) - \frac{1}{4} \begin{pmatrix} 1 \\ 0 \end{pmatrix} \exp(-2t)$$
$$+ \frac{1}{4} \begin{pmatrix} 1 \\ -1 \end{pmatrix} t \exp(-2t)$$

Example 9

Find the GS to the following InHomo system.

$$\begin{pmatrix} x \\ y \end{pmatrix}' = \begin{pmatrix} 1 & -1 \\ 1 & 3 \end{pmatrix} \begin{pmatrix} x \\ y \end{pmatrix} + \begin{pmatrix} \sin t \\ \cos t \end{pmatrix} \tag{4.56}$$

Solution

We use the E-method. The coefficient matrix is

$$A = \begin{pmatrix} 1 & -1 \\ 1 & 3 \end{pmatrix}$$

and its e-value equation is

$$\det(A - \lambda I) = \det \begin{pmatrix} 1 - \lambda & -1 \\ 1 & 3 - \lambda \end{pmatrix} = \lambda^2 - 4\lambda + 4 = 0$$

whose roots, or the A's e-values, are

$$\lambda_1 = 2, \qquad \lambda_2 = 2$$

For $\lambda_1 = 1$, we get the e-vector by solving

$$\begin{pmatrix} -1 & -1 \\ 1 & 1 \end{pmatrix} V_1 = \begin{pmatrix} 0 \\ 0 \end{pmatrix}$$

and the e-vector is

$$V_1 = \begin{pmatrix} 1 \\ -1 \end{pmatrix}$$

So, the first solution is

$$\begin{pmatrix} 1 \\ -1 \end{pmatrix} \exp(2t)$$

The second solution is

$$(V_1 t + V_2) \exp(2t)$$

where V_2 satisfies the following equations

$$(A - \lambda_1 I)^2 V_2 = 0$$

$$(A - \lambda_1 I)^2 = \begin{pmatrix} -1 & -1 \\ 1 & 1 \end{pmatrix} \begin{pmatrix} -1 & -1 \\ 1 & 1 \end{pmatrix} = \begin{pmatrix} 0 & 0 \\ 0 & 0 \end{pmatrix}$$

Essentially, any V_2 will work. For convenience, we select

$$V_2 = \begin{pmatrix} 0 \\ -1 \end{pmatrix}$$

So, the GS to the Homo system is

$$X_C(t) = C_1 \begin{pmatrix} 1 \\ -1 \end{pmatrix} \exp(2t) + C_2 \left(\begin{pmatrix} 1 \\ -1 \end{pmatrix} t + \begin{pmatrix} 0 \\ -1 \end{pmatrix} \right) \exp(2t)$$

The InHomo term

$$F(t) = \begin{pmatrix} 1 \\ 0 \end{pmatrix} \sin t + \begin{pmatrix} 0 \\ 1 \end{pmatrix} \cos t$$

The TS is

$$X_P(t) = E_s \sin t + E_c \cos t$$

where E_s and E_c are vectors. The derivative of the above is

$$X_P'(t) = E_s \cos t - E_c \sin t$$

Plugging them into the original InHomo system, we get

$$E_s \cos t - E_c \sin t = A(E_s \sin t + E_c \cos t) + \begin{pmatrix} 1 \\ 0 \end{pmatrix} \sin t + \begin{pmatrix} 0 \\ 1 \end{pmatrix} \cos t$$

Matching the $\sin t$ terms, we get

$$-E_c = A E_s + \begin{pmatrix} 1 \\ 0 \end{pmatrix}$$

Matching the $\cos t$ terms, we get

$$E_s = A\,E_c + \begin{pmatrix} 0 \\ 1 \end{pmatrix}$$

Combining these two equations, we get

$$-E_c = A\left(AE_c + \begin{pmatrix} 0 \\ 1 \end{pmatrix}\right) + \begin{pmatrix} 1 \\ 0 \end{pmatrix}$$

$$= A^2 E_c + A\begin{pmatrix} 0 \\ 1 \end{pmatrix} + \begin{pmatrix} 1 \\ 0 \end{pmatrix}$$

$$= A^2 E_c + \begin{pmatrix} 0 \\ 3 \end{pmatrix}$$

We know

$$A^2 = \begin{pmatrix} 0 & -4 \\ 4 & 8 \end{pmatrix}$$

Finally,

$$\begin{pmatrix} 1 & -4 \\ 4 & 9 \end{pmatrix} E_c = \begin{pmatrix} 0 \\ -3 \end{pmatrix}$$

Using the Cramer's rule that may be found in a linear algebra book

$$\begin{pmatrix} a_{11} & a_{12} \\ a_{21} & a_{22} \end{pmatrix}^{-1} = \frac{\begin{pmatrix} a_{22} & -a_{12} \\ -a_{21} & a_{11} \end{pmatrix}}{\det\begin{pmatrix} a_{11} & a_{12} \\ a_{21} & a_{22} \end{pmatrix}}$$

we get

$$\begin{pmatrix} 1 & -4 \\ 4 & 9 \end{pmatrix}^{-1} = \frac{1}{25}\begin{pmatrix} 9 & 4 \\ -4 & 1 \end{pmatrix}$$

Thus,

$$E_c = \frac{1}{25}\begin{pmatrix} 9 & 4 \\ -4 & 1 \end{pmatrix}\begin{pmatrix} 0 \\ -3 \end{pmatrix}$$

$$= -\frac{1}{25}\begin{pmatrix} 12 \\ 3 \end{pmatrix}$$

$$E_s = A\,E_c + \begin{pmatrix} 0 \\ 1 \end{pmatrix}$$

$$= \begin{pmatrix} 1 & -1 \\ 1 & 3 \end{pmatrix}\left(-\frac{1}{25}\begin{pmatrix} 12 \\ 3 \end{pmatrix}\right) + \begin{pmatrix} 0 \\ 1 \end{pmatrix}$$

$$= \frac{1}{25}\begin{pmatrix} -9 \\ 4 \end{pmatrix}$$

Thus, we have

$$E_c = -\frac{1}{25}\begin{pmatrix} 12 \\ 3 \end{pmatrix}$$

$$E_s = \frac{1}{25}\begin{pmatrix} -9 \\ 4 \end{pmatrix}$$

Thus, the PS is

$$X_P(t) = \frac{1}{25}\begin{pmatrix} -9 \\ 4 \end{pmatrix}\sin t - \frac{1}{25}\begin{pmatrix} 12 \\ 3 \end{pmatrix}\cos t$$

Then, the GS to the original InHomo system is

$$X(t) = X_C(t) + X_P(t)$$

$$= C_1 \begin{pmatrix} 1 \\ -1 \end{pmatrix} \exp(2t) + C_2 \left(\begin{pmatrix} 1 \\ -1 \end{pmatrix} t + \begin{pmatrix} 0 \\ -1 \end{pmatrix} \right) \exp(2t) + \frac{1}{25} \begin{pmatrix} -9 \\ 4 \end{pmatrix} \sin t$$

$$- \frac{1}{25} \begin{pmatrix} 12 \\ 3 \end{pmatrix} \cos t$$

To corroborate, we re-solve the InHomo system using the S-method. The original InHomo system can be written as

$$\begin{cases} x' = x - y + \sin t \\ y' = x + 3y + \cos t \end{cases}$$

From the first DE, we have

$$y = x - x' + \sin t$$
$$y' = x' - x'' + \cos t$$

Plugging them into the Second DE, we get

$$x'' - 4x' + 4x = -3 \sin t$$

Solving the above 2nd.O InHomo DE, we get

$$x_C(t) = C_1 \exp(2t) + C_2 t \exp(2t)$$

The TS is

$$x_P(t) = A \sin t + B \cos t$$

We have

$$x_P'(t) = A \cos t - B \sin t$$
$$x_P''(t) = -A \sin t - B \cos t$$

Plugging into the InHomo DE and matching coefficients for $\sin t$ and $\cos t$, we get

$$4B + 3A = -3$$
$$3B - 4A = 0$$

Thus,

$$A = -\frac{9}{25}, \quad B = -\frac{12}{25}$$

Thus,

$$x_P(t) = -\frac{9}{25} \sin t - \frac{12}{25} \cos t$$

The GS is

$$x(t) = C_1 \exp(2t) + C_2 t \exp(2t) - \frac{9}{25} \sin t - \frac{12}{25} \cos t$$

Taking the derivative wrt t of the above, we get

$$x'(t) = 2C_1 \exp(2t) + C_2 \exp(2t) + 2C_2 t \exp(2t) - \frac{9}{25} \cos t + \frac{12}{25} \sin t$$

Plugging back, we get

$$y(t) = -C_1 \exp(2t) - C_2 \exp(2t) - C_2 t \exp(2t) + \frac{4}{25} \sin t - \frac{3}{25} \cos t$$

If we wish to express it in vector form, it is

$$\begin{pmatrix} x(t) \\ y(t) \end{pmatrix} = C_1 \begin{pmatrix} 1 \\ -1 \end{pmatrix} \exp(2t) + C_2 \left(\begin{pmatrix} 1 \\ -1 \end{pmatrix} t + \begin{pmatrix} 0 \\ -1 \end{pmatrix} \right) \exp(2t) + \frac{1}{25} \begin{pmatrix} -9 \\ 4 \end{pmatrix} \sin t$$
$$- \frac{1}{25} \begin{pmatrix} 12 \\ 3 \end{pmatrix} \cos t$$

One may express the GS for the Homo system as

$$X_C(t) = C_1 \begin{pmatrix} 1 \\ -1 \end{pmatrix} \exp(2t) + C_2 \left(\begin{pmatrix} 1 \\ -1 \end{pmatrix} t + \begin{pmatrix} 0 \\ -1 \end{pmatrix} \right) \exp(2t)$$

One may express the PS to the InHomo system as

$$X_P(t) = \frac{1}{25} \begin{pmatrix} -9 \\ 4 \end{pmatrix} \sin t - \frac{1}{25} \begin{pmatrix} 12 \\ 3 \end{pmatrix} \cos t$$

Example 10

Find the GS to the following InHomo system.

$$\begin{pmatrix} x \\ y \end{pmatrix}' = \begin{pmatrix} 2 & 1 \\ 1 & 2 \end{pmatrix} \begin{pmatrix} x \\ y \end{pmatrix} + \begin{pmatrix} \exp(t) \\ t \end{pmatrix} \qquad (4.57)$$

Solution

We use the E-method. The coefficient matrix is

$$A = \begin{pmatrix} 2 & 1 \\ 1 & 2 \end{pmatrix}$$

The e-value equation is

$$\det(A - \lambda I) = \det \begin{pmatrix} 2-\lambda & 1 \\ 1 & 2-\lambda \end{pmatrix} = (2-\lambda)^2 - 1 = 0$$

whose roots, or the A's e-values, are

$$\lambda_{1,2} = 1, 3$$

For $\lambda_1 = 1$, we get the e-vector by solving

$$\begin{pmatrix} 1 & 1 \\ 1 & 1 \end{pmatrix} V_1 = \begin{pmatrix} 0 \\ 0 \end{pmatrix}$$

and the e-vector is

$$V_1 = \begin{pmatrix} -1 \\ 1 \end{pmatrix}$$

For $\lambda_2 = 3$, we get the e-vector by solving

$$\begin{pmatrix} -1 & 1 \\ 1 & -1 \end{pmatrix} V_2 = \begin{pmatrix} 0 \\ 0 \end{pmatrix}$$

and the e-vector is

$$V_2 = \begin{pmatrix} 1 \\ 1 \end{pmatrix}$$

So, the GS to the Homo system is

$$X(t) = C_1 \begin{pmatrix} -1 \\ 1 \end{pmatrix} \exp(t) + C_2 \begin{pmatrix} 1 \\ 1 \end{pmatrix} \exp(3t)$$

The InHomo term

$$F(t) = \begin{pmatrix} \exp(t) \\ t \end{pmatrix} = \begin{pmatrix} 0 \\ 1 \end{pmatrix} t + \begin{pmatrix} 1 \\ 0 \end{pmatrix} \exp(t)$$

The TS is

$$X_P(t) = G_0 + G_1 t + E_0 \exp(t) + E_1 t \exp(t)$$

where G_0, G_1, E_0, E_1 are vectors.

The term $E_1 t \exp(t)$ was chosen because $V_1 \exp(t)$ was part of the complementary solution to the Homo system. The term $E_0 \exp(t)$ was chosen because $\begin{pmatrix} 1 \\ 0 \end{pmatrix} \exp(t)$ is an InHomo term and because E_0 is not a multiple of V_1, i. e., $E_0 \nparallel V_1$, in general.

The derivative of the above is

$$X'_P(t) = G_1 + E_0 \exp(t) + E_1 \exp(t) + E_1 t \exp(t)$$

Making $X_P(t)$ satisfy the original system $X'_P(t) = AX_P(t) + F(t)$, we get

$$G_1 + E_0 \exp(t) + E_1 \exp(t) + E_1 t \exp(t)$$
$$= A(G_0 + G_1 t + E_0 \exp(t) + E_1 t \exp(t)) + \begin{pmatrix} 0 \\ 1 \end{pmatrix} t$$
$$+ \begin{pmatrix} 1 \\ 0 \end{pmatrix} \exp(t)$$

Matching the corresponding terms, we get

$$G_1 = AG_0$$
$$0 = AG_1 + \begin{pmatrix} 0 \\ 1 \end{pmatrix}$$
$$E_0 + E_1 = AE_0 + \begin{pmatrix} 1 \\ 0 \end{pmatrix}$$
$$E_1 = AE_1$$

Solving the above four equations, we get 4 vectors

$$G_0 = \frac{1}{9} \begin{pmatrix} 4 \\ -5 \end{pmatrix}$$
$$G_1 = \frac{1}{3} \begin{pmatrix} 1 \\ -2 \end{pmatrix}$$
$$E_0 = -\frac{1}{2} \begin{pmatrix} 1 \\ 0 \end{pmatrix}$$
$$E_1 = \frac{1}{2} \begin{pmatrix} 1 \\ -1 \end{pmatrix}$$

Then, the GS to the original InHomo system is

$$X(t) = C_1 \begin{pmatrix} -1 \\ 1 \end{pmatrix} \exp(t) + C_2 \begin{pmatrix} 1 \\ 1 \end{pmatrix} \exp(3t) + \frac{1}{9} \begin{pmatrix} 4 \\ -5 \end{pmatrix} + \frac{1}{3} \begin{pmatrix} 1 \\ -2 \end{pmatrix} t$$
$$- \frac{1}{2} \begin{pmatrix} 1 \\ 0 \end{pmatrix} \exp(t) + \frac{1}{2} \begin{pmatrix} 1 \\ -1 \end{pmatrix} t \exp(t)$$

To corroborate, we re-solve the InHomo system using the S-method.
$$\begin{cases} x' = 2x + y + \exp(t) \\ y' = x + 2y + t \end{cases}$$
From the Second DE, we have
$$x = y' - 2y - t$$
$$x' = y'' - 2y' - 1$$
Plugging into the first DE, we get
$$y'' - 4y' + 3y = 1 - 2t + \exp(t)$$
Solving the above 2nd.O InHomo DE, we get
$$y_C(t) = C_1 \exp(t) + C_2 \exp(3t)$$
The TS is
$$y_P(t) = A_0 + A_1 t + E t \exp(t)$$
and
$$y_P'(t) = A_1 + E \exp(t) + E t \exp(t)$$
$$y_P''(t) = 2E \exp(t) + E t \exp(t)$$
Plugging into the original DE, we get
$$2E \exp(t) + E t \exp(t) - 4(A_1 + E \exp(t) + E t \exp(t))$$
$$+ 3(A_0 + A_1 t + E t \exp(t)) = 1 - 2t + \exp(t)$$
Matching t^0: $-4A_1 + 3A_0 = 1$
Matching t^1: $3A_1 = -2$
Matching $\exp(t)$: $-2E = 1$
Solving these, we get
$$A_0 = -\frac{5}{9}$$
$$A_1 = -\frac{2}{3}$$
$$E = -\frac{1}{2}$$
Thus,
$$y_P(t) = -\frac{5}{9} - \frac{2}{3}t - \frac{1}{2}t \exp(t)$$
The GS is
$$y(t) = C_1 \exp(t) + C_2 \exp(3t) - \frac{5}{9} - \frac{2}{3}t - \frac{1}{2}t \exp(t)$$
Plugging back, we get
$$x(t) = -C_1 \exp(t) + C_2 \exp(3t) + \frac{4}{9} + \frac{1}{3}t - \frac{1}{2}\exp(t) + \frac{1}{2}t \exp(t)$$
If we wish to express it in vector form, it is

$$\begin{pmatrix} x(t) \\ y(t) \end{pmatrix} = C_1 \begin{pmatrix} -1 \\ 1 \end{pmatrix} \exp(t) + C_2 \begin{pmatrix} 1 \\ 1 \end{pmatrix} \exp(3t) + \frac{1}{9} \begin{pmatrix} 4 \\ -5 \end{pmatrix} + \frac{1}{3} \begin{pmatrix} 1 \\ -2 \end{pmatrix} t$$

$$- \frac{1}{2} \begin{pmatrix} 1 \\ 0 \end{pmatrix} \exp(t) + \frac{1}{2} \begin{pmatrix} 1 \\ -1 \end{pmatrix} t \exp(t)$$

One may express the GS for the Homo system as

$$X_C(t) = C_1 \begin{pmatrix} -1 \\ 1 \end{pmatrix} \exp(t) + C_2 \begin{pmatrix} 1 \\ 1 \end{pmatrix} \exp(3t)$$

One may express the PS to the InHomo system as

$$X_P(t) = \frac{1}{9} \begin{pmatrix} 4 \\ -5 \end{pmatrix} + \frac{1}{3} \begin{pmatrix} 1 \\ -2 \end{pmatrix} t - \frac{1}{2} \begin{pmatrix} 1 \\ 0 \end{pmatrix} \exp(t) + \frac{1}{2} \begin{pmatrix} 1 \\ -1 \end{pmatrix} t \exp(t)$$

Example 11

Find the GS to the following InHomo system.

$$X'(t) = \begin{pmatrix} 2 & 4 \\ 1 & -1 \end{pmatrix} X(t) + \begin{pmatrix} \exp(t) \\ -t \end{pmatrix} \tag{4.58}$$

Solution

We use the E-method. The coefficient matrix is

$$A = \begin{pmatrix} 2 & 4 \\ 1 & -1 \end{pmatrix}$$

The e-value equation is

$$\det(A - \lambda I) = \det \begin{pmatrix} 2 - \lambda & 4 \\ 1 & -1 - \lambda \end{pmatrix} = (\lambda + 2)(\lambda - 3) = 0$$

whose roots, or the A's e-values, are

$$\lambda_{1,2} = -2, 3$$

The e-vector for $\lambda_1 = -2$ is

$$V_1 = \begin{pmatrix} -1 \\ 1 \end{pmatrix} \text{ by solving } \begin{pmatrix} 2 & 4 \\ 1 & -1 \end{pmatrix} V_1 = -2V_1$$

The e-vector for $\lambda_2 = 3$ is

$$V_2 = \begin{pmatrix} 4 \\ 1 \end{pmatrix} \text{ by solving } \begin{pmatrix} 2 & 4 \\ 1 & -1 \end{pmatrix} V_2 = 3V_2$$

So, the GS to the Homo system is

$$X(t) = C_1 \begin{pmatrix} -1 \\ 1 \end{pmatrix} \exp(-2t) + C_2 \begin{pmatrix} 4 \\ 1 \end{pmatrix} \exp(3t)$$

The InHomo term

$$F(t) = \begin{pmatrix} \exp(t) \\ -t \end{pmatrix} = - \begin{pmatrix} 0 \\ 1 \end{pmatrix} t + \begin{pmatrix} 1 \\ 0 \end{pmatrix} \exp(t)$$

The TS is

$$X_P(t) = G_0 + G_1 t + E \exp(t)$$

where G_0, G_1, E are vectors. The derivative of the above is

$$X'_P(t) = G_1 + E \exp(t)$$

Making $X_P(t)$ satisfy the original system $X'_P(t) = AX_P(t) + F(t)$, we get

$$G_1 + E \exp(t) = A(G_0 + G_1 t + E \exp(t)) - \begin{pmatrix} 0 \\ 1 \end{pmatrix} t + \begin{pmatrix} 1 \\ 0 \end{pmatrix} \exp(t)$$

Matching the corresponding terms, we get

$$G_1 = AG_0$$

$$0 = AG_1 - \binom{0}{1}$$

$$E = AE + \binom{1}{0}$$

Solving the above 4 equations, we get the vectors

$$G_0 = \frac{1}{9}\binom{-1}{2}$$

$$G_1 = \frac{1}{3}\binom{2}{-1}$$

$$E = -\frac{1}{6}\binom{2}{1}$$

Then, the GS to the original InHomo system is

$$X(t) = C_1 \binom{-1}{1} \exp(-2t) + C_2 \binom{4}{1} \exp(3t) + \frac{1}{9}\binom{-1}{2} + \frac{1}{3}\binom{2}{-1}t$$
$$- \frac{1}{6}\binom{2}{1} \exp(t)$$

Example 12
Find the GS to the following DE.Syst

$$\begin{pmatrix} x \\ y \\ z \end{pmatrix}' = \begin{pmatrix} 5 & 5 & 2 \\ -6 & -6 & -5 \\ 6 & 6 & 5 \end{pmatrix} \begin{pmatrix} x \\ y \\ z \end{pmatrix} \qquad (4.59)$$

Solution
Using the operator notation, we can write the Homo system as

$$\begin{pmatrix} D-5 & -5 & -2 \\ 6 & D+6 & 5 \\ -6 & -6 & D-5 \end{pmatrix} \begin{pmatrix} x \\ y \\ z \end{pmatrix} = 0$$

$$\begin{pmatrix} D-5 & -5 & -2 \\ 6 & D+6 & 5 \\ -6 & -6 & D-5 \end{pmatrix} \Rightarrow \begin{pmatrix} 6 & D+6 & 5 \\ D-5 & -5 & -2 \\ -6 & -6 & D-5 \end{pmatrix}$$

$$\Rightarrow \begin{pmatrix} 6 & D+6 & 5 \\ 6(D-5) & -30 & -12 \\ 0 & D & D \end{pmatrix}$$

$$\Rightarrow \begin{pmatrix} 6 & D+6 & 5 \\ 0 & -30+(D+6)(5-D) & -12+5(5-D) \\ 0 & D & D \end{pmatrix}$$

$$\Rightarrow \begin{pmatrix} 6 & D+6 & 5 \\ 0 & D(-D-1) & 13-5D \\ 0 & D & D \end{pmatrix}$$

$$\Rightarrow \begin{pmatrix} 6 & D+6 & 5 \\ 0 & D & D \\ 0 & D(-D-1) & 13-5D \end{pmatrix}$$

$$\Rightarrow \begin{pmatrix} 6 & D+6 & 5 \\ 0 & D & D \\ 0 & 0 & D(D+1)+13-5D \end{pmatrix}$$

$$\Rightarrow \begin{pmatrix} 6 & D+6 & 5 \\ 0 & D & D \\ 0 & 0 & D^2-4D+13 \end{pmatrix}$$

Thus,

$$(D^2 - 4D + 13)z(t) = 0$$

whose C-Eq is

$$r^2 - 4r + 13 = 0$$
$$r_{1,2} = 2 \pm 3i$$

Thus,

$$z(t) = c_1 \exp(2t)\cos(3t) + c_2 \exp(2t)\sin(3t)$$
$$z'(t) = (2c_1 + 3c_2)\exp(2t)\cos(3t) + (2c_2 - 3c_1)\exp(2t)\sin(3t)$$

From the second and third rows of the matrix, we have

$$y' + z' = 0$$

$$y'(t) = -z'(t)$$
$$= (-2c_1 - 3c_2)\exp(2t)\cos(3t) + (-2c_2 + 3c_1)\exp(2t)\sin(3t)$$

So,

$$y(t) = \int y'(t)dt = -\int z'(t)dt = -z(t) + c_3$$
$$= -c_1\exp(2t)\cos(3t) - c_2\exp(2t)\sin(3t) + c_3$$

From the second row of the matrix, we have

$$x(t) = \frac{1}{6}\left(-y'(t) - 6y(t) - 5z(t)\right)$$
$$= \frac{1}{2}(c_1 + c_2)\exp(2t)\cos(3t) - \frac{1}{2}(c_1 - c_2)\exp(2t)\sin(3t) - c_3$$

Then, the GS is

$$x(t) = \frac{1}{2}(c_1 + c_2)\exp(2t)\cos(3t) - \frac{1}{2}(c_1 - c_2)\exp(2t)\sin(3t) - c_3$$
$$y(t) = -c_1\exp(2t)\cos(3t) - c_2\exp(2t)\sin(3t) + c_3$$
$$z(t) = c_1\exp(2t)\cos(3t) + c_2\exp(2t)\sin(3t)$$

Example 13

Find the GS to the following DE.Syst $X'(t) = AX(t)$ for

$$A = \begin{pmatrix} 1 & 0 & 0 \\ 2 & 2 & 0 \\ 3 & 3 & 3 \end{pmatrix} \tag{4.60}$$

Solution

The Eigen equation for this Homo system is

$$\det\begin{pmatrix} 1-\lambda & 0 & 0 \\ 2 & 2-\lambda & 0 \\ 3 & 3 & 3-\lambda \end{pmatrix} = 0$$

whose e-values are

$$\lambda_{1,2,3} = 1, 2, 3$$

and the solution should be

$$X(t) = c_1 V_1 \exp(\lambda_1 t) + c_2 V_2 \exp(\lambda_2 t) + c_3 V_3 \exp(\lambda_3 t)$$

where $V_{1,2,3}$ are the corresponding e-vectors of $\lambda_{1,2,3}$. Now, let's calculate them.

For e-value $\lambda_1 = 1$, we have

$$\begin{pmatrix} 1-\lambda_1 & 0 & 0 \\ 2 & 2-\lambda_1 & 0 \\ 3 & 3 & 3-\lambda_1 \end{pmatrix} V_1 = \begin{pmatrix} 0 & 0 & 0 \\ 2 & 1 & 0 \\ 3 & 3 & 2 \end{pmatrix} V_1 = 0$$

$$V_1 = \begin{pmatrix} 2 \\ -4 \\ 3 \end{pmatrix}$$

For e-value $\lambda_2 = 2$, we have

$$\begin{pmatrix} 1-\lambda_2 & 0 & 0 \\ 2 & 2-\lambda_2 & 0 \\ 3 & 3 & 3-\lambda_2 \end{pmatrix} V_2 = \begin{pmatrix} -1 & 0 & 0 \\ 2 & 0 & 0 \\ 3 & 3 & 1 \end{pmatrix} V_2 = 0$$

$$V_2 = \begin{pmatrix} 0 \\ -1 \\ 3 \end{pmatrix}$$

For e-value $\lambda_3 = 3$, we have

$$\begin{pmatrix} 1-\lambda_3 & 0 & 0 \\ 2 & 2-\lambda_3 & 0 \\ 3 & 3 & 3-\lambda_3 \end{pmatrix} V_3 = \begin{pmatrix} -2 & 0 & 0 \\ 2 & -1 & 0 \\ 3 & 3 & 0 \end{pmatrix} V_3 = 0$$

$$V_3 = \begin{pmatrix} 0 \\ 0 \\ 1 \end{pmatrix}$$

Thus, the solution vector is

$$X(t) = c_1 \begin{pmatrix} 2 \\ -4 \\ 3 \end{pmatrix} \exp(t) + c_2 \begin{pmatrix} 0 \\ -1 \\ 3 \end{pmatrix} \exp(2t) + c_3 \begin{pmatrix} 0 \\ 0 \\ 1 \end{pmatrix} \exp(3t)$$

Example 14

Find the GS to the following DE.Syst $X'(t) = AX(t)$ for

$$A = \begin{pmatrix} 2 & 2 & -3 \\ 5 & 1 & -5 \\ -3 & 4 & 0 \end{pmatrix} \tag{4.61}$$

Solution

The Eigen equation for this Homo system is

$$\det \begin{pmatrix} 2-\lambda & 2 & -3 \\ 5 & 1-\lambda & -5 \\ -3 & 4 & 0-\lambda \end{pmatrix} = 0$$

whose e-values are

$$\lambda_{1,2,3} = 1, 1, 1$$

and the multiplicity of the e-value is 3 and

$$(A - 1I)^3 = \begin{pmatrix} 1 & 2 & -3 \\ 5 & 0 & -5 \\ -3 & 4 & -1 \end{pmatrix} \begin{pmatrix} 1 & 2 & -3 \\ 5 & 0 & -5 \\ -3 & 4 & -1 \end{pmatrix} \begin{pmatrix} 1 & 2 & -3 \\ 5 & 0 & -5 \\ -3 & 4 & -1 \end{pmatrix}$$

$$= \begin{pmatrix} 0 & 0 & 0 \\ 0 & 0 & 0 \\ 0 & 0 & 0 \end{pmatrix}$$

For V_3, we have

$$(A - 1I)^3 V_3 = 0$$

The vector V_3 is arbitrary and, for simplicity, we select

$$V_3 = \begin{pmatrix} 1 \\ 0 \\ 0 \end{pmatrix}$$

Thus,

$$V_2 = (A - 1I)V_3 = \begin{pmatrix} 1 & 2 & -3 \\ 5 & 0 & -5 \\ -3 & 4 & -1 \end{pmatrix} \begin{pmatrix} 1 \\ 0 \\ 0 \end{pmatrix} = \begin{pmatrix} 1 \\ 5 \\ -3 \end{pmatrix}$$

$$V_1 = (A - 1I)V_2 = \begin{pmatrix} 1 & 2 & -3 \\ 5 & 0 & -5 \\ -3 & 4 & -1 \end{pmatrix} \begin{pmatrix} 1 \\ 5 \\ -3 \end{pmatrix} = 20 \begin{pmatrix} 1 \\ 1 \\ 1 \end{pmatrix}$$

Therefore, the GS is

$$X(t) = \left(c_1 V_1 + c_2 (V_1 t + v_2) + c_3 \left(\frac{1}{2} V_1 t^2 + V_2 t + V_3 \right) \right) \exp(t)$$

$$= \left(c_1 20 \begin{pmatrix} 1 \\ 1 \\ 1 \end{pmatrix} + c_2 \left(20 \begin{pmatrix} 1 \\ 1 \\ 1 \end{pmatrix} t + \begin{pmatrix} 1 \\ 5 \\ -3 \end{pmatrix} \right) \right.$$

$$\left. + c_3 \left(10 \begin{pmatrix} 1 \\ 1 \\ 1 \end{pmatrix} t^2 + \begin{pmatrix} 1 \\ 5 \\ -3 \end{pmatrix} t + \begin{pmatrix} 1 \\ 0 \\ 0 \end{pmatrix} \right) \right) \exp(t)$$

The coefficient $c_1 20$ can be combined.

Problems

Problem 4.5.1 Solve the following Homo system
$$X'(t) = \begin{pmatrix} 1 & -3 \\ 3 & 7 \end{pmatrix} X(t)$$

Problem 4.5.2 Solve the following Homo system
$$\begin{cases} x' = 4x - 3y \\ y' = 3x + 4y \\ x(0) = 2 \\ y(0) = 3 \end{cases}$$

Problem 4.5.3 Solve the following InHomo system
$$X'(t) = \begin{pmatrix} 1 & 2 \\ 2 & 1 \end{pmatrix} X(t) + \begin{pmatrix} 3\exp(t) \\ -t^2 \end{pmatrix}$$

Problem 4.5.4 Solve the following InHomo system
$$\begin{pmatrix} x \\ y \end{pmatrix}' = \begin{pmatrix} 1 & 3 \\ 1 & -1 \end{pmatrix} \begin{pmatrix} x \\ y \end{pmatrix} + \begin{pmatrix} 1 \\ 0 \end{pmatrix} \exp(-2t)$$

Problem 4.5.5 Solve the following InHomo system
$$\begin{pmatrix} x \\ y \end{pmatrix}' = \begin{pmatrix} 1 & 2 \\ 4 & -1 \end{pmatrix} \begin{pmatrix} x \\ y \end{pmatrix} + \begin{pmatrix} -1 \\ 1 \end{pmatrix} t$$

Problem 4.5.6 Solve the following Homo system
$$\begin{cases} x' - 4x + 2y = 0 \\ y' + 4x - 4y + 2z = 0 \\ z' + 4y - 4z = 0 \end{cases}$$

4.6　Examples of Systems of DEs

There are broad applications of DE systems. Many DE systems, linear or nonlinear, Homo or InHomo, first-order or higher-order, can find applications in science, engineering, and finance. Examples include electric networks, predator-prey, home heating, liquid rheostat, biomass transfer, *etc*. The list is simply too long to compile.

4.6.1　Predator-prey DEs

A prey-predator model involving two competing and, sometimes cooperating, species can be expressed by the following system of two 1st.O nonlinear DEs with ICs

$$\begin{cases} \dfrac{dx}{dt} = ax + bxy \\[2mm] \dfrac{dy}{dt} = cx + exy \\[2mm] x(t = 0) = x_0 \\[1mm] y(t = 0) = y_0 \end{cases} \tag{4.62}$$

where $x(t)$ and $y(t)$ are the populations of the two species at time t. This DE system was initially introduced by the US mathematician A. J. Lotka (1880-1949) for the theory of autocatalytic chemical reactions in 1920 and was independently investigated by the Italian mathematician V. Volterra (1860-1940). Thus, this model is also called Lokta-Volterra DE System. In fact, this Lokta-Volterra model is a special example of the Kolmogorov model, a general framework for modeling the dynamics of ecological systems with predator-prey interactions, competition, disease, and mutualism.

Obtaining analytical solutions to many DEs and their systems is a dream mostly unrealizable while approximate solutions can be obtained by a series of approximation methods including the power series methods, successive approximation methods, and multiple methods originating from the perturbation theory. These approximation methods are still limiting for realistic applications. Since the 1940s when digital computing entered the scene, numerical methods gained attention and have been growing to become the mainstream and dominant solution methods. Euler methods, the implicit Backward Euler method, the family of linear multistep methods, as well as the family of the widely-used Runge-Kutta methods developed around 1900 by German mathematicians C. Runge (1856-1927) and M. W. Kutta (1867-1944).

Because the prey-predator model we described is nonlinear, we are unable to obtain its analytical solution. To gain insights, we solve it numerically by using the Runge-Kutta method, after assigning some appropriate values for the parameters and ICs.

$$
\begin{cases}
\dfrac{dx}{dt} = ax + bxy \\[2mm]
\dfrac{dy}{dt} = cy + exy \\[2mm]
x(t = 0) = 199 \\
y(t = 0) = 21 \\
\quad a = 0.222 \\
\quad b = -0.0011 \\
\quad c = -1.999 \\
\quad e = 0.010
\end{cases}
\tag{4.63}
$$

where we have selected the values for parameters, a, b, c and e, and given the ICs for the two DVs $x(t)$ and $y(t)$.

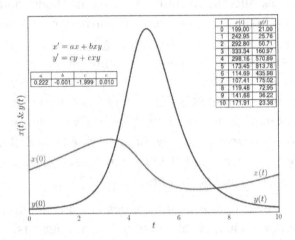

Figure 4.9 The solutions, in time series, for a predator-prey system.

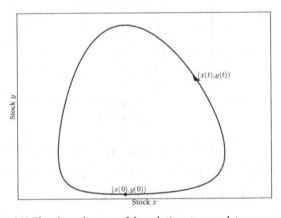

Figure 4.10 The phase diagram of the solutions to a predator-prey system.

4.6.2   Cascade of Tanks

Consider a Brine cascade of three tanks of volumes V_1, V_2, V_3 with the same flow rate r for flowing to the first tank (with water containing no salt), from first tank to the second and-then-from the second to the third. We assume the salt concentration throughout each tank is uniform due to stirring in each tank. We establish a Homo system to compute the salt concentration in each tank at time t: $x_1(t)$, $x_2(t)$, $x_3(t)$. According to chemical balance law that the change of the concentration rate in each tank is the difference between the input and output rates. For convenience, we set $V_1 = 2$, $V_2 = 4$, $V_3 = 6$ and $r = 1$. We have the following DEs.

For the first tank

$$\frac{dx_1}{dt} = 0 - \frac{1}{2}x_1 \tag{4.64}$$

For the second tank

$$\frac{dx_2}{dt} = \frac{1}{2}x_1 - \frac{1}{4}x_2 \tag{4.65}$$

For the third tank

$$\frac{dx_3}{dt} = \frac{1}{4}x_2 - \frac{1}{6}x_3 \tag{4.66}$$

If the initial concentrations are c_1, c_2 and c_3 for the three tanks respectively, we can form the following DE.Syst

$$\begin{cases} \dfrac{dx_1}{dt} = 0 - \dfrac{1}{2}x_1 \\[2mm] \dfrac{dx_2}{dt} = \dfrac{1}{2}x_1 - \dfrac{1}{4}x_2 \\[2mm] \dfrac{dx_3}{dt} = \dfrac{1}{4}x_2 - \dfrac{1}{6}x_3 \\[2mm] x_1(0) = c_1 \\[1mm] x_2(0) = c_2 \\[1mm] x_3(0) = c_3 \end{cases} \qquad (4.67)$$

Solving the above linear DE.Syst, we get the following solutions

$$\begin{cases} x_1(t) = c_1 \exp\left(-\dfrac{t}{2}\right) \\[3mm] x_2(t) = -2c_1 \exp\left(-\dfrac{t}{2}\right) + (c_2 + 2c_1) \exp\left(-\dfrac{t}{4}\right) \\[3mm] x_3(t) = \dfrac{3}{2}c_1 \exp\left(-\dfrac{t}{2}\right) - (3c_2 + 6c_1) \exp\left(-\dfrac{t}{4}\right) \\[3mm] \qquad + \left(c_3 - \dfrac{3}{2}c_1 + 3c_2 + 6c_1\right) \exp\left(-\dfrac{t}{6}\right) \end{cases} \qquad (4.68)$$

shown in Figure 4.11 if we set $c_1 = 1$, $c_2 = 2$ and $c_3 = 3$.

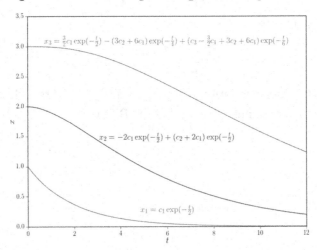

Figure 4.11 The solutions, salt concentrations of each tank, of the DE system.

Problem

Problem 4.6.1 Convert the following DE into a DE.Syst
$$x''' - 5x'' + 9x = t \sin 2t$$

Chapter 5
Laplace Transforms

5.1   Laplace Transforms

Laplace Transform (LT) is an integral transform, named after French scholar Pierre-Simon Laplace (1749-1827) who made major contributions in mathematics, statistics, physics and astronomy. The LT is widely used in engineering, physics, as well as mathematics. We focus on learning its uses for solving DEs, especially, linear DEs and DE systems.

In Chapters 3 and 4, we discussed how to solve linear DEs and DE systems. In certain cases, the methods in Chapter 3 can be quite tedious while other methods including the LT method can be much less cumbersome. The Differentiation Operator D can be viewed as a transform which, when applied to the function $f(t)$, yields a new function $D\{f(t)\} = f'(t)$. Similarly, the LT operator $\mathcal{L}$ involves the operation of integration and yields a new function $\mathcal{L}\{f(t)\} = F(s)$ of a new variable s. This new function is usually easier to manipulate and it is one of the reasons we study LT.

After learning how to compute the LT $F(s)$ of a function $f(t)$, we will learn how the LT converts a DE with the unknown function $f(t)$ and its derivatives to an AE in $F(s)$. AEs are usually easier to solve than DEs and this method simplifies the problem of finding the solution $f(t)$.

Definition

Given a function $f(t) \, \forall t \geq 0$, the LT of function $f(t)$ is another function $F(s)$ defined as

$$F(s) = \mathcal{L}\{f(t)\} = \int_0^\infty \exp(-st) \, f(t) dt \qquad (5.1)$$

for all values of s for which the improper integral converges.

Problems

Problem 5.1.1 Use the definition of the LT to prove
$$\mathcal{L}\{f(t) + g(t)\} = \mathcal{L}\{f(t)\} + \mathcal{L}\{g(t)\}$$

Problem 5.1.2 Use the definition of the LT to prove
$$\mathcal{L}\{cf(t)\} = c\mathcal{L}\{f(t)\}$$
where c is an arbitrary constant.

5.2   Properties of Laplace Transforms

Computing LTs requires understanding of the key properties of the transforms. In the following, we present a few such properties.

The first and the most important property is that of linearity.

Linearity Property

$$\mathcal{L}\{f(t)\} = \int_0^\infty f(t) \exp(-st)\, dt = F(s) \qquad (5.2)$$

and

$$\mathcal{L}\{g(t)\} = \int_0^\infty g(t) \exp(-st)\, dt = G(s) \qquad (5.3)$$

The linearity of the LT states that,

$$\mathcal{L}\{\alpha f(t) + \beta g(t)\} = \alpha F(s) + \beta G(s) \qquad (5.4)$$

where α and β are constants.

Proof of the Linearity Property

$$
\begin{aligned}
\text{LHS} &= \mathcal{L}\{\alpha f(t) + \beta g(t)\} \\
&= \int_0^\infty \big(\alpha f(t) + \beta g(t)\big) \exp(-st)\, dt \\
&= \int_0^\infty \alpha f(t) \exp(-st)\, dt \\
&\quad + \int_0^\infty \beta g(t) \exp(-st)\, dt \\
&= \alpha F(s) + \beta G(s) \\
&= \text{RHS}
\end{aligned}
\qquad (5.5)
$$

5.2.1   Laplace Transforms for Polynomials

Polynomials are a class of functions of form

$$p_n(t) = a_0 + a_1 t + a_2 t^2 + \cdots + a_n t^n \tag{5.6}$$

From the linearity, we know the LT $P_n(s) = \mathcal{L}\{p_n(t)\}$ has form

$$P_n(s) = a_0 \mathcal{L}\{1\} + a_1 \mathcal{L}\{t\} + \cdots + a_n \mathcal{L}\{t^n\} \tag{5.7}$$

Let's now compute $\mathcal{L}\{t^n\}$. We start from the simplest case $\mathcal{L}\{1\}$

$$
\begin{aligned}
\mathcal{L}\{1\} &= \int_0^\infty 1 \times \exp(-st)\, dt \\
&= \int_0^\infty \exp(-st)\, dt \\
&= -\frac{1}{s}\exp(-st)\Big|_0^\infty \\
&= -\frac{1}{s}\exp(-st)\Big|_{t\to\infty} + \frac{1}{s}\exp(-st)\Big|_{t\to 0} \\
&= \lim_{t\to 0}\frac{1}{s}\exp(-st)
\end{aligned}
\tag{5.8}
$$

It is obvious that the integral diverges if $s \leq 0$, resulting in a meaningless LT. We only consider $s > 0$ for which we have

$$\mathcal{L}\{1\} = \frac{1}{s} \tag{5.9}$$

We can also easily prove $\mathcal{L}\{0\} = 0$ where the two 0's are defined in two different domains, *i.e.*, the first 0 is in the t-domain while the second 0 is in the s-domain.

Next, we consider $\mathcal{L}\{t\}$ starting from the definition

$$\mathcal{L}\{t\} = \int_0^\infty t \exp(-st)\, dt \tag{5.10}$$

Using integration by parts, we have

$$\mathcal{L}\{t\} = -\frac{1}{s}\int_0^\infty t(\exp(-st))\, dt \tag{5.11}$$

$$= -\frac{1}{s} t \exp(-st) \Big|_0^\infty$$

$$-\frac{1}{s}\left(-\int_0^\infty \exp(-st)\, dt\right)$$

$$= -\frac{1}{s}\left(-\int_0^\infty \exp(-st)\, dt\right)$$

$$= \frac{1}{s} L\{1\}$$

Using (5.9) $\mathcal{L}\{1\} = \frac{1}{s}$ $\forall s > 0$, we have

$$\mathcal{L}\{t\} = \frac{1}{s^2}\ \forall\, s > 0 \tag{5.12}$$

Alternatively, we can derive this directly by taking derivative of

$$\mathcal{L}\{1\} = \frac{1}{s} \tag{5.13}$$

That is

$$\frac{d}{ds}\mathcal{L}\{1\} = \frac{d}{ds}\left(\frac{1}{s}\right) \tag{5.14}$$

$$\text{LHS} = \frac{d}{ds}\mathcal{L}\{1\} = \frac{d}{ds}\int_0^\infty \exp(-st)\, dt$$

$$= \int_0^\infty \frac{\partial}{\partial s}(\exp(-st))\, dt \tag{5.15}$$

$$= -\int_0^\infty t \exp(-st)\, dt = -L\{t\}$$

$$\text{RHS} = \frac{d}{ds}\left(\frac{1}{s}\right) = -\frac{1}{s^2} \tag{5.16}$$

By equating both sides, we get

$$-\mathcal{L}\{t\} = -\frac{1}{s^2} \tag{5.17}$$

Finally, the intended LT is

$$\mathcal{L}\{t\} = \frac{1}{s^2} \tag{5.18}$$

Using this method recursively, we can derive

$$\frac{d}{ds}\mathcal{L}\{t\} = \frac{d}{ds}\left(\frac{1}{s^2}\right)$$

$$\int_0^\infty \frac{\partial}{\partial s}(t\exp(-st))dt = -\frac{2}{s^3} \tag{5.19}$$

$$\int_0^\infty t^2\exp(-st)\,dt = \frac{2}{s^3}$$

Thus,

$$\mathcal{L}\{t^2\} = \frac{2}{s^3} \quad \forall s > 0 \tag{5.20}$$

It is easy to derive the general formula for a power function (and naturally for a polynomial)

$$\mathcal{L}\{t^n\} = \frac{n!}{s^{n+1}} \quad \forall s > 0 \tag{5.21}$$

which can be proven by induction as follows.

Proof

For $n = 0$, we have already discussed.

Suppose the formula is true for n. This means

$$\mathcal{L}\{t^n\} = \frac{n!}{s^{n+1}} \quad \forall s > 0$$

from which we have

$$\frac{d}{ds}\mathcal{L}\{t^n\} = \frac{d}{ds}\left(\frac{n!}{s^{n+1}}\right)$$

$$\int_0^\infty \frac{\partial}{\partial s}(t^n\exp(-st))dt = -\frac{n!\cdot(n+1)}{s^{n+2}} \tag{5.22}$$

$$\int_0^\infty t^{n+1} \exp(-st)\, dt = \frac{(n+1)!}{s^{n+2}}$$

That is

$$\mathcal{L}\{t^{n+1}\} = \frac{(n+1)!}{s^{n+2}} \quad \forall s > 0 \tag{5.23}$$

QED.

Example 1

If $T(\alpha) = \mathcal{L}\{t^\alpha\}$, prove

$$\frac{d}{ds}\big(T(\alpha)\big) = -T(\alpha + 1) \tag{5.24}$$

Solution

To prove this, we simply derive the formula directly

$$\text{LHS} = \frac{d}{ds}\mathcal{L}\{t^\alpha\}$$

$$= \frac{d}{ds}\int_0^\infty t^\alpha \exp(-st)\, dt$$

$$= \int_0^\infty \frac{\partial}{\partial s}(t^\alpha \exp(-st))\, dt$$

$$= -\int_0^\infty t^{\alpha+1} \exp(-st)\, dt$$

$$= -\mathcal{L}\{t^{\alpha+1}\} = \text{RHS}$$

5.2.2 The Translator Property

Given $\mathcal{L}\{f(t)\} = F(s)$, compute $\mathcal{L}\{f(t)\exp(at)\}$.

$$\mathcal{L}\{f(t)\exp(at)\} = \int_0^\infty f(t)\exp(at - st)\, dt$$

$$= \int_0^\infty f(t)\exp((a - s)t)\, dt \tag{5.25}$$

By substituting $\rho = s - a$, we have

$$\mathcal{L}\{f(t) \exp(at)\} = \int_0^\infty f(t) \exp(-\rho t) \, dt = F(\rho) \qquad (5.26)$$

That is

$$\mathcal{L}\{f(t) \exp(at)\} = F(s - a) \qquad (5.27)$$

Similarly, we get

$$\mathcal{L}\{f(t) \exp(-at)\} = F(s + a) \qquad (5.28)$$

Combining the two, we have

$$\mathcal{L}\{f(t) \exp(\pm at)\} = F(s \mp a) \qquad (5.29)$$

This formula is the translator property of LT.

Example 1

Compute $\mathcal{L}\{\exp(at)\}$

Solution

$$\mathcal{L}\{\exp(at)\} = \mathcal{L}\{\exp(at) \times 1\}$$

Let $f(t) = 1$, we know

$$F(s) = \mathcal{L}\{f(t)\} = \frac{1}{s}, \qquad s > 0$$

Thus, using the translator property, we have

$$\mathcal{L}\{\exp(at)\} = F(s - a) = \frac{1}{s - a}, \qquad s > a$$

Example 2

Compute $\mathcal{L}\{t \exp(at)\}$

Solution

Let $f(t) = t$, we have

$$F(s) = \mathcal{L}\{f(t)\} = \mathcal{L}\{t\} = \frac{1}{s^2}, \qquad s > a$$

Using the translator property, we have

$$\mathcal{L}\{t \exp(at)\} = F(s - a) = \frac{1}{(s - a)^2}, \qquad s > a$$

Example 3

Compute $\mathcal{L}\{t^2 \exp(at)\}$

Solution

Let $f(t) = t^2$, we have

$$F(s) = \mathcal{L}\{f(t)\} = \mathcal{L}\{t^2\} = \frac{2}{s^3}, \qquad s > a$$

Using the translator property, we have

$$\mathcal{L}\{t^2 \exp(at)\} = F(s - a) = \frac{2}{(s - a)^3}, \qquad s > a$$

In general, we can prove

$$\mathcal{L}\{t^n \exp(at)\} = \frac{n!}{(s - a)^{n+1}}, \qquad s > a \qquad (5.30)$$

The above Example 1 can be generalized to the complex plane:

$$\mathcal{L}\{\exp(i\omega t)\} = \frac{1}{s - i\omega} \qquad (5.31)$$

and

$$\mathcal{L}\{\exp(-i\omega t)\} = \frac{1}{s + i\omega} \qquad (5.32)$$

Example 4

Compute $\mathcal{L}\{\cos \omega t\}$ and $\mathcal{L}\{\sin \omega t\}$

Solution

From Euler's formulas

$$\begin{cases} \exp(i\omega t) = \cos \omega t + i \sin \omega t \\ \exp(-i\omega t) = \cos \omega t - i \sin \omega t \end{cases}$$

we get

$$\cos \omega t = \frac{1}{2}(\exp(i\omega t) + \exp(-i\omega t))$$

Thus,

$$\mathcal{L}\{\cos \omega t\} = \mathcal{L}\left\{\frac{1}{2}(\exp(i\omega t) + \exp(-i\omega t))\right\}$$

$$= \frac{1}{2}\mathcal{L}\{\exp(i\omega t)\} + \frac{1}{2}\mathcal{L}\{\exp(-i\omega t)\}$$

$$= \frac{1}{2}\left(\frac{1}{s - i\omega}\right) + \frac{1}{2}\left(\frac{1}{s + i\omega}\right)$$

$$= \frac{1}{2}\left(\frac{1}{s - i\omega} + \frac{1}{s + i\omega}\right)$$

$$= \frac{s}{s^2 + \omega^2}$$

Therefore, we have obtained one of the most important LTs,

$$\mathcal{L}\{\cos \omega t\} = \frac{s}{s^2 + \omega^2}$$

Similarly, for $\sin \omega t$, we have

$$\sin \omega t = \frac{1}{2i}(\exp(i\omega t) - \exp(-i\omega t))$$

Then,

$$\mathcal{L}\{\sin \omega t\} = \frac{1}{2i}(\mathcal{L}\{\exp(i\omega t)\} - \mathcal{L}\{\exp(-i\omega t)\})$$

$$= \frac{1}{2i}\left(\frac{1}{s - i\omega} - \frac{1}{s + i\omega}\right)$$

$$= \frac{\omega}{s^2 + \omega^2}$$

Therefore, we have obtained another one of the most important LTs,

$$\mathcal{L}\{\sin \omega t\} = \frac{\omega}{s^2 + \omega^2}$$

Example 5
Compute $\mathcal{L}\{\cosh \omega t\}$

Solution
Because

$$\cosh \omega t = \frac{1}{2}(\exp(\omega t) + \exp(-\omega t))$$

we get

$$\mathcal{L}\{\cosh \omega t\} = \mathcal{L}\left\{\frac{1}{2}(\exp(\omega t) + \exp(-\omega t))\right\}$$

$$= \frac{1}{2}\left(\frac{1}{s - \omega} + \frac{1}{s + \omega}\right)$$

$$= \frac{s}{s^2 - \omega^2}$$

Similarly, we have

$$\mathcal{L}\{\sinh \omega t\} = \frac{\omega}{s^2 - \omega^2}$$

Example 6
Compute $\mathcal{L}\{\exp(at) \cos \omega t\}$

Solution

Let $f(t) = \cos \omega t$. From the previous discussion, we know that

$$F(s) = L\{\cos \omega t\} = \frac{s}{s^2 + \omega^2}$$

Therefore, using the translator property, we get

$$L\{\exp(at) \cos \omega t\} = F(s - a)$$
$$= \frac{s - a}{(s - a)^2 + \omega^2}$$

Example 7

Compute $L\{3 \exp(2t) + 2 \sin^2 3t\}$

Solution

$$L\{3 \exp(2t) + 2 \sin^2 3t\} = L\{3 \exp(2t) + 1 - \cos 6t\}$$
$$= \frac{3}{s - 2} + \frac{1}{s} - \frac{s}{s^2 + 36}$$
$$= \frac{3s^3 + 144s - 72}{s(s - 2)(s^2 + 36)}$$

5.2.3   Transforms of Step and Delta Functions

Unit Function: Given a unit step function, aka, Heaviside function for honoring the self-taught electric engineer Oliver Heaviside (1850-1925), shown by Figure 5.1, defined as

$$u(t) = \begin{cases} 0, & t < 0 \\ 1, & t \geq 0 \end{cases}$$

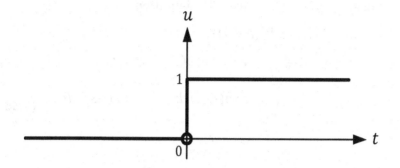

Figure 5.1 The unit step function $u(t)$.

Since $u(t) = 1$ for $t \geq 0$ and, because the LT involves only the values of a function for $t \geq 0$, we see immediately that

$$\mathcal{L}\{u(t)\} = \frac{1}{s}, \qquad s \geq 0 \tag{5.33}$$

Furthermore, for any function $f(t)$, $F(s) = \mathcal{L}\{f(t)\}$, we know that

$$\mathcal{L}\{f(t)u(t)\} = F(s) \tag{5.34}$$

A more generalized step function $u_a(t) = u(t - a)$ shown by Figure 5.2 sets its jump at $t = a$ rather than at $t = 0$,

$$u_a(t) = u(t - a) = \begin{cases} 0, & t < a \\ 1, & t \geq a \end{cases} \tag{5.35}$$

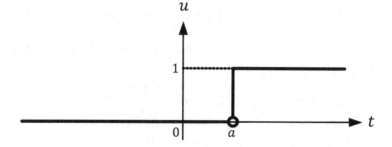

Figure 5.2 The step function $u(t - a)$.

To perform LT on $u_a(t)$, we use LT's definition, *i.e.*,

$$
\begin{aligned}
\mathcal{L}\{u_a(t)\} &= \mathcal{L}\{u(t-a)\} \\
&= \int_0^\infty u(t-a)\exp(-st)\,dt \\
&= \int_0^a 0 \cdot \exp(-st)\,dt + \int_a^\infty \exp(-st)\,dt \\
&= -\frac{1}{s}\exp(-st)\Big|_{t=a}^\infty \\
&= \frac{1}{s}\exp(-as), \qquad s>0, a>0
\end{aligned}
\tag{5.36}
$$

Thus, we obtain a more general LT,

$$
\mathcal{L}\{u_a(t)\} = \frac{1}{s}\exp(-as) \qquad \forall s, a>0
\tag{5.37}
$$

Example 1

Suppose $F(s) = \mathcal{L}\{f(t)\}$, compute $\mathcal{L}\{u(t-a)f(t-a)\}$

Solution

From the definition of the LT, we have

$$
F(s) = \mathcal{L}\{f(t)\} = \int_0^\infty f(t)\exp(-st)\,dt
$$

On the other hand, we have

$$
\begin{aligned}
\mathcal{L}\{u(t-a)f(t-a)\} &= \int_0^\infty u(t-a)f(t-a)\exp(-st)\,dt \\
&= \int_0^a 0\,dt + \int_a^\infty f(t-a)\exp(-st)\,dt \\
&= \exp(-as)\int_a^\infty f(t-a)\exp\big(-s(t-a)\big)\,dt
\end{aligned}
$$

Let $\tau = t - a$, we have

$$
\begin{aligned}
\mathcal{L}\{u(t-a)f(t-a)\} &= \exp(-as)\int_0^\infty f(\tau)\exp(-s\tau)\,d\tau \\
&= \exp(-as)\,\mathcal{L}\{f(\tau)\} \\
&= \exp(-as)\,F(s)
\end{aligned}
$$

Example 2

Compute $\mathcal{L}\{(t-a)u(t-a)\}$

Solution

Let $f(t) = t$, we know that

$$F(s) = \mathcal{L}\{f(t)\} = \frac{1}{s^2}$$

Using the result from Example 1, we have

$$\mathcal{L}\{(t-a)u(t-a)\} = \mathcal{L}\{f(t-a)u(t-a)\}$$
$$= \exp(-as)\,F(s)$$
$$= \frac{1}{s^2}\exp(-as)$$

Delta Function: Another generalized function, the so-called Delta function, aka, Dirac Delta function for the introduction of the function by the English theoretical physicist Paul Dirac (1902-1984), is defined as

$$\delta(t-a) = \begin{cases} 0, & t \neq a \\ \infty, & t = a \end{cases}$$

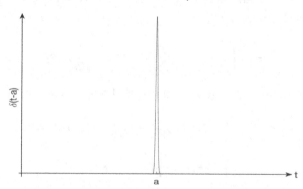

Figure 5.3 The Delta function $\delta(t-a)$.

Mathematically speaking, $\delta(t-a)$ or $\delta(t)$, is not strictly a function. But, it serves the "mathematical" purposes of engineering nicely, *e.g.*, it is used to express the unit impulse in signal processing, it is used to express the point charge in electromagnetism, and it is also used to express point mass in study of gravity.

The Delta function has a few important properties including

$$\int_{-\infty}^{\infty} \delta(t - a)dt = 1$$

and

$$\int_{-\infty}^{\infty} f(t)\delta(t - a)dt = f(a)$$

Using these properties, we can find the LT of $\delta(t - a)$ as

$$\mathcal{L}\{\delta(t - a)\} = \int_{0}^{\infty} \delta(t - a)\exp(-st)\,dt$$

$$= \exp(-sa)$$

5.2.4   The *t*-multiplication Property

The *t*-multiplication property, also called the frequency differentiation, is another important property. Given $\mathcal{L}\{f(t)\} = F(s)$, when taking derivative wrt s, we have

$$\frac{d}{ds}\left(\int_{0}^{\infty} f(t)\exp(-st)\,dt\right) = \frac{d}{ds}F(s) \tag{5.38}$$

By exchanging the derivative and integral operation, we have

$$\int_{0}^{\infty} \frac{d}{ds}(f(t)\exp(-st))dt = F'(s)$$

$$-\int_{0}^{\infty} tf(t)\exp(-st)\,dt = F'(s) \tag{5.39}$$

Thus,

$$\mathcal{L}\{tf(t)\} = \int_{0}^{\infty} tf(t)\exp(-st)\,dt = -F'(s) \tag{5.40}$$

Similarly, we have

$$\mathcal{L}\{t^2 f(t)\} = F''(s) \tag{5.41}$$

Generally,

$$\mathcal{L}\{t^n f(t)\} = (-1)^n F^{(n)}(s) \tag{5.42}$$

Example 1

Compute $\mathcal{L}\{t \exp(at)\}$

Solution

Method 1: By t-multiplication property.

Let $f(t) = \exp(at)$, we know from shifting property,

$$F(s) = \mathcal{L}\{\exp(at)\} = \frac{1}{s-a}$$

By applying the t-multiplication property, we have

$$\mathcal{L}\{tf(t)\} = -F'(s)$$
$$= -\left(\frac{1}{s-a}\right)'$$
$$= \frac{1}{(s-a)^2}$$

Method 2: Using the shifting property.

Let $f(t) = t$, we know that

$$F(s) = \mathcal{L}\{t\} = \frac{1}{s^2}$$

By applying the shifting property, we have

$$\mathcal{L}\{\exp(at) f(t)\} = F(s-a) = \frac{1}{(s-a)^2}$$

Method 3: Using the differentiation property.

Considering $f(t) = t \exp(at)$ and denoting $F(s) = \mathcal{L}\{f(t)\}$, we have

$$f'(t) = \exp(at) + at \exp(at)$$

Performing LT on both sides, we have

$$\mathcal{L}\{f'(t)\} = \mathcal{L}\{\exp(at)\} + a\mathcal{L}\{f(t)\}$$

Applying the differentiation property, we have

$$sF(s) - f(0) = \frac{1}{s-a} + aF(s)$$

Solving this AE, we get

$$F(s) = \frac{1}{(s-a)^2} + \frac{f(0)}{s-a}$$

Since $f(0) = 0$, we have

$$F(s) = \frac{1}{(s-a)^2}$$

Example 2

Compute $L\{t \sin \omega t\}$

Solution

Let $f(t) = \sin \omega t$ and thus,

$$F(s) = \mathcal{L}\{f(t)\} = \frac{\omega}{s^2 + \omega^2}$$

Applying the t-multiplication property, we have

$$\mathcal{L}\{tf(t)\} = -F'(s)$$
$$= -\left(\frac{\omega}{s^2 + \omega^2}\right)'$$
$$= \frac{2\omega s}{(s^2 + \omega^2)^2}$$

That is

$$\mathcal{L}\{t \sin \omega t\} = \frac{2\omega s}{(s^2 + \omega^2)^2}$$

Remarks

Similarly, we may calculate

$$\mathcal{L}\{t^2 \sin \omega t\} = \frac{d^2}{ds^2}\left(\frac{\omega}{s^2 + \omega^2}\right) = \frac{6\omega s^2 - 2\omega^3}{(s^2 + \omega^2)^3} \tag{5.43}$$

Example 3

Compute $\mathcal{L}\{t \cos \omega t\}$ and $\mathcal{L}\{t^2 \cos \omega t\}$

Solution

Similar to Example 2, we have

$$\mathcal{L}\{t \cos \omega t\} = -\left(\frac{s}{s^2 + \omega^2}\right)' = \frac{s^2 - \omega^2}{(s^2 + \omega^2)^2}$$

and

$$\mathcal{L}\{t^2 \cos \omega t\} = \left(\frac{s}{s^2 + \omega^2}\right)'' = \frac{2s^3 - 6s\omega^2}{(s^2 + \omega^2)^3}$$

5.2.5   Periodic Functions

For a periodic function $f(t)$ with period T that can be defined as

$$f(t) = f(t + T) \quad \forall\, t \geq 0 \tag{5.44}$$

the LT can be obtained using the original definition of the LT

$$\mathcal{L}\{f(t)\} = \int_0^\infty f(t) \exp(-st)\, dt \tag{5.45}$$

$$= \int_0^T f(t) \exp(-st)\, dt + \int_T^{2T} f(t) \exp(-st)\, dt$$

$$+ \int_{2T}^{3T} f(t) \exp(-st)\, dt \,...$$

$$= \int_0^T f(t) \exp(-st)\, dt + \int_0^T f(t + T) \exp\bigl(-s(t + T)\bigr)\, dt$$

$$+ \int_0^T f(t + 2T) \exp\bigl(-s(t + 2T)\bigr)\, dt \,...$$

$$= (1 + \exp(-sT) + \exp(-s2T) + \cdots) \int_0^T f(t) \exp(-st)\, dt$$

$$= \frac{1}{1 - \exp(-sT)} \int_0^T f(t) \exp(-st)\, dt$$

Therefore, for periodic functions, one only needs to compute the integral for the first period and multiply it by a period-dependent factor

$$\mathcal{L}\{f(t)\} = \frac{1}{1 - \exp(-sT)} \int_0^T f(t) \exp(-st)\, dt \tag{5.46}$$

5.2.6   Differentiation and Integration Property

In order to utilize LTs to solve DEs, we need to be able to perform LTs on differentiation and integration.

Let's first consider $\mathcal{L}\{f'(t)\}$ using the definition $F(s) = \mathcal{L}\{f(t)\}$. Using integration by parts, we have

$$
\begin{aligned}
\mathcal{L}\{f'(t)\} &= \int_0^\infty f'(t)\exp(-st)\,dt \\
&= \exp(-st)f(t)|_0^\infty \\
&\quad - \int_0^\infty f(t)\,d(\exp(-st)) \\
&= \exp(-s \times \infty)f(\infty) \\
&\quad - \exp(-s \times 0)f(0) \\
&\quad + s\int_0^\infty f(t)\exp(-st)\,dt
\end{aligned}
\tag{5.47}
$$

Since $F(s) = \int_0^\infty f(t)\exp(-st)\,dt$, we have

$$
\mathcal{L}\{f'(t)\} = sF(s) - f(0) \tag{5.48}
$$

from which we can derive the LTs for higher order derivatives, *e.g.*,

$$
\begin{aligned}
\mathcal{L}\{f''(t)\} &= \mathcal{L}\left\{\left(f'(t)\right)'\right\} \\
&= s\mathcal{L}\{f'(t)\} - f'(0) \\
&= s\left(sF(s) - f(0)\right) - f'(0)
\end{aligned}
\tag{5.49}
$$

Thus,

$$
\mathcal{L}\{f''(t)\} = s^2 F(s) - sf(0) - f'(0) \tag{5.50}
$$

Similarly,

$$
\begin{aligned}
\mathcal{L}\{f'''(t)\} &= s\mathcal{L}\{f''(t)\} - f''(0) \\
&= s^3 F(s) - s^2 f(0) - sf'(0) \\
&\quad - f''(0)
\end{aligned}
\tag{5.51}
$$

In the same manner, we find the general LT for $f^{(n)}(t)$

$$\mathcal{L}\{f^{(n)}(t)\} = s^n F(s) - s^{n-1}f(0) - \cdots - f^{(n-1)}(0) \qquad (5.52)$$

However, computing the LT for an integration $\mathcal{L}\left\{\int_0^t f(\tau)d\tau\right\}$ is a little more involved. Consider

$$g(t) = \int_0^t f(\tau)d\tau \qquad (5.53)$$

and, denoting $G(s) = \mathcal{L}\{g(t)\}$ and $F(s) = \mathcal{L}\{f(t)\}$ and using (5.48), we get

$$\mathcal{L}\{g'(t)\} = sG(s) - g(0) \qquad (5.54)$$

Since $g'(t) = f(t)$, the above can be written as

$$F(s) = sG(s) - g(0) \qquad (5.55)$$

Thus,

$$G(s) = \frac{1}{s}(F(s) + g(0)) \qquad (5.56)$$

It is easy to recognize $g(0) = \int_0^0 f(\tau)d\tau = 0$

Finally, we get $G(s) = \frac{1}{s}F(s)$ or

$$\mathcal{L}\left\{\int_0^t f(\tau)d\tau\right\} = \frac{1}{s}F(s) \qquad (5.57)$$

Similarly, we have

$$\mathcal{L}\left\{\int^t \left(\int^\tau f(t_1)dt_1\right)d\tau\right\} = \left(\frac{1}{s}\right)^2 F(s) \qquad (5.58)$$

and

$$\mathcal{L}\left\{\int^t \left(\int^\tau \left(\int^{t_1} f(t_2)dt_2\right)dt_1\right)d\tau\right\} = \left(\frac{1}{s}\right)^3 F(s) \qquad (5.59)$$

Example 4

Re-compute $\mathcal{L}\{t \sin \omega t\}$

Solution

Let $f(t) = t \sin \omega t$ and denote $F(s) = \mathcal{L}\{f(t)\}$. By taking derivative, we have

$$f'(t) = \sin \omega t + \omega t \cos \omega t$$
$$f''(t) = 2\omega \cos \omega t - \omega^2 t \sin \omega t$$

Performing LT on $f''(t)$, we get

$$\mathcal{L}\{f''(t)\} = 2\omega \mathcal{L}\{\cos \omega t\} - \omega^2 \mathcal{L}\{t \sin \omega t\}$$

Using (5.50), we get

$$s^2 F(s) - sf(0) - f'(0) = \frac{2\omega s}{s^2 + \omega^2} - \omega^2 F(s)$$

Obviously, $f'(0) = f(0) = 0$. Thus,

$$F(s) = \frac{2\omega s}{(s^2 + \omega^2)^2}$$

Problems

Problem 5.2.1 The square wave function $g(t)$ is shown in Figure 5.4 and its mathematical expression can be constructed by Step functions as

$$g(t) = 2 \sum_{n=0}^{\infty} (-1)^n u(t - n) - u(t - 0)$$

Prove

$$\mathcal{L}\{g(t)\} = \frac{1 - \exp(-s)}{s(1 + \exp(-s))} = \frac{1}{s} \tanh \frac{s}{2}$$

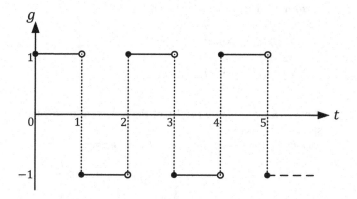

Figure 5.4 The square wave function for Problem 5.2.1.

Problem 5.2.2 Compute the LT
$$\mathcal{L}\left\{\frac{d}{dt}\left(t^2 \exp(\alpha t) \sin(\omega t)\right)\right\}$$

Problem 5.2.3 Compute the LT
$$\mathcal{L}\left\{\exp(t)\int_0^{\sqrt{t}} \exp(-\tau^2)\, d\tau\right\}$$

Problem 5.2.4 Compute the LT of the following function.
$$f(t) = t^2 \sin(\omega_1 t) + \exp(\alpha t)\cos(\omega_2 t)$$

Problem 5.2.5 Compute the LTs of the following un-related functions.
$$f_1(t) = \sum_{n=0}^{\infty} u(t-n)$$
$$f_2(t) = t - \lfloor t \rfloor$$
In both cases above, $t > 0$ and $\lfloor t \rfloor$ is the floor function of t and $u(t-n)$ is the usual step function.

Problem 5.2.6 LT is defined as
$$F(s) = \int_0^{\infty} f(t)\exp(-st)\, dt$$

Perform LT on the following functions:

    (1) $f(t) = t^5$

    (2) $f(t) = \exp(at)$

    (3) $f(t) = \sin(\omega t)$

    (4) $f(t) = \exp(at)\sin(\omega t)$

Problem 5.2.7 Compute the LT of the following triangular wave function shown in Figure 5.5.

Hint: the function during the 1st period can be written as

$$f(t) = \begin{cases} 2A\left(\dfrac{t}{T}\right), & 0 \le t \le \dfrac{T}{2} \\ 2A\left(1 - \dfrac{t}{T}\right), & \dfrac{T}{2} \le t \le T \end{cases}$$

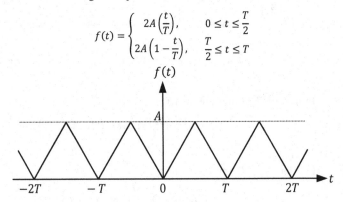

Figure 5.5 The triangular wave function for Problem 5.2.7.

5.3   Inverse Laplace Transforms

We introduce the inverse LT

$$\mathcal{L}^{-1}\{F(s)\} = f(t) \qquad (5.60)$$

where

$$F(s) = \mathcal{L}\{f(t)\} \qquad (5.61)$$

Theoretically, we have to calculate the Bromwich integral (also called the Fourier-Mellin integral) to find the inverse LT.

$$f(t) = \mathcal{L}^{-1}\{F(s)\} = \frac{1}{2\pi i} \int_{\gamma - i\infty}^{\gamma + i\infty} \exp(st) \, F(s) ds \qquad (5.62)$$

where $\gamma > R(s_p)$ for every singularity s_p of $F(s)$.

To actually calculate this integral, one needs to use the Cauchy residual theorem. However, for most cases, we can find out the inverse LT by simply looking up in the LT table, with some basic manipulations. Like the *forward* LT $\mathcal{L}\{f(t)\}$, the inverse LT $\mathcal{L}^{-1}\{F(s)\}$, is also linear, *i.e.*,

$$\mathcal{L}^{-1}\{\alpha F(s) + \beta G(s)\} = \alpha \mathcal{L}^{-1}\{F(s)\} + \beta \mathcal{L}^{-1}\{G(s)\}$$
$$= \alpha f(t) + \beta g(t)$$

Example 1
Compute

$$\mathcal{L}^{-1}\left\{\frac{1}{s^2} - \frac{1}{s-a}\right\} \qquad (5.63)$$

Solution

$$\mathcal{L}^{-1}\left\{\frac{1}{s^2} - \frac{1}{s-a}\right\} = \mathcal{L}^{-1}\left\{\frac{1}{s^2}\right\} - \mathcal{L}^{-1}\left\{\frac{1}{s-a}\right\}$$
$$= t - \exp(at)$$

Example 2
Compute

$$\mathcal{L}^{-1}\left\{\tan^{-1}\frac{1}{s}\right\} \tag{5.64}$$

Solution

Using two important relationships:

(1) The derivative of $\tan^{-1}\frac{1}{s}$ is a simple rational function

$$\frac{d}{ds}\left(\tan^{-1}\frac{1}{s}\right) = -\frac{1}{s^2+1}$$

(2) The t-multiplication property

$$\mathcal{L}^{-1}\{F'(s)\} = -t\mathcal{L}^{-1}\{F(s)\}$$

Or

$$\mathcal{L}^{-1}\{F(s)\} = \left(-\frac{1}{t}\right)\mathcal{L}^{-1}\{F'(s)\}$$

Thus,

$$\mathcal{L}^{-1}\left\{\tan^{-1}\frac{1}{s}\right\} = \left(-\frac{1}{t}\right)\mathcal{L}^{-1}\left\{\frac{d}{ds}\left(\tan^{-1}\frac{1}{s}\right)\right\}$$

$$= -\frac{1}{t}\mathcal{L}^{-1}\left\{-\frac{1}{s^2+1}\right\}$$

$$= \frac{\sin t}{t}$$

Therefore,

$$\mathcal{L}^{-1}\left\{\tan^{-1}\frac{1}{s}\right\} = \frac{\sin t}{t}$$

Example 3

Compute

$$\mathcal{L}^{-1}\left\{\frac{1}{s^2(s-a)}\right\} \tag{5.65}$$

Solution

Method 1: Using the integration property.

Given

$$\mathcal{L}\left\{\int_0^t f(\tau)d\tau\right\} = \frac{1}{s}F(s)$$

Conversely,

$$\mathcal{L}^{-1}\left\{\frac{1}{s}F(s)\right\} = \int_0^t f(\tau)d\tau$$

Now, we can proceed

$$\mathcal{L}^{-1}\left\{\frac{1}{s-a}\right\} = \exp(at)$$

$$\mathcal{L}^{-1}\left\{\frac{1}{s}\frac{1}{s-a}\right\} = \int_0^t \exp(a\tau)d\tau$$

$$= \frac{1}{a}(\exp(at) - 1)$$

$$\mathcal{L}^{-1}\left\{\frac{1}{s^2}\frac{1}{s-a}\right\} = \int_0^t \frac{1}{a}(\exp(a\tau) - 1)d\tau$$

$$= \frac{1}{a^2}(\exp(at) - at - 1)$$

Method 2: Using partial fractions.
Let

$$\frac{1}{s^2(s-a)} = \frac{A}{s} + \frac{B}{s^2} + \frac{C}{s-a}$$

We have

$$1 = (As + B)(s - a) + Cs^2$$
$$= (A + C)s^2 + (B - Aa)s - Ba$$

Matching coefficients, we have

$$A + C = 0$$
$$B - Aa = 0$$
$$-Ba = 1$$

Thus,

$$A = -\frac{1}{a^2}, \qquad B = -\frac{1}{a}, \qquad C = \frac{1}{a^2}$$

which means

$$\frac{1}{s^2(s-a)} = \left(-\frac{1}{a^2}\right)\frac{1}{s} + \left(-\frac{1}{a}\right)\frac{1}{s^2} + \left(\frac{1}{a^2}\right)\frac{1}{s-a}$$

By linearity, we have

$$\mathcal{L}^{-1}\left\{\frac{1}{s^2(s-a)}\right\} = -\frac{1}{a^2}\mathcal{L}^{-1}\left\{\frac{1}{s}\right\} - \frac{1}{a}\mathcal{L}^{-1}\left\{\frac{1}{s^2}\right\} + \frac{1}{a^2}\mathcal{L}^{-1}\left\{\frac{1}{s-a}\right\}$$

$$= -\frac{1}{a^2} - \frac{1}{a}t + \frac{1}{a^2}\exp(at)$$

$$= \frac{1}{a^2}(\exp(at) - at - 1)$$

Problems

Problem 5.3.1 Prove that

$$\mathcal{L}^{-1}\left\{\frac{1}{(s^2 + l^2)^2}\right\} = \frac{1}{2k^3}\left(\sin(kt) - kt\cos(kt)\right)$$

If given

$$\mathcal{L}\{k\cos(kt)\} = \frac{s^2 - k^2}{(s^2 + k^2)^2}$$

$$\mathcal{L}\{\sin(kt)\} = \frac{k}{s^2 + k^2}$$

Problem 5.3.2 Show the relationship

$$\mathcal{L}^{-1}\left\{\frac{\exp\left(-\frac{1}{s}\right)}{s}\right\} = J_0\left(2\sqrt{t}\right)$$

where $J_0(t)$ is the so-called Bessel's function defined as

$$J_0\left(2\sqrt{t}\right) = \sum_{n=0}^{\infty} \frac{(-1)^n}{(n!)^2}\left(\sqrt{t}\right)^{2n}$$

Problem 5.3.3 Compute the following inverse LT

$$\mathcal{L}^{-1}\left\{\frac{2s}{(s^2 - 1)^2}\right\}$$

Problem 5.3.4 Compute the following inverse LT

$$\mathcal{L}^{-1}\left\{\frac{\exp(-as)}{s(1 - \exp(-as))}\right\}$$

Problem 5.3.5 The following equation is one of the so-called Volterra Integro-Differential Equations (IDE) containing the unknown function $X(t)$

$$x(t) = \sin(t) + 2\int_0^t \cos(t - \tau)x(\tau)d\tau$$

Use LT to solve this equation by completing the following steps:
(1) Use LT to transform the above IDE into an AE containing unknown function in Laplace space $X(s)$ with given formula

$$\mathcal{L}\left\{\int_0^t \cos(t-\tau)\,x(\tau)d\tau\right\} = \frac{s}{s^2+1}X(s)$$

(2) Solve the AE for $X(s)$.

(3) Inverse LT to obtain the unknown function $x(t)$ as a function of t.

Problem 5.3.6 Solve the following Volterra IDE.

$$x(t) = 2\exp(3t) - \int_0^t \exp\big(2(t-\tau)\big)\,x(\tau)d\tau$$

5.4   The Convolution of Two Functions

Consider two functions $f(t)$ and $g(t)$, a binary operation defined as

$$f(t) \otimes g(t) = \int_0^t f(\tau)g(t-\tau)d\tau \qquad (5.66)$$

is called the convolution of these functions. This operation has elegant and useful properties associated with LTs and one of the properties is called Convolution Theorem.

Convolution Theorem

Denoting $F(s) = \mathcal{L}\{f(t)\}$ and $G(s) = \mathcal{L}\{g(t)\}$, we have

$$\mathcal{L}\{f(t) \otimes g(t)\} = \mathcal{L}\{f(t)\}\mathcal{L}\{g(t)\} = F(s)G(s) \qquad (5.67)$$

Proof

$$
\begin{aligned}
f(t) \otimes g(t) &= \int_0^t f(\tau)g(t-\tau)d\tau \\
&= \int_0^t f(\tau)g(t-\tau)d\tau \\
&\quad + \int_t^\infty 0 \times f(\tau)g(t-\tau)d\tau \\
&= \int_0^\infty u(t-\tau)f(\tau)g(t-\tau)d\tau
\end{aligned}
\qquad (5.68)
$$

where $u(t-\tau)$ is the unit step function. Thus, we have

$$
\begin{aligned}
\mathcal{L}\{f(t) \otimes g(t)\} &= \int_0^\infty \exp(-st)\,dt \int_0^\infty u(t-\tau)f(\tau)g(t-\tau)d\tau \\
&= \int_0^\infty \int_0^\infty \exp(-st)\,u(t-\tau)f(\tau)g(t-\tau)dtd\tau
\end{aligned}
\qquad (5.69)
$$

Introducing a new variable $t_1 = t - \tau$ to replace t, we have

$$\mathcal{L}\{f(t)\otimes g(t)\} = \int_0^\infty \int_0^\infty \exp[-s(t_1 + \tau)]\, u(t_1)f(\tau)g(t_1)d\tau dt_1$$

$$= \int_0^\infty \exp(-st_1)\, u(t_1)g(t_1)dt_1$$

$$= \int_0^\infty \exp(-st_1)\, g(t_1)dt_1 \int_0^\infty \exp(-s\tau)\, f(\tau)d\tau$$

$$= F(s) \times G(s) \tag{5.70}$$

Conversely,

$$\mathcal{L}^{-1}\{F(s) \times G(s)\} = f(t)\otimes g(t) \tag{5.71}$$

Thus, we can find the inverse LT of the usual product of two functions $F(s)G(s)$ as long as we can evaluate the convolution of $f(t)$ and $g(t)$. The binary operation, convolution, satisfies the following simple rules:

(1) Commutative: $f_1 \otimes f_2 = f_2 \otimes f_1$
(2) Distributive: $f_1 \otimes (f_A + f_B) = f_1 \otimes f_A + f_1 \otimes f_B$
(3) Associative: $f_1 \otimes (f_2 \otimes f_3) = (f_1 \otimes f_2) \otimes f_3$
(4) Zero: $f_1 \otimes 0 = 0$

Example 1

Compute the following simple convolutions

$$f_1(t) = 1 \otimes 1$$
$$f_2(t) = 1 \otimes t = t \otimes 1$$
$$f_3(t) = t \otimes t \tag{5.72}$$
$$f_4(t) = \exp(\alpha t) \otimes \exp(\beta t)$$

Solution

$$f_1(t) = 1 \otimes 1 = \int_0^t 1 \times 1 \, d\tau = t$$

$$f_2(t) = 1 \otimes t = \int_0^t 1 \times (t - \tau) \, d\tau = \frac{1}{2}t^2$$

$$f_3(t) = t \otimes t = \int_0^t \tau(t - \tau) \, d\tau = \frac{1}{6}t^3$$

$$f_4(t) = \exp(\alpha t) \otimes \exp(\beta t)$$

$$= \int_0^t \exp(\alpha \tau) \exp\big(\beta(t - \tau)\big)\, d\tau$$

$$= \exp(\beta t) \int_0^t \exp\big((\alpha - \beta)\tau\big)\, d\tau$$

$$= \frac{\exp(\alpha t) - \exp(\beta t)}{\alpha - \beta}$$

where

$$\lim_{\alpha \to \beta} \frac{\exp(\alpha t) - \exp(\beta t)}{\alpha - \beta} = t \exp(\alpha t) = t \exp(\beta t)$$

Example 2

Compute $\cos t \otimes \sin t$ and $\sin \omega t \otimes \sin \omega t$ and $\cos \omega t \otimes \cos \omega t$

Solution

Recognizing the trigonometric identity

$$\sin(A \pm B) = \sin A \cos B \pm \cos A \sin B$$
$$\cos(A \pm B) = \cos A \cos B \mp \sin A \sin B$$

These can be written as

$$\cos A \sin B = \frac{1}{2}[\sin(A + B) - \sin(A - B)]$$
$$\sin A \sin B = \frac{1}{2}[\cos(A - B) - \cos(A + B)]$$

Thus,

$$\cos t \otimes \sin t = \int_0^t \cos(t - \tau) \sin \tau\, d\tau$$

$$= \frac{1}{2} \int_0^t (\sin t - \sin(t - 2\tau))\, d\tau$$

$$= \frac{1}{2} \left(\tau \sin t - \frac{1}{2} \cos(t - 2\tau) \right) \Big|_{\tau=0}^{t}$$

$$= \frac{1}{2} t \sin t$$

So,

$$\cos t \otimes \sin t = \frac{1}{2} t \sin t$$
$$\sin t \otimes \cos t = \frac{1}{2} t \sin t$$

Obviously,

$$\cos t \otimes \sin t \neq \cos t \times \sin t$$

Next, we compute

$$\sin \omega t \otimes \sin \omega t = \int_0^t \sin \omega(t - \tau) \sin \omega \tau \, d\tau$$

$$= \frac{1}{2} \int_0^t \big(\cos(\omega(t - 2\tau)) - \cos(\omega t)\big) d\tau$$

$$= \frac{1}{2} \int_0^t \big(\cos(\omega(t - 2\tau)) - \cos(\omega t)\big) d\tau$$

$$= \frac{1}{2\omega} (\sin \omega t - \omega t \cos \omega t)$$

Similarly,

$$\cos \omega t \otimes \cos \omega t = \int_0^t \cos \omega(t - \tau) \cos \omega \tau \, d\tau$$

$$= \frac{1}{2} \int_0^t \big(\cos(\omega(t - 2\tau)) + \cos(\omega t)\big) d\tau$$

$$= \frac{1}{2} \int_0^t \big(\cos(\omega(t - 2\tau)) + \cos(\omega t)\big) d\tau$$

$$= \frac{1}{2\omega} (\sin \omega t + \omega t \cos \omega t)$$

Thus,

$$\sin \omega t \otimes \sin \omega t = \frac{1}{2\omega} (\sin \omega t - \omega t \cos \omega t)$$

$$\cos \omega t \otimes \cos \omega t = \frac{1}{2\omega} (\sin \omega t + \omega t \cos \omega t)$$

If $\omega = 1$, we get

$$\sin t \otimes \sin t = \frac{1}{2} (\sin t - t \cos t)$$

$$\cos t \otimes \cos t = \frac{1}{2} (\sin t + t \cos t)$$

Example 3

Compute

$$\mathcal{L}^{-1}\left\{\frac{1}{s^2(s - a)}\right\} \tag{5.73}$$

Solution

$$\mathcal{L}^{-1}\left\{\frac{1}{s^2(s - a)}\right\} = \mathcal{L}^{-1}\left\{\frac{1}{s^2} \cdot \frac{1}{s - a}\right\}$$

$$= \mathcal{L}^{-1}\left\{\frac{1}{s^2}\right\} \otimes \mathcal{L}^{-1}\left\{\frac{1}{s-a}\right\}$$

$$= t \otimes \exp(at)$$

$$= \int_0^t \exp(a\tau)\,(t-\tau)d\tau$$

$$= \int_0^t (t\exp(a\tau) - \tau\exp(a\tau))d\tau$$

$$= t\int_0^t \exp(a\tau)\,d\tau - \int_0^t \frac{\tau}{a}d(\exp(a\tau))$$

$$= \left.\left(\frac{t}{a}\exp(a\tau) - \frac{\tau}{a}\exp(a\tau) + \frac{1}{a^2}\exp(a\tau)\right)\right|_{\tau=0}^{t}$$

$$= \frac{1}{a^2}(\exp(at) - at - 1)$$

Consistent with results obtained earlier in Example 3 of the previous section.

Problems

Problem 5.4.1 Use Convolution Theorem to prove the following identity

$$\mathcal{L}^{-1}\left\{\frac{1}{(s-1)\sqrt{s}}\right\} = \exp(t)\,\text{erf}(\sqrt{t})$$

where $\mathcal{L}^{-1}$ denotes an inverse LT and $\text{erf}(x)$ is the "error function" defined by

$$\text{erf}(x) \equiv \frac{2}{\sqrt{\pi}}\int_0^x \exp(-u^2)\,du$$

(Hint: Use substitution $u = \sqrt{t}$.)

Problem 5.4.2 Find the GS to the following DE

$$y(x) = x\cos(3x) - \int_0^x \exp(\tau)\,y(x-\tau)\,d\tau$$

5.5 Applications

LTs can help solve some of the linear (and, in special circumstances, nonlinear) DEs much more conveniently than the other methods we learned earlier. Starting from Section 5.2.4, we demonstrate such values of the LT methods.

The most essential step of applying LT to solving DEs is to use the following formulas

$$\mathcal{L}\{f'(t)\} = s\mathcal{L}\{f(t)\} - f(0)$$
$$\mathcal{L}\{f''(t)\} = s^2\mathcal{L}\{f(t)\} - sf(0) - f'(0)$$

$$\cdots$$

$$\mathcal{L}\{f^{(n)}(t)\} = s^n\mathcal{L}\{f(t)\} - s^{n-1}f(0) - \cdots - f^{(n-1)}(0) \qquad (5.74)$$

$$= s^n\mathcal{L}\{f(t)\} - \sum_{i=1}^{n} s^{n-i} f^{(i-1)}(0)$$

Thus, for a linear DE of the following form

$$\sum_{j=0}^{m} a_j f^{(j)}(t) = \phi(t) \qquad (5.75)$$

applying LTs on both sides yields

$$\sum_{j=0}^{m} a_j \mathcal{L}\{f^{(j)}(t)\} = \mathcal{L}\{\phi(t)\} \qquad (5.76)$$

Applying the differentiation property, we write the equation as

$$\sum_{j=0}^{m} a_j \left(s^j \mathcal{L}\{f(t)\} - \sum_{i=1}^{j} s^{j-i} f^{(i-1)}(0) \right) = \mathcal{L}\{\phi(t)\} \qquad (5.77)$$

$$\sum_{j=0}^{m} a_j s^j \mathcal{L}\{f(t)\} - \sum_{j=0}^{m}\left(a_j \sum_{i=1}^{j} s^{j-i} f^{(i-1)}(0)\right) = \mathcal{L}\{\phi(t)\}$$

$$\mathcal{L}\{f(t)\} = \frac{1}{\sum_{j=0}^{m} a_j s^j}\left(\mathcal{L}\{\phi(t)\} + \sum_{j=0}^{m}\left(a_j \sum_{i=1}^{j} s^{j-i} f^{(i-1)}(0)\right)\right)$$

where $f^{(k)}(0)$ are ICs.

Finally, applying inverse LT, we get the solution to the DE,

$$f(t) = \mathcal{L}^{-1}\left\{\frac{1}{\sum_{j=0}^{m} a_j s^j}\left(\mathcal{L}\{\phi(t)\}\right.\right.$$
$$\left.\left. + \sum_{j=0}^{m}\left(a_j \sum_{i=1}^{j} s^{j-i} f^{(i-1)}(0)\right)\right)\right\}$$

$$(5.78)$$

Example 1

Use LTs method to solve the following IVP (Homo DE with given IC's)

$$\begin{cases} x'' - x' - 6x = 0 \\ x(0) = 2 \\ x'(0) = -1 \end{cases} \qquad (5.79)$$

Solution

Applying LTs to both sides of the DE, we get

$$\mathcal{L}\{x'' - x' - 6x\} = L\{0\}$$
$$\mathcal{L}\{x''\} - L\{x'\} - 6L\{x\} = 0$$
$$\left(s^2 X(s) - sx(0) - x'(0)\right) - \left(sX(s) - x(0)\right) - 6X(s) = 0$$

Plugging in the IC's, we get

$$s^2 X(s) - 2s + 1 - sX(s) + 2 - 6X(s) = 0$$
$$X(s)(s^2 - s - 6) - 2s + 3 = 0$$
$$X(s) = \frac{2s - 3}{s^2 - s - 6}$$

By partial fraction decomposition, we have,

$$X(s) = \frac{7}{5}\left(\frac{1}{s+2}\right) + \frac{3}{5}\left(\frac{1}{s-3}\right)$$

Thus, we can find the PS to the original DE by applying inverse LT

$$x(t) = \mathcal{L}^{-1}\{X(s)\}$$
$$= \mathcal{L}^{-1}\left\{\frac{7}{5}\left(\frac{1}{s+2}\right) + \frac{3}{5}\left(\frac{1}{s-3}\right)\right\}$$
$$= \frac{7}{5}\exp(-2t) + \frac{3}{5}\exp(3t)$$

Example 2

Use LTs method to solve the following IVP (InHomo DE with given IC's)

$$\begin{cases} x'' + 4x = \sin 3t \\ \quad x(0) = 0 \\ \quad x'(0) = 0 \end{cases} \tag{5.80}$$

Solution

Such a problem arises in the motion of a mass attached to a spring with external force, as shown in Figure 5.6.

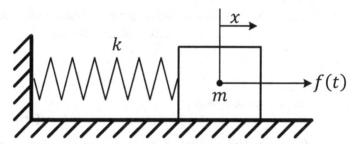

Figure 5.6 A block-spring system with an external force $f(t)$.

Apply LTs on both sides of the DE

$$\mathcal{L}\{x'' + 4x\} = \mathcal{L}\{\sin 3t\}$$

we get

$$\left(s^2 X(s) - sx(0) - x'(0)\right) + 4X(s) = \frac{3}{s^2 + 3^2}$$

Plugging in the IC's, we get

$$X(s)(s^2 + 4) = \frac{3}{s^2 + 9}$$
$$X(s) = \frac{3}{(s^2 + 4)(s^2 + 9)}$$

By partial fraction decomposition

$$X(s) = \frac{A}{s^2 + 4} + \frac{B}{s^2 + 9}$$

we get

$$A = \frac{3}{5}, \qquad B = -\frac{3}{5}$$

Thus,

$$X(s) = \frac{3}{5}\left(\frac{1}{s^2+4}\right) - \frac{3}{5}\left(\frac{1}{s^2+9}\right)$$

$$= \frac{3}{10}\left(\frac{2}{s^2+2^2}\right) - \frac{1}{5}\left(\frac{3}{s^2+3^2}\right)$$

Finally, applying inverse LT generates the PS to the original DE

$$x(t) = \mathcal{L}^{-1}\{X(s)\}$$

$$= \mathcal{L}^{-1}\left\{\frac{3}{10}\left(\frac{2}{s^2+2^2}\right) - \frac{1}{5}\left(\frac{3}{s^2+3^2}\right)\right\}$$

$$= \frac{3}{10}\mathcal{L}^{-1}\left\{\frac{2}{s^2+2^2}\right\} - \frac{1}{5}\mathcal{L}^{-1}\left\{\frac{3}{s^2+3^2}\right\}$$

$$= \frac{3}{10}\sin 2t - \frac{1}{5}\sin 3t$$

Example 3

Use LTs method to solve the following IVP (InHomo DE with given IC's)

$$\begin{cases} x'' - x' - 12x = \cos 3t + \exp(4t) \\ \qquad\qquad x(0) = 2 \\ \qquad\qquad x'(0) = 1 \end{cases} \qquad (5.81)$$

Solution

With the given IC's, we get

$$\mathcal{L}\{x''\} - \mathcal{L}\{x'\} - 12\mathcal{L}\{x\} = \mathcal{L}\{\cos 3t + \exp(4t)\}$$

$$(s^2 X(s) - sx(0) - x'(0)) - (sX(s) - x(0)) - 12X(s) = \frac{s}{s^2+9} + \frac{1}{s-4}$$

$$(s-4)(s+3)X(s) = \frac{s}{s^2+9} + \frac{1}{s-4} + 2s - 1$$

$$X(s) = \frac{1}{7}\left(\frac{1}{s-4} - \frac{1}{s+3}\right)\left(\frac{s}{s^2+9} + \frac{1}{s-4} + 2s - 1\right)$$

$$= \frac{667}{650}\left(\frac{1}{s+3}\right) + \frac{199}{195}\left(\frac{1}{s-4}\right) - \frac{7}{150}\left(\frac{s}{s^2+3^2}\right) - \frac{1}{150}\left(\frac{3}{s^2+3^2}\right)$$

$$+ \frac{1}{4}\left(\frac{1}{s-4}\right)^2$$

Applying inverse LT of the above, we get

$$x(t) = \frac{667}{650}\exp(-3t) + \frac{199}{195}\exp(4t) - \frac{7}{150}\cos 3t - \frac{1}{150}\sin 3t$$

$$+ \frac{1}{4}t\exp(4t)$$

Example 4

Find the GS to the following DE.Syst

$$\begin{cases} 2x'' = -6x + 2y \\ \quad y'' = 2x - 2y + 40\sin 3t \\ x(0) = x'(0) = y(0) = y'(0) = 0 \end{cases} \tag{5.82}$$

which models the following spring system. (Figure 5.7)

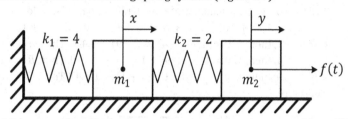

Figure 5.7 A system of two blocks connected by two springs and an external force $f(t)$.

Solution

With the given IC's, we have

$$\mathcal{L}\{x''(t)\} = s^2 X(s)$$
$$\mathcal{L}\{y''(t)\} = s^2 Y(s)$$

Performing LT on both sides of the original DEs, we have,

$$2s^2 X(s) = -6X(s) + 2Y(s)$$

$$s^2 Y(s) = 2X(s) - 2Y(s) + 40\left(\frac{3}{s^2 + 9}\right)$$

where we have used

$$\mathcal{L}\{\sin 3t\} = \frac{3}{s^2 + 9}$$

Resulting two AEs defined in the s-space

$$\begin{cases} (s^2 + 3)X(s) - Y(s) = 0 \\ -2X(s) + (s^2 + 2)Y(s) = \dfrac{120}{s^2 + 9} \end{cases}$$

Substituting the first equation

$$Y(s) = (s^2 + 3)X(s)$$

to the second, we get

$$X(s) = \frac{120}{(s^2 + 9)\big((s^2 + 2)(s^2 + 3) - 2\big)}$$

$$= \frac{120}{(s^2 + 1)(s^2 + 4)(s^2 + 9)}$$

Plugging this back into the first equation, we have

$$Y(s) = \frac{120(s^2 + 3)}{(s^2 + 1)(s^2 + 4)(s^2 + 9)}$$

Next, express $X(s)$ and $Y(s)$ using partial fractions, for example for $X(s)$ we let

$$\frac{120}{(s^2 + 1)(s^2 + 4)(s^2 + 9)} = \frac{A}{s^2 + 1} + \frac{B}{s^2 + 4} + \frac{C}{s^2 + 9}$$

The following steps demonstrate the details of partial fractions.

$$120 = A(s^2 + 4)(s^2 + 9) + B(s^2 + 1)(s^2 + 9) + C(s^2 + 1)(s^2 + 4)$$

If $s^2 = -1$, then, $120 = A(-1 + 4)(-1 + 9) + 0 + 0$. Thus, $A = 5$.
If $s^2 = -4$, then, $120 = 0 + B(-4 + 1)(-4 + 9) + 0$. Thus, $B = -8$.
If $s^2 = -9$, then, $120 = 0 + 0 + C(-9 + 1)(-9 + 4)$. Thus, $C = -3$.

Therefore,

$$X(s) = \frac{5}{s^2 + 1} - \frac{8}{s^2 + 4} + \frac{3}{s^2 + 9}$$

$$= 5\frac{1}{s^2 + 1} - 4\frac{2}{s^2 + 4} + \frac{3}{s^2 + 9}$$

Applying inverse LT produces

$$x(t) = \mathcal{L}^{-1}\{X(s)\}$$

$$= 5\sin t - 4\sin 2t + \sin 3t$$

Similarly, for $Y(s)$

$$\frac{120(s^2 + 3)}{(s^2 + 1)(s^2 + 4)(s^2 + 9)} = \frac{E}{s^2 + 1} + \frac{F}{s^2 + 4} + \frac{G}{s^2 + 9}$$

Thus,

$$120(s^2 + 3) = E(s^2 + 4)(s^2 + 9) + F(s^2 + 1)(s^2 + 9)$$
$$+ G(s^2 + 1)(s^2 + 4)$$

If $s^2 = -1$, then $240 = E(-1 + 4)(-1 + 9) + 0 + 0$. Thus, $E = 10$.
If $s^2 = -4$, then $-120 = 0 + F(-4 + 1)(-4 + 9) + 0$. Thus, $F = 8$.
If $s^2 = -9$, then $-720 = 0 + 0 + G(-9 + 1)(-9 + 4)$. Thus, $G = -18$.

Therefore,

$$Y(s) = \frac{10}{s^2 + 1} + \frac{8}{s^2 + 4} - \frac{18}{s^2 + 9}$$

$$= 10\frac{1}{s^2 + 1} + 4\frac{2}{s^2 + 4} - 6\frac{3}{s^2 + 9}$$

Finally,

$$y(t) = 10\sin t + 4\sin 2t - 6\sin 3t$$

Example 5

Solve IVP for the Bessel's equation of order 0.

$$\begin{cases} tx'' + x' + tx = 0 \\ \qquad\qquad x(0) = 1 \\ \qquad\qquad x'(0) = 0 \end{cases} \qquad (5.83)$$

Solution

Using the LT of derivatives and the IC's, we have

$$\mathcal{L}\{x'(t)\} = sX(s) - 1$$
$$\mathcal{L}\{x''(t)\} = s^2X(s) - s$$

Because x and x'' are each multiplied by t, by applying the t-multiplication property, we get the transformed equation

$$-\frac{d}{ds}(s^2X(s) - s) + (sX(s) - 1) - \frac{d}{ds}(X(s)) = 0$$

The result of differentiation and simplification is the DE

$$(s^2 + 1)X'(s) + sX(s) = 0$$

Since the DE is separable we have,

$$\frac{X'(s)}{X(s)} = -\frac{s}{s^2 + 1}$$

and its GS is

$$X(s) = \frac{C}{\sqrt{s^2 + 1}}$$

Remarks

(1) Here the constant C in the GS of $X(s)$ is actually not an arbitrary number. Let $s = 0$, we have

$$X(0) = C \qquad (5.84)$$

On the other hand, from the definition of the LT,

$$X(0) = \int_0^\infty x(t)\,dt \qquad (5.85)$$

This means C is a normalization factor.

(2) The solution to the original DE, *i.e.*, the inverse LT for

$$X(s) = \frac{1}{\sqrt{s^2 + 1}} \qquad (5.86)$$

is a function called 0th-order Bessel function of first kind. The general form of α order Bessel function is defined as

$$J_0(t) = \sum_{m=0}^{\infty} \frac{(-1)^m}{m!\,\Gamma(m + \alpha + 1)} \left(\frac{x}{2}\right)^{2m+\alpha} \qquad (5.87)$$

Example 6

Find the PS to the following IVP

$$\begin{cases} x'' + 4x = 2\exp(t) \\ \quad x(0) = x'(0) = 0 \end{cases} \qquad (5.88)$$

Solution

Applying LT on both sides of the DE, we get

$$\mathcal{L}\{x'' + 4x\} = \mathcal{L}\{2\exp(t)\}$$

From the IC, we have

$$(s^2 + 4)X(s) = \frac{2}{s - 1}$$

$$X(s) = \left(\frac{2}{s - 1}\right)\left(\frac{1}{s^2 + 4}\right)$$

Therefore, the PS to the original DE is

$$x(t) = \mathcal{L}^{-1}\{X(s)\}$$

$$= \mathcal{L}^{-1}\left\{\left(\frac{1}{s - 1}\right)\left(\frac{2}{s^2 + 4}\right)\right\}$$

$$= \mathcal{L}^{-1}\left\{\frac{1}{s - 1}\right\} \otimes \mathcal{L}^{-1}\left\{\frac{2}{s^2 + 4}\right\}$$

$$= \exp(t) \otimes \sin 2t$$

Since

$$\exp(t) \otimes \sin 2t = \int_0^t \exp(t - \tau)\sin 2\tau\, d\tau$$

$$= \exp(t)\int_0^t \exp(-\tau)\sin 2\tau\, d\tau$$

Let

$$I = \int_0^t \exp(-\tau)\sin 2\tau\, d\tau$$

Using integration by parts repeatedly, we have

$$I = -\int_0^t \sin 2\tau\, d(\exp(-\tau))$$

$$= -\exp(-\tau)\sin 2\tau\big|_0^t + 2\int_0^t \exp(-\tau)\cos 2\tau\, d\tau$$

$$= -\exp(-\tau)\sin 2t - 2\int_0^t \cos 2\tau\, d(\exp(-\tau))$$

$$= -\exp(-\tau)\sin 2t - 2\exp(-\tau)\cos 2\tau\big|_0^t + 4\int_0^t \exp(-\tau)\sin 2\tau\, d\tau$$

$$= -\exp(-t)\sin 2t - 2\exp(-t)\cos 2t + 2 - 4I$$

This gives

$$I = -\frac{1}{5}\exp(-t)\sin 2t - \frac{2}{5}\exp(-t)\cos 2t + \frac{2}{5}$$

Finally,

$$x(t) = \exp(-t)I = \frac{2}{5}\exp(-t) - \frac{1}{5}\sin 2t - \frac{2}{5}\cos 2t$$

Example 7

Find the PS to the following IVP
$$\begin{cases} x'' - \alpha^2 x = \exp(\beta t) \\ \qquad x(0) = x'(0) = 0 \end{cases} \tag{5.89}$$
where α and β are constants.

Solution

Applying LT on both sides of the InHomo DE, we get
$$(s^2 - \alpha^2)X(s) = \mathcal{L}\{\exp(\beta t)\}$$

Thus,
$$X(s) = \frac{1}{2\alpha}\left(\frac{1}{s-\alpha} - \frac{1}{s+\alpha}\right)\mathcal{L}\{\exp(\beta t)\}$$

Applying inverse LT of the above, we get
$$x(t) = \mathcal{L}^{-1}\left(\frac{1}{2\alpha}\left(\frac{1}{s-\alpha} - \frac{1}{s+\alpha}\right)\mathcal{L}\{\exp(\beta t)\}\right)$$
$$= \frac{1}{2\alpha}(\exp(\alpha t) - \exp(-\alpha t)) \otimes \exp(\beta t)$$

Let's now compute the convolutions as we did in previous section.
$$\exp(\alpha t) \otimes \exp(\beta t) = \int_0^t \exp(\alpha(t-\tau))\exp(\beta\tau)\,d\tau$$
$$= \exp(\alpha t)\int_0^t \exp((\beta - \alpha)\tau)\,d\tau$$
$$= \begin{cases} t\exp(\alpha t) & \text{if } \alpha - \beta = 0 \\ \dfrac{\exp(\alpha t) - \exp(\beta t)}{\alpha - \beta} & \text{if } \alpha - \beta \neq 0 \end{cases}$$

$$\exp(-\alpha t) \otimes \exp(\beta t) = \int_0^t \exp(-\alpha(t-\tau))\exp(\beta\tau)\,d\tau$$
$$= \exp(-\alpha t)\int_0^t \exp((\alpha + \beta)\tau)\,d\tau$$
$$= \begin{cases} t\exp(-\alpha t) & \text{if } \alpha + \beta = 0 \\ \dfrac{-\exp(-\alpha t) + \exp(\beta t)}{\alpha + \beta} & \text{if } \alpha + \beta \neq 0 \end{cases}$$

Thus, there are four cases to examine.

Case 1: $\alpha - \beta = 0$ and $\alpha + \beta = 0$.

This case is equivalent to $\alpha = \beta = 0$ for which the PS is
$$x(t) = \frac{1}{2}t^2$$
a special case where the original IVP reduces to

$$\begin{cases} x'' = 1 \\ x(0) = x'(0) = 0 \end{cases}$$

Case 2: $\alpha - \beta = 0$ and $\alpha + \beta \neq 0$.

This case is equivalent to $\alpha = \beta \neq 0$ for which the PS is

$$x(t) = \frac{1}{2\alpha}\left(t\exp(\alpha t) - \frac{-\exp(-\alpha t) + \exp(\beta t)}{\alpha + \beta} \right)$$

$$= \frac{1}{2\alpha}\left(t\exp(\alpha t) - \frac{1}{\alpha}\left(\frac{-\exp(-\alpha t) + \exp(\alpha t)}{2} \right) \right)$$

$$= \frac{1}{2\alpha}\left(t\exp(\alpha t) - \frac{1}{\alpha}\sinh(\alpha t) \right)$$

Case 3: $\alpha - \beta \neq 0$ and $\alpha + \beta = 0$.

This case is equivalent to $\alpha = -\beta \neq 0$ for which the PS is

$$x(t) = \frac{1}{2\alpha}\left(\frac{\exp(\alpha t) - \exp(\beta t)}{\alpha - \beta} - t\exp(-\alpha t) \right)$$

$$= \frac{1}{2\alpha}\left(\frac{1}{\alpha}\left(\frac{\exp(\alpha t) - \exp(-\alpha t)}{2} \right) - t\exp(-\alpha t) \right)$$

$$= \frac{1}{2\alpha}\left(\frac{1}{\alpha}\sinh(\alpha t) - t\exp(-\alpha t) \right)$$

Case 4: $\alpha - \beta \neq 0$ and $\alpha + \beta \neq 0$.

This case is equivalent to $\alpha \neq \pm\beta$ for which the PS is

$$x(t) = \frac{1}{2\alpha}\left(\frac{\exp(\alpha t) - \exp(\beta t)}{\alpha - \beta} - \frac{-\exp(-\alpha t) + \exp(\beta t)}{\alpha + \beta} \right)$$

$$= \frac{1}{2\alpha}\left(\frac{\exp(\alpha t) - \exp(\beta t)}{\alpha - \beta} + \frac{\exp(-\alpha t) - \exp(\beta t)}{\alpha + \beta} \right)$$

Example 8

Find the PS to the following IVP

$$\begin{cases} x'' + \omega^2 x = \cos(\omega_1 t) \\ x(0) = x'(0) = 0 \end{cases} \tag{5.90}$$

where ω and ω_1 are given constants and they may be equal or they may not be.

Solution

Applying LT on both sides of the DE, we get

$$\mathcal{L}\{x''\} + \omega^2 \mathcal{L}\{x\} = \mathcal{L}\{\cos(\omega_1 t)\}$$

$$s^2 X(s) - sx(0) - x'(0) + \omega^2 X(s) = \mathcal{L}\{\cos(\omega_1 t)\}$$

Thus,

$$X(s) = \frac{\mathcal{L}\{\cos(\omega_1 t)\}}{s^2 + \omega^2}$$

Applying inverse LT, we get

$$x(t) = \mathcal{L}^{-1}\left\{\frac{1}{s^2 + \omega^2}\right\} \otimes \mathcal{L}^{-1}\{\mathcal{L}\{\cos(\omega_1 t)\}\}$$

$$= \frac{1}{\omega}\mathcal{L}^{-1}\left\{\frac{\omega}{s^2 + \omega^2}\right\} \otimes \cos(\omega_1 t) = \frac{1}{\omega}\sin(\omega t) \otimes \cos(\omega_1 t)$$

$$= \frac{1}{\omega}\int_0^t \sin(\omega \tau)\cos(\omega_1(t - \tau))d\tau$$

We have three cases to examine.

Case 1: $\omega = \omega_1 = 0$. For this case, the PS is

$$x(t) = \int_0^t \left(\lim_{\omega \to 0} \frac{\sin(\omega \tau)}{\omega}\right)d\tau$$

$$= \int_0^t \tau \, d\tau$$

$$= \frac{1}{2}t^2$$

Remark: One may also obtain this PS by solving the IVP with $\omega = \omega_1 = 0$

$$\begin{cases} x'' = 1 \\ x(0) = x'(0) = 0 \end{cases}$$

Case 2: $\omega = \omega_1 \neq 0$. For this case, the PS is

$$x(t) = \frac{1}{2\omega}\int_0^t \sin(\omega \tau + \omega(t - \tau))d\tau + \frac{1}{2\omega}\int_0^t \sin(\omega \tau) - \omega(t - \tau))d\tau$$

$$= \frac{1}{2\omega}t\sin \omega t - \frac{1}{2\omega}\int_0^t \sin(2\omega \tau - \omega t)\,d\tau$$

$$= \frac{1}{2\omega}t\sin \omega t$$

Remark: Taking the limit of $\omega \to 0$, this Case 2 recovers nicely the Case 1

$$\lim_{\omega \to 0}\frac{1}{2\omega}t\sin \omega t = \frac{1}{2}t^2$$

Case 3: $\omega \neq \omega_1$ and $\omega \neq 0$. For this case, the PS is

$$x(t) = \frac{1}{2\omega}\int_0^t \sin(\omega \tau + \omega_1(t - \tau))d\tau + \frac{1}{2\omega}\int_0^t \sin(\omega \tau - \omega_1(t - \tau))d\tau$$

$$= \frac{1}{2\omega}\int_0^t \sin((\omega - \omega_1)\tau + \omega_1 t)d\tau + \frac{1}{2\omega}\int_0^t \sin((\omega - \omega_1)\tau - \omega_1 t)d\tau$$

$$= \frac{1}{2\omega}\left(\frac{1}{\omega - \omega_1}\right)(2\cos(\omega_1 t) - \cos(\omega t) - \cos(\omega t - 2\omega_1 t))$$

Example 9

Find the PS to the following InHomo system

$$\begin{pmatrix} x \\ y \end{pmatrix}' = \begin{pmatrix} 2 & 4 \\ 1 & -1 \end{pmatrix}\begin{pmatrix} x \\ y \end{pmatrix} + \begin{pmatrix} \cos t \\ \sin t \end{pmatrix}$$
$$\begin{pmatrix} x(0) \\ y(0) \end{pmatrix} = \begin{pmatrix} 0 \\ 0 \end{pmatrix} \qquad\qquad (5.91)$$

Solution

Applying LT on both sides of the InHomo system, we get

$$s\begin{pmatrix} X(s) \\ Y(s) \end{pmatrix} = \begin{pmatrix} 2 & 4 \\ 1 & -1 \end{pmatrix}\begin{pmatrix} X(s) \\ Y(s) \end{pmatrix} + \begin{pmatrix} \mathcal{L}\{\cos t\} \\ \mathcal{L}\{\sin t\} \end{pmatrix}$$

Now, we can solve this system in terms of $X(s)$ and $Y(s)$

$$\begin{pmatrix} s-2 & -4 \\ -1 & s+1 \end{pmatrix}\begin{pmatrix} X(s) \\ Y(s) \end{pmatrix} = \begin{pmatrix} \mathcal{L}\{\cos t\} \\ \mathcal{L}\{\sin t\} \end{pmatrix}$$

Thus, we get

$$X(s) = \frac{1}{(s+2)(s-3)}\left((s+1)\mathcal{L}\{\cos t\} - 4\mathcal{L}\{\sin t\}\right)$$

$$Y(s) = \frac{1}{(s+2)(s-3)}\left(\mathcal{L}\{\cos t\} + (s-2)\mathcal{L}\{\sin t\}\right)$$

Using partial fraction decomposition, we get

$$X(s) = \frac{1}{(s+2)(s-3)}\left((s+1)\mathcal{L}\{\cos t\} - 4\mathcal{L}\{\sin t\}\right)$$
$$= \frac{1}{5}\left(\frac{4}{s-3} + \frac{1}{s+2}\right)\mathcal{L}\{\cos t\} - \frac{4}{5}\left(\frac{1}{s-3} - \frac{1}{s+2}\right)\mathcal{L}\{\sin t\}$$

$$Y(s) = \frac{1}{(s+2)(s-3)}\left(\mathcal{L}\{\cos t\} + (s-2)\mathcal{L}\{\sin t\}\right)$$
$$= \frac{1}{5}\left(\frac{1}{s-3} - \frac{1}{s+2}\right)\mathcal{L}\{\cos t\} + \frac{1}{5}\left(\frac{1}{s-3} + \frac{4}{s+2}\right)\mathcal{L}\{\sin t\}$$

Thus,

$$x(t) = \mathcal{L}^{-1}\left\{\frac{1}{5}\left(\frac{4}{s-3} + \frac{1}{s+2}\right)\mathcal{L}\{\cos t\} - \frac{4}{5}\left(\frac{1}{s-3} - \frac{1}{s+2}\right)\mathcal{L}\{\sin t\}\right\}$$
$$= \frac{1}{5}(4\exp(3t) + \exp(-2t)) \otimes \cos t - \frac{4}{5}(\exp(3t) - \exp(-2t)) \otimes \sin t$$

$$y(t) = \mathcal{L}^{-1}\left\{\frac{1}{5}\left(\frac{1}{s-3} - \frac{1}{s+2}\right)\mathcal{L}\{\cos t\} + \frac{1}{5}\left(\frac{1}{s-3} + \frac{4}{s+2}\right)\mathcal{L}\{\sin t\}\right\}$$
$$= \frac{1}{5}(\exp(3t) - \exp(-2t)) \otimes \cos t + \frac{1}{5}(\exp(3t) + 4\exp(-2t)) \otimes \sin t$$

Example 10

Find the PS to the following IDE

$$\begin{cases} x'(t) + 2x(t) - 4 \int_0^t e^{(t-\tau)} x(\tau) d\tau = \exp(t) \\ x(0) = 0 \end{cases} \tag{5.92}$$

Solution

Applying LT on both sides of the equation and using the Convolution Theorem, we find

$$(sX(s) - x(0)) + 2X(s) - 4\mathcal{L}\{e^t \otimes x(t)\} = \mathcal{L}(\exp(t))$$

$$sX(s) + 2X(s) - 4\left(\frac{1}{s-1}\right)X(s) = \mathcal{L}(\exp(t))$$

Solving for $X(s)$ and performing partial fraction, we obtain

$$X(s) = \left(\frac{s-1}{s^2 + s - 6}\right)\mathcal{L}(\exp(t))$$

$$= \frac{1}{5}\left(\frac{1}{s-2} + \frac{4}{s+3}\right)\mathcal{L}(\exp(t))$$

Applying inverse LT, we get

$$x(t) = \mathcal{L}^{-1}\{X(s)\}$$

$$= \frac{1}{5}\mathcal{L}^{-1}\left\{\left(\frac{1}{s-2} + \frac{4}{s+3}\right)\mathcal{L}(\exp(t))\right\}$$

$$= \frac{1}{5}(\exp(2t) + \exp(-3t)) \otimes \exp(t)$$

Using the convolution formula obtained in the previous section,

$$\exp(\alpha t) \otimes \exp(\beta t) = \frac{\exp(\alpha t) - \exp(\beta t)}{\alpha - \beta}$$

we get

$$x(t) = \frac{1}{5}\left(\frac{\exp(2t) - \exp(t)}{2 - 1} + \frac{\exp(-3t) - \exp(t)}{-3 - 1}\right)$$

$$= \frac{1}{20}(-3\exp(t) + 4\exp(2t) - \exp(-3t))$$

Problems

Problem 5.5.1 Apply the convolution theorem to derive the indicated solution $x(t)$ of the given DE with IC $x(0) = x'(0) = 0$.

$$\begin{cases} x'' + 4x = f(t) \\ x(t) = \dfrac{1}{2}\displaystyle\int_0^t f(t-\tau)\sin 2\tau \, d\tau \end{cases}$$

Problem 5.5.2 Find the GS to the following IDE and you may not leave the solution in its convolution form.

$$\begin{cases} x'(t) + 2x(t) - 4\displaystyle\int_0^t \exp(t-\tau)x(\tau)d\tau = \sin t \\ x(0) = 0 \end{cases}$$

Problem 5.5.3 Find the GS to the following DE system

$$\begin{cases} x'' = -4x + \sin t \\ y'' = 4x - 8y \end{cases}$$

Problem 5.5.4 Use LT to find the PS to the following IVP

$$\begin{cases} x''' + x'' - 6x' = 0 \\ x(0) = 0 \\ x'(0) = x''(0) = 1 \end{cases}$$

Problem 5.5.5 Find the PS to the following IVP

$$\begin{cases} x'' + 4x = -5\delta(t-3) \\ x(0) = 1 \\ x'(0) = 0 \end{cases}$$

Problem 5.5.6 Find the PS to the following IVP

$$\begin{cases} x'' + 6x' + 8x = -\delta(t-2) \\ x(0) = 1 \\ x'(0) = 0 \end{cases}$$

Problem 5.5.7 Find the PS to the following IVP
$$\begin{cases} x'' + 2x' + x = \delta(t) - \delta(t-2) \\ \quad x(0) = 0 \\ \quad x'(0) = 0 \end{cases}$$

Problem 5.5.8 Find the PS to the following IVP
$$\begin{cases} x'' + \omega^2 x = \displaystyle\sum_{n=0}^{\infty} \delta(t - 2nt_0) \\ \quad x(0) = 0 \\ \quad x'(0) = 0 \end{cases}$$

Problem 5.5.9 Find the PS to the following IVP
$$\begin{cases} x'' + 2x' + x = u(t-a) + \delta(t-b) \\ \quad x(0) = 0 \\ \quad x'(0) = 0 \end{cases}$$

Problem 5.5.10 Find the PS to the following system of IVP.
$$\begin{cases} \quad x' = x + 2y \\ \quad y' = 2x - 2y \\ x(0) = 1 \\ y(0) = 0 \end{cases}$$

Problem 5.5.11 Find the PS to the following system of IVP.
$$\begin{cases} \quad x_1'' = -2x_1 + x_2 + \delta(t-\tau) \\ \quad x_2'' = x_1 - x_2 + \delta(t-2\tau) \\ x_1(0) = x_1'(0) = 0 \\ x_2(0) = x_2'(0) = 0 \end{cases}$$

Problem 5.5.12 Find the PS to the following system of IVP
$$\begin{cases} x'(t) + 4x(t) + 6x(t) \otimes \exp(t) = \sin(\omega t) \\ \qquad\qquad\qquad\qquad\qquad x(0) = 0 \end{cases}$$

Problem 5.5.13 Find the PS to the following DE.Syst
$$\begin{pmatrix} x \\ y \end{pmatrix}' = \begin{pmatrix} 2 & 1 \\ 1 & 2 \end{pmatrix} \begin{pmatrix} x \\ y \end{pmatrix} + \begin{pmatrix} t \\ \exp(t) \end{pmatrix}$$

Problem 5.5.14 Use LT to find the PS to the following IVP
$$\begin{cases} x'' - 6x' + 8x = 2 \\ x(0) = x'(0) = 0 \end{cases}$$

Problem 5.5.15 Find the PS to the following IVP
$$\begin{cases} tx'' + 2(t-1)x' - 2x = 0 \\ \qquad\qquad x(0) = x'(0) = 0 \end{cases}$$

Problem 5.5.16 Find the PS to the following IVP
$$\begin{cases} tx'' + x' + tx = 0 \\ \qquad x(0) = \alpha = \text{constant} \\ \qquad x'(0) = 0 \end{cases}$$

Problem 5.5.17 Find the PS to the following IVP
$$\begin{cases} x'' + 2x' + x = f(t) \\ x(0) = x'(0) = 0 \end{cases}$$

Problem 5.5.18 Use LT to find the PS to the following IVP
$$\begin{cases} x'' + 4x' + 13x = t\exp(-t) \\ \qquad\qquad x(0) = 0 \\ \qquad\qquad x'(0) = 2 \end{cases}$$

Problem 5.5.19 Use LT and another method to find the PS to the following IVP and compare the results.
$$\begin{cases} x'' - x' - 12x = \sin 4t + \exp(3t) \\ \qquad\qquad x(0) = 2 \\ \qquad\qquad x'(0) = 1 \end{cases}$$

Problem 5.5.20 Find the PS to the following IVP (where x_0, v_0 and ω are given constants) by: (1) VOP; (2) LT.
$$\begin{cases} x'' + \omega^2 x = f(t) \\ \qquad x(0) = x_0 \\ \qquad x'(0) = v_0 \end{cases}$$

Problem 5.5.21 Find the PS to the following IVP
$$\begin{cases} x'' + \omega_1^2 x = \sin(\omega_2 t) \\ x(0) = x'(0) = 0 \end{cases}$$

by
(1) Any method of your choice except LT.
(2) The method of LT.

Problem 5.5.22 Find the PS to the following IVP using two different methods

$$\begin{cases} x' - x = 1 - (t-1)u(t-1) \\ x(0) = 0 \end{cases}$$

where $u(t-1)$ is the so-called Step Function defined as

$$u(t-1) = \begin{cases} 0, t \leq 1 \\ 1, t \geq 1 \end{cases}$$

(1) LT method.
(2) Any other method of your choice.

Problem 5.5.23 Find the PS to the following IVP

$$\begin{cases} x''(t) + 4x(t) = \left(1 - u(t-2\pi)\right)\cos 2t \\ x(0) = 0 \\ x'(0) = 0 \end{cases}$$

Problem 5.5.24 Find the PS to the following IVP

$$\begin{cases} x''(t) + x(t) = (-1)^{[\![t]\!]} \\ x(0) = 0 \\ x'(0) = 0 \end{cases}$$

where $[\![t]\!]$ denotes the greatest integer not exceeding $[\![t]\!]$, e.g., $[\![0.918]\!] = 0$; $[\![1.234]\!] = 1$; $[\![1989.64]\!] = 1989$. Your final answer may not contain the convolution operator.

Problem 5.5.25 The motion of a particle in the plane can be described by

$$\begin{cases} x'' + \omega^2 x = b_0 \sin\omega_0 t \\ y'' + \omega^2 y = b_0 \cos\omega_0 t \end{cases}$$

where ω, ω_0, and b_0 are all constants. Initially, the particle is placed at the origin at rest. Find the trajectories of the particle for
(1) $\omega = \omega_0$
(2) $\omega \neq \omega_0$

Problem 5.5.26 A block of mass m is attached to a massless spring of spring constant k, and they are placed on a horizontal and perfectly smooth bench. During the first $\tau/2$ time, we add a constant force f to the block from left to right. During the second $\tau/2$ time, the force direction is reversed but its constant magnitude retains. Repeat this process until eternity. Find the displacement of the block as a function of time. The initial displacement and speed can be set to zero.

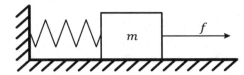

Figure 5.8 The block-spring system for Problem 5.5.26 & Problem 5.5.27.

Problem 5.5.27 A block is attached to a massless spring of spring and they are placed on a horizontal and perfectly smooth bench. We add a force $f(t) = \cos 2t$ to the block during $t \in [0, 2\pi]$ and remove it at all other times (the spring is still there). The EoM of the block is
$$\begin{cases} x'' + 4x = f(t) \\ x(0) = x'(0) = 0 \end{cases}$$
Solve the equation.

Problem 5.5.28 A block of mass m is attached to two massless springs of spring constants k_1 and k_2 and they are placed on a horizontal and perfectly smooth bench. During $\tau/2$ time, we add a constant force f to the block from left to right. At the end of the $\tau/2$ time, immediately, we reverse the direction of the force (but keep the magnitude) and act on the block for another $\tau/2$ time. Then, we reverse the force and keep repeating this process forever. Find the displacement of the block as a function of time.

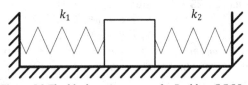

Figure 5.9 The block-spring system for Problem 5.5.28.

Problem 5.5.29 Two blocks (A & B) of the same mass m are attached to three identical massless springs (as shown). The assembly is placed on a horizontal and perfectly smooth bench. Initially, springs stay at their natural lengths and both blocks are at rest. During two brief moments, two forces were added to the blocks respectively

$$f_A(t) = \begin{cases} f_0 & t \in [0, \tau/2] \\ 0 & \text{O.W.} \end{cases} \quad \text{and} \quad f_B(t) = \begin{cases} f_0 & t \in [\tau/2, \tau] \\ 0 & \text{O.W.} \end{cases}$$

Compute the displacements of the blocks as a function of time. All mentioned parameters, including the spring constant k, are given. You may not leave the solution in its convolution form.

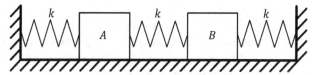

Figure 5.10 The block-spring system for Problem 5.5.29.

Problem 5.5.30 Two blocks (A & B) of the same mass m are attached to two identical massless springs (as shown). The assembly is placed on a horizontal and perfectly smooth bench. Initially, springs stay at their natural lengths and both blocks are at rest. During two brief moments, two forces were added to the blocks respectively

$$f_A(t) = \begin{cases} f_0 & t \in [0, \tau/2] \\ 0 & \text{O.W.} \end{cases} \quad \text{and} \quad f_B(t) = \begin{cases} f_0 & t \in [\tau/2, \tau] \\ 0 & \text{O.W.} \end{cases}$$

Compute the displacements of the blocks as a function of time. All mentioned parameters, including the spring constant, are given. You may not leave the solution in its convolution form.

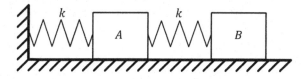

Figure 5.11 The block-spring system for Problem 5.5.30 & Problem 5.5.31.

Problem 5.5.31 Two blocks (A & B) of the same mass m are attached to two identical massless springs (as shown). The assembly is placed on a horizontal and perfectly smooth bench. Initially, springs stay at their natural lengths, and both blocks are at rest. During two instances $t = t_0$ and $2t_0$, two forces were added to the blocks respectively. The EoM of the two blocks are

$$\begin{cases} mx_1'' = -kx_1 + k(x_2 - x_1) + f_0\delta(t - t_0) \\ mx_2'' = -k(x_2 - x_1) + f_0\delta(t - 2t_0) \\ x_1(0) = x_1'(0) = 0 \\ x_2(0) = x_2'(0) = 0 \end{cases}$$

Compute the displacements of the blocks as a function of time. All parameters, including m, k, f_0, t_0, are given. You may leave the solution in its convolution form if you need to.

Problem 5.5.32 Consider a system of two masses m_1 and m_2 (from left to right) connected to three massless springs whose spring constants are k_1, k_2, and k_3 (from left to right). The entire space is placed on a frictionless leveled surface as shown in Figure 5.12.

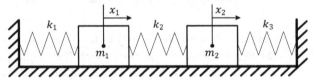

Figure 5.12 The block-spring system for Problem 5.5.32.

(1) Derive the EoMs for the two masses.

(2) Solve the DEs you derived above with the following IC and simplified parameters $m_1 = m_2 = 1, k_1 = 1, k_2 = 2, k_3 = 3$

$$\begin{cases} x_1(0) = x_2(0) = 0 \\ x_1'(0) = 0 \\ x_2'(0) = v \end{cases}$$

Appendix A
Solutions to Problems

Chapter 1 First-Order DEs

1.1 Definition of DEs

Problem 1.1.1

$$y_1 = \cos x - \cos 2x \Rightarrow \begin{cases} y_1' = -\sin x + 2\sin 2x \\ y_1'' = -\cos x + 4\cos 2x \end{cases}$$

$$\text{LHS} = y_1'' + y_1 = -\cos x + 4\cos 2x + \cos x - \cos 2x$$
$$= 3\cos 2x = \text{RHS}$$

y_1 is a solution to the DE.

$$y_2 = \sin x - \cos 2x \Rightarrow \begin{cases} y_2' = \cos x + 2\sin 2x \\ y_2'' = -\sin x + 4\cos 2x \end{cases}$$

$$\text{LHS} = y_2'' + y_2 = -\sin x + 4\cos 2x + \sin x - \cos 2x$$
$$= 3\cos 2x = \text{RHS}$$

y_2 is a solution to the DE.

Problem 1.1.2

$$y_1 = x\cos(\ln x)$$
$$y_1' = \cos(\ln x) - \sin(\ln x)$$
$$y_1'' = -\frac{\sin(\ln x) + \cos(\ln x)}{x}$$

$$\text{LHS} = x^2 y_1'' - x y_1' + 2 y_1$$

$$= x^2 \left(-\frac{\sin(\ln x) + \cos(\ln x)}{x} \right) - x(\cos(\ln x) - \sin(\ln x))$$
$$+ 2x \cos(\ln x)$$
$$= 0 = \text{RHS}$$

y_1 is a solution to the DE.

$$y_2 = x \sin(\ln x)$$
$$y_2' = \sin(\ln x) + \cos(\ln x)$$
$$y_2'' = \frac{\cos(\ln x) - \sin(\ln x)}{x}$$

$$\text{LHS} = x^2 y_2'' - x y_2' + 2 y_2$$
$$= x^2 \left(\frac{\cos(\ln x) - \sin(\ln x)}{x} \right) - x(\sin(\ln x) + \cos(\ln x)) + 2x \sin(\ln x)$$
$$= 0 = \text{RHS}$$

y_2 is a solution to the DE.

Problem 1.1.3

From the orthogonality relation, for two curves $f(x)$ and $g(x)$ that are orthogonal to each other, we have
$$f'g' = -1$$

Here we have
$$f(x) = x^2 + k$$
$$f'(x) = 2x$$

Thus, we have
$$g'(x) = -\frac{1}{2x}$$
$$g(x) = \int \left(-\frac{1}{2x} \right) dx$$
$$= -\frac{1}{2} \ln x + C$$

Problem 1.1.4

By observation, we can find that
$$(xy)' = xy' + x'y = xy' + y = \text{LHS}$$

and
$$(x^3)' = 3x^2 = \text{RHS}$$

This gives
$$xy = x^3$$

That means $y = x^2$ is a solution.

Problem 1.1.5

Guess

$$y = \cos x$$
$$y' = -\sin x$$
$$y'' = -\cos x$$
$$\text{LHS} = y'' + y = -\cos x + \cos x = 0 = \text{RHS}$$

$y = \cos x$ is a solution to the DE.

Problem 1.1.6

$$y = \frac{1}{1 + x^2} \Rightarrow y' = -\frac{2x}{(1 + x^2)^2}$$
$$\text{LHS} = y' + 2xy^2 = 0 = \text{RHS}$$
$$y = \frac{1}{1 + x^2}$$

is a solution to the DE.

Problem 1.1.7

$$y(x) = C \exp(-x^3) \Rightarrow y'(x) = -3Cx^2 \exp(-x^3)$$
$$\text{LHS} = y' + 3x^2 y = -3Cx^2 \exp(-x^3) + 3x^2(C \exp(-x^3))$$
$$= 0 = \text{RHS}$$

$y = C \exp(-x^3)$ is the GS to the DE.

Plugging $y(0) = 7$ into the GS, we have

$$C \exp(0) = 7$$

gives $C = 7$. Thus,

$$y(x) = 7 \exp(-x^3)$$

Problem 1.1.8

$$y = (x + C) \cos x \Rightarrow y' = \cos x - (x + C) \sin x$$
$$\text{LHS} = \cos x - (x + C) \sin x + (x + C) \cos x \tan x$$
$$= \cos x - (x + C) \sin x + (x + C) \sin x$$
$$= \cos x = \text{RHS}$$

Plugging $y(\pi) = 0$ into the GS, we have

$$y(\pi) = (\pi + C) \cos \pi$$
$$= -(\pi + C) = 0$$

gives $C = -\pi$. Thus,

$$y(x) = (x - \pi) \cos x$$

Problem 1.1.9

$$y' = (\tan(x^3 + C))'$$

Using

$$(\tan x)' = \frac{1}{\cos^2 x}$$

and the chain rule, we get

$$y' = \frac{3x^2}{\cos^2(x^3 + C)}$$

On the other hand,

$$y' = 3x^2(y^2 + 1)$$
$$= 3x^2\left(\frac{\sin^2(x^3 + C)}{\cos^2(x^3 + C)} + 1\right)$$
$$= 3x^2\frac{\sin^2(x^3 + C) + \cos^2(x^3 + C)}{\cos^2(x^3 + C)}$$
$$= 3x^2\frac{1}{\cos^2(x^3 + C)}$$

They match. Thus, the given function satisfies the DE.

Plugging the IC $y(0) = 1$ into the GS $y(x) = \tan(x^3 + C)$, we get $\tan C = 1$ whose root is $C = \pi/4$. One may express the PS as

$$y(x) = \tan(x^3 + \pi/4)$$

Problem 1.1.10

$$y(x) = \frac{1}{4}x^5 + \frac{C}{x^3}$$
$$y'(x) = \frac{5}{4}x^4 - \frac{3C}{x^4}$$
$$\text{LHS} = \frac{5}{4}x^5 - \frac{3C}{x^3} + \frac{3}{4}x^5 + \frac{3C}{x^3} = 2x^5 = \text{RHS}$$

Plugging $y(2) = 1$ into the GS, we have

$$y(2) = \frac{1}{4}(2^5) + \frac{C}{2^3} = 1$$

Thus,

$$C = -56$$

Problem 1.1.11

$$y_1 = \exp(3x)$$
$$y_1' = 3\exp(3x)$$
$$y_1'' = 9\exp(3x) = 9y_1$$
$$y_2 = \exp(-3x)$$

$$y_2' = -3\exp(-3x)$$
$$y_2'' = 9\exp(-3x) = 9y_2$$

Problem 1.1.12

$$y = \ln(x + C)$$
$$y' = \frac{1}{x + C}$$

$$\text{LHS} = \exp(\ln(x + C))\frac{1}{x + C} = (x + C)\frac{1}{x + C} = 1 = \text{RHS}$$

Plugging $y(0) = 0$ into the GS, we have

$$y(0) = \ln C = 0$$

Thus,

$$C = 1$$

Problem 1.1.13

$$y = C\exp(-x^n)$$
$$y' = -nx^{n-1}C\exp(-x^n)$$
$$\text{LHS} = -nx^{n-1}C\exp(-x^n) + nx^{n-1}C\exp(-x^n)$$
$$= 0 = \text{RHS}$$

Plugging $y(0) = 2014$ into the GS, we have

$$y(0) = C = 2014$$

1.2 Slope Fields and Solution Curves

Problem 1.2.1

The slope field can be plotted by using the DE $y' = x^2 - y$ while the solution curve by the solution of the DE $y = x^2 - 2x + 2 + C\exp(-x)$:

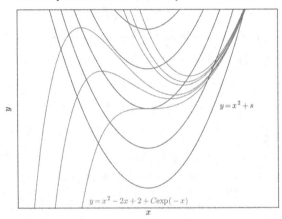

Problem 1.2.2

The slope field can be plotted by using the DE $y' + y = x + 2$ while the solution curve by the solution of the DE $y = x + 1 + C\exp(-x)$:

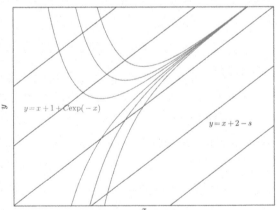

1.3 Separation of Variables

Problem 1.3.1

The DE is separable and, using the SOV method, we have

$$\frac{3xdx}{x^2 + 1} = -\frac{dy}{y - 2}$$

$$\int \frac{3xdx}{x^2 + 1} = -\int \frac{dy}{y - 2}$$

$$\frac{3}{2}\ln(x^2 + 1) = -\ln(y - 2) + C_1$$

$$y = 2 + \frac{C}{(x^2 + 1)^{\frac{3}{2}}}$$

Problem 1.3.2

$$\frac{dy}{dx} = 4x^{\frac{1}{3}}y^{\frac{1}{3}}$$

$$y^{-\frac{1}{3}}dy = 4x^{\frac{1}{3}}dx$$

$$\int y^{-\frac{1}{3}}dy = \int 4x^{\frac{1}{3}}dx$$

$$\frac{3}{2}y^{\frac{2}{3}} = 3x^{\frac{4}{3}} + C$$

$$y = \left(2x^{\frac{4}{3}} + C\right)^{\frac{3}{2}}$$

Problem 1.3.3

The DE is separable. Thus, we have

$$\frac{dy}{dx} = (1 + x)(1 + y)$$

$$\frac{dy}{1 + y} = (1 + x)dx$$

$$\int \frac{dy}{1 + y} = \int (1 + x)dx$$

$$\ln|1 + y| = x + \frac{1}{2}x^2 + C$$

Problem 1.3.4

Since $\frac{dy}{dx} = x\exp(-x)$, we have $dy = x\exp(-x)\,dx$. Thus,

$$y = \int x \exp(-x)\, dx$$

$$= -\int x\, d(\exp(-x))$$

$$= -x \exp(-x) + \int \exp(-x)\, dx$$

$$= -x \exp(-x) - \exp(-x) + C$$

Applying the IC $y(0) = 1$,

$$-1 + C = 1$$

we get $C = 2$ and the PS to the DE

$$y(x) = -x \exp(-x) - \exp(-x) + 2$$

Problem 1.3.5

$$\frac{dy}{dx} = -2 \cos 2x$$

$$y = \int -2 \cos 2x\, dx$$

$$= -\sin 2x + C$$

IC $y(0) = 2014$ gives

$$-\sin 0 + C = 2014$$

$$C = 2014$$

Problem 1.3.6

$$\sec^2 y \cdot y' = \frac{1}{2\sqrt{x}}$$

$$\int \sec^2 y\, dy = \int \frac{dx}{2\sqrt{x}}$$

$$\tan y = \sqrt{x} + C$$

The GS is

$$y = \tan^{-1}(\sqrt{x} + C)$$

Plugging the IC $y(4) = \frac{\pi}{4}$ into the GS, we get

$$\frac{\pi}{4} = \tan^{-1}(2 + C)$$

$$C = -1$$

The PS to the DE is

$$y = \tan^{-1}(\sqrt{x} - 1)$$

Problem 1.3.7

$$\frac{y'}{y} = 2x + 3x^2 \exp(x^3)$$

$$\int \frac{dy}{y} = \int (2x + 3x^2 \exp(x^3)) dx$$

$$\ln y = x^2 + \exp(x^3) + C$$

The GS is

$$y = \exp(x^2 + \exp(x^3) + C)$$

Given the IC $y(0) = 5$, we have

$$5 = \exp(0 + 1 + C) \Rightarrow C = \ln 5 - 1$$

Thus, the PS to the DE is

$$y = \exp(x^2 + \exp(x^3) + \ln 5 - 1)$$
$$= 5 \exp(x^2 + \exp(x^3) - 1)$$

Problem 1.3.8

$$\frac{y^3}{y^4 + 1} y' = \cos x$$

$$\int \frac{y^3}{y^4 + 1} dy = \int \cos x \, dx$$

$$\frac{1}{4} \int \frac{d(y^4 + 1)}{y^4 + 1} = \int \cos x \, dx$$

$$\frac{1}{4} \ln(y^4 + 1) = \sin x + C$$

Problem 1.3.9

The DE is a separable DE. Thus,

$$\frac{y'}{y^2} = 5x^4 - 4x$$

$$\int \frac{dy}{y^2} = \int (5x^4 - 4x) dx$$

$$-\frac{1}{y} = x^5 - 2x^2 + C$$

Problem 1.3.10

$$\frac{dy}{dx} = 2xy^2 + 3x^2 y^2 = y^2(2x + 3x^2)$$

which is a separable DE:

$$\int \frac{dy}{y^2} = \int (2x + 3x^2)\, dx$$

$$-\frac{1}{y} = x^2 + x^3 + C$$

$$y = -\frac{1}{x^2 + x^3 + C}$$

Applying the IC $y(1) = -1$

$$-1 = -\frac{1}{1 + 1 + C}$$

we get $C = -1$ and

$$y = -\frac{1}{x^2 + x^3 - 1}$$

Problem 1.3.11

$$y' \cos y = 2x$$

$$\int \cos y\, dy = \int 2x\, dx$$

$$\sin y = x^2 + C$$

Problem 1.3.12

$$\sin x \frac{dy}{dx} = y \cos x$$

$$\int \frac{dy}{y} = \int \frac{\cos x}{\sin x}\, dx$$

$$\ln |y| = \ln |\sin x| + C_1$$

$$y = C \sin x$$

Applying the IC $y\left(\frac{\pi}{2}\right) = \frac{\pi}{2}$ to the GS, we have $\frac{\pi}{2} = C \sin \frac{\pi}{2}$ resulting in $C = \frac{\pi}{2}$ and thus the PS is

$$y = \frac{\pi}{2} \sin x$$

Problem 1.3.13

$$(x^2 + 1) \tan y \frac{dy}{dx} = x$$

$$\int \tan y\, dy = \int \frac{x}{x^2 + 1}\, dx$$

$$-\ln|\cos y| = \frac{1}{2}\ln(x^2 + 1) + C$$

Problem 1.3.14

$$(x+3)^3 \frac{dy}{dx} = (y-2)^2$$

$$(y-2)^{-2}dy = (x+3)^{-3}dx$$

$$\int (y-2)^{-2}dy = \int (x+3)^{-3}dx$$

$$-(y-2)^{-1} = \left(-\frac{1}{2}\right)(x+3)^{-2} - C_1$$

$$(y-2)^{-1} = \left(\frac{1}{2}\right)(x+3)^{-2} + C_1$$

$$(y-2)^{-1} = \frac{\frac{1}{2} + C_1(x+3)^2}{(x+3)^2}$$

$$y-2 = \frac{(x+3)^2}{\frac{1}{2} + C_1(x+3)^2}$$

$$y = 2 + \frac{(x+3)^2}{\frac{1}{2} + C_1(x+3)^2}$$

Problem 1.3.15

$$y' - xy = 3y$$

$$\left(\frac{1}{y}\right)\left(\frac{dy}{dx}\right) = x+3$$

$$\int \frac{1}{y}dy = \int (x+3)dx$$

$$\ln y = \frac{(x+3)^2}{2} + C_1$$

$$y = \exp\left(\frac{(x+3)^2}{2} + C_1\right)$$

$$= C \exp\left(\frac{(x+3)^2}{2}\right)$$

Applying IC $y(1) = 1$, we have $1 = C \exp\left(\frac{(1+3)^2}{2}\right) = c\exp(8)$ resulting in $C = \exp(-8)$. Thus, the PS is

$$y = \exp\left(\frac{(x+3)^2}{2} - 8\right)$$

Problem 1.3.16

The DE can be solved as

$$((x+1)y)' = \cos x$$
$$(x+1)y = \int \cos x \, dx$$
$$= \sin x + C$$
$$y = \frac{\sin x + C}{x+1}$$

Applying IC, we find $C = 1$ and the PS

$$y = \frac{\sin x + 1}{x+1}$$

Problem 1.3.17

$$\frac{dy}{\beta y(\alpha - \ln y)} = dt$$
$$\frac{d(\ln y)}{\beta(\alpha - \ln y)} = dt$$
$$\frac{-d(\alpha - \ln y)}{\beta(\alpha - \ln y)} = dt$$
$$\frac{-d(\ln(\alpha - \ln y))}{\beta} = dt$$
$$-\int \frac{d(\ln(\alpha - \ln y))}{\beta} = \int dt$$
$$-\frac{\ln(\alpha - \ln y)}{\beta} = t + C$$
$$\alpha - \ln y = \exp(-(\beta t + C_1))$$

The GS is

$$y = \exp(\alpha - \exp(-(\beta t + C_1)))$$

If $y(0) = y_0$

$$\alpha - \ln y_0 = \exp(-C_1)$$

If $y(\infty) = y_\infty$

$$y_\infty = \exp(\alpha), \qquad \beta > 0$$
$$\alpha = \ln y_\infty$$

If $\beta < 0$, $y_\infty = 0$.

1.4 First-Order Linear DEs

Problem 1.4.1

Dividing $(x^2 + 1)$ on both sides of the DE, we have

$$y' + \frac{3x^3}{x^2 + 1}y = \frac{6x \exp\left(-\frac{3}{2}x^2\right)}{x^2 + 1}$$

Let

$$\rho(x) = \exp\left(\int \frac{3x^3}{x^2 + 1}dx\right)$$

$$= \exp\left(\int \frac{3x(x^2 + 1) - 3x}{x^2 + 1}dx\right)$$

$$= \exp\left(\frac{3}{2}x^2 - \frac{3}{2}\ln(x^2 + 1)\right)$$

$$= (x^2 + 1)^{-\frac{3}{2}}\exp\left(\frac{3}{2}x^2\right)$$

We have

$$(\rho(x)y)' = 6x(x^2 + 1)^{-\frac{5}{2}}$$

$$y = \frac{1}{\rho(x)}\int 6x(x^2 + 1)^{-\frac{5}{2}}dx$$

$$= (x^2 + 1)^{\frac{3}{2}}\exp\left(-\frac{3}{2}x^2\right)\left(-2(x^2 + 1)^{-\frac{3}{2}} + C\right)$$

$$= -2\exp\left(-\frac{3}{2}x^2\right) + C(x^2 + 1)^{\frac{3}{2}}\exp\left(-\frac{3}{2}x^2\right)$$

which is the GS. Given $y(0) = 1$, we have

$$1 = -2 + C \Rightarrow C = 3$$

Thus, the PS to the DE is

$$y = -2\exp\left(-\frac{3}{2}x^2\right) + 3(x^2 + 1)^{\frac{3}{2}}\exp\left(-\frac{3}{2}x^2\right)$$

Problem 1.4.2

(a) We know that

$$y'_c = \left(C\exp\left(-\int P(x)dx\right)\right)'$$

$$= C\exp\left(-\int P(x)dx\right)\left(-\int P(x)dx\right)'$$

$$= -C\exp\left(-\int P(x)dx\right)P(x)$$

$$= -P(x)y_c$$

Obviously,
$$y'_C + P(x)y_C = -P(x)y_C + P(x)y_C = 0$$
This proves that it is a GS to the original DE.

(b) We have that

$$y'_P(x) = \left(\left(\exp\left(-\int P(x)dx \right) \right) \left(\int Q(x) \exp\left(\int P(x)dx \right) dx \right) \right)'$$

$$= \left(\exp\left(-\int P(x)dx \right) \right)' \left(\int Q(x) \exp\left(\int P(x)dx \right) dx \right)$$

$$\qquad + \exp\left(-\int P(x)dx \right) \left(\int Q(x) \exp\left(\int P(x)dx \right) dx \right)'$$

$$= -P(x) \exp\left(-\int P(x)dx \right) \left(\int Q(x) \exp\left(\int P(x)dx \right) dx \right)$$

$$\qquad + \exp\left(-\int P(x)dx \right) Q(x) \exp\left(\int P(x)dx \right)$$

$$= -P(x)y_P(x) + Q(x)$$

Obviously,

$$\begin{aligned} \text{LHS} &= y'_P + P(x)y_P \\ &= -P(x)y_P + Q(x) + P(x)y_P \\ &= Q(x) = \text{RHS} \end{aligned}$$

(c) Since $y_C(x)$ is a GS to $y' + P(x)y = 0$, we get

$$y'_C = -P(x)y_C$$

On the other hand, we know that

$$y'_P = Q(x) - P(x)y_P$$

Then, for $y = y_C + y_P$, we have

$$\begin{aligned} y' &= (y_C + y_P)' \\ &= y'_C + y'_P \\ &= -P(x)y_C + Q(x) - P(x)y_P \\ &= -P(x)(y_C + y_P) + Q(x) \\ &= -P(x)y + Q(x) \end{aligned}$$

That means

$$y' + P(x)y = Q(x)$$

which means that it is a GS to the DE.

Problem 1.4.3

The DE can be written as

$$y' + \frac{2}{x}y = 7x$$

Let

$$\rho(x) = \exp\left(\int \frac{2}{x} dx\right)$$
$$= \exp(2\ln x)$$
$$= x^2$$

We have that

$$(\rho(x)y)' = 7x^3$$

That gives

$$y = x^{-2}\left(\int 7x^3 dx\right)$$
$$= x^{-2}\left(\frac{7}{4}x^4 + C\right)$$
$$= \frac{7}{4}x^2 + Cx^{-2}$$

Given $y(2) = 5$, we have

$$5 = 7 + \frac{C}{4} \Rightarrow C = -8$$

Thus, the PS is

$$y = \frac{7}{4}x^2 - 8x^{-2}$$

Problem 1.4.4

Let

$$\rho(x) = \exp\left(\int \cot x \, dx\right)$$
$$= \exp(\ln \sin x)$$
$$= \sin x$$

We have

$$(\rho(x)y)' = \sin x \cos x$$

which gives

$$y = \frac{1}{\sin x}\left(\int \sin x \cos x \, dx\right)$$
$$= \frac{\sin^2 x + C}{2\sin x}$$

Problem 1.4.5

Method 1: The 1st.O linear DE method.

This DE can be written as

$$y' - 3x^2 y = 21x^2$$

which is a 1st.O linear DE. Let the IF

$$\rho(x) = \exp\left(\int -3x^2 dx\right)$$
$$= \exp(-x^3)$$

We know that

$$(\exp(-x^3)\,y)' = 21x^2 \exp(-x^3)$$

Thus,

$$y = \exp(x^3)\int 21x^2 \exp(-x^3)\,dx$$
$$= -7 + C\exp(x^3)$$

Method 2: The SOV method.
The DE can be written as

$$\frac{y'}{y+7} = 3x^2$$

which is separable. Thus, we have

$$\int \frac{dy}{y+7} = \int 3x^2 dx$$
$$\ln(y+7) = x^3 + C$$
$$y = C\exp(x^3) - 7$$

Problem 1.4.6

The DE can be written as

$$y' - \frac{2}{x}y = x^2 \cos x$$

Let

$$\rho(x) = \exp\left(\int -\frac{2}{x}dx\right)$$
$$= \exp(-2\ln x)$$
$$= x^{-2}$$

We have

$$(\rho(x)y)' = x^2 \cos x\,\rho(x)$$

That gives

$$y = \frac{1}{\rho(x)}\int x^2 \cos x\,\rho(x)dx$$
$$= x^2 \int \cos x\,dx$$
$$= x^2(\sin x + C)$$

Thus, the GS to the original DE is

$$y = x^2 \sin x + Cx^2$$

Problem 1.4.7

$$y' + \frac{1}{2x+1}y = (2x+1)^{\frac{1}{2}}$$

Let

$$\rho(x) = \exp\left(\int \frac{dx}{2x+1}\right)$$

$$= (2x+1)^{\frac{1}{2}}$$

We have

$$\left((2x+1)^{\frac{1}{2}}y\right)' = 2x+1$$

Thus,

$$y = (2x+1)^{-\frac{1}{2}} \int (2x+1)dx$$

$$= (2x+1)^{-\frac{1}{2}}(x^2 + x + C)$$

Problem 1.4.8

Method 1: The SOV method.

The DE can be written as

$$y' = \frac{2x(y+1)}{x^2+1}$$

$$\frac{y'}{y+1} = \frac{2x}{x^2+1}$$

$$\ln(y+1) = \ln(x^2+1) + C_1$$

$$y = Cx^2 + C - 1$$

Method 2: The 1st.O linear DE method.

Rewriting the DE, there emerges a linear DE

$$y' - \frac{2x}{x^2+1}y = \frac{2x}{x^2+1}$$

whose IF is

$$\rho(x) = \exp\left(\int -\frac{2x}{x^2+1}dx\right)$$

$$= \frac{1}{x^2+1}$$

Thus, we know,

$$\left(\frac{y}{x^2+1}\right)' = \frac{2x}{(x^2+1)^2}$$

$$y = (x^2+1) \int \frac{2xdx}{(x^2+1)^2}$$

$$= (x^2 + 1)\left(-\frac{1}{x^2+1} + C\right)$$
$$= -1 + C(x^2 + 1)$$

Problem 1.4.9

To solve this DE, we can regard x as the DV and y as the IV. Thus, we have

$$(1 + 2xy)\frac{dy}{dx} = 1 + y^2$$
$$\frac{dy}{dx} = \frac{1+y^2}{1+2xy}$$
$$\frac{dx}{dy} = \frac{1+2xy}{1+y^2}$$
$$\frac{dx}{dy} - \frac{2y}{1+y^2}x = \frac{1}{1+y^2}$$

a 1st.O linear DE of x wrt y whose IF is

$$\rho(y) = \exp\left(\int\left(-\frac{2y}{1+y^2}\right)dy\right)$$
$$= \exp\left(-\int \frac{d(1+y^2)}{1+y^2}\right)$$
$$= \exp[-\ln(1+y^2)]$$
$$= \frac{1}{1+y^2}$$

Thus,

$$(\rho(y)x)' = \frac{1}{1+y^2}\rho(y)$$
$$x = \frac{1}{\rho(y)}\int \frac{1}{1+y^2}\rho(y)dy$$
$$= (1+y^2)\int \frac{1}{(1+y^2)^2}dy$$
$$= (1+y^2)\left(\frac{1}{2}\left(\arctan y + \frac{y}{1+y^2}\right) + C\right)$$

Note: The integral $\int \frac{1}{(1+y^2)^2}dy$ can be evaluated as follows. Let $y = \tan\theta$, we have $1 + y^2 = \sec^2\theta$ and $dy = \sec^2\theta\, d\theta$.

$$\int \frac{1}{(1+y^2)^2}dy = \int \frac{1}{\sec^4\theta}\sec^2\theta\, d\theta$$
$$= \int \cos^2\theta\, d\theta$$
$$= \frac{1}{2}\int(1 - \cos 2\theta)d\theta$$

$$= \frac{\theta}{2} - \frac{\sin 2\theta}{4} + C$$

$$= \frac{1}{2}(\theta - \sin\theta\cos\theta) + C$$

$$= \frac{1}{2}\left(\theta - \frac{\tan\theta}{\sec^2\theta}\right) + C$$

$$= \frac{1}{2}\left(\arctan y + \frac{y}{1+y^2}\right) + C$$

Problem 1.4.10

The DE can be written as

$$y' + \frac{1}{2x}y = 5x^{-\frac{1}{2}}$$

whose IF is

$$\rho(x) = \exp\left(\int \frac{1}{2x}dx\right)$$

$$= \exp\left(\frac{1}{2}\ln x\right)$$

$$= x^{\frac{1}{2}}$$

Thus,

$$(\rho(x)y)' = 5x^{-\frac{1}{2}}\rho(x)$$

This gives

$$y = \frac{1}{\rho(x)}\int 5x^{-\frac{1}{2}}\rho(x)dx$$

$$= x^{-\frac{1}{2}}(5x + C)$$

$$= 5\sqrt{x} + \frac{C}{\sqrt{x}}$$

Problem 1.4.11

The DE can be written as

$$\frac{dx}{dy} - x = y\exp(y)$$

which can be considered as a DE of x wrt y in the following new form

$$\frac{dx}{dy} - P(y)x = Q(y)$$

Thus, the IF for the DE is

$$\rho(y) = \exp\left(\int -1dy\right)$$

$$= \exp(-y)$$

Thus,

$$(\rho(y)x)' = y \exp(y)\rho(y)$$

which gives

$$
\begin{aligned}
x &= \frac{1}{\rho(y)} \int y \exp(y)\rho(y)dy \\
&= \exp(y) \int y dy \\
&= \exp(y)\left(\frac{1}{2}y^2 + C\right) \\
&= \frac{1}{2}y^2 \exp(y) + C \exp(y)
\end{aligned}
$$

Problem 1.4.12

$$y' + \frac{2}{x+1}y = 3$$

Let

$$
\begin{aligned}
\rho(x) &= \exp\left(\int \frac{2}{x+1}dx\right) \\
&= (x+1)^2
\end{aligned}
$$

Thus,

$$
\begin{aligned}
((x+1)^2 y)' &= 3(x+1)^2 \\
y &= (x+1)^{-2}((x+1)^3 + C) \\
&= x+1+\frac{C}{(x+1)^2}
\end{aligned}
$$

Problem 1.4.13

Converting the DE

$$y' + \frac{1}{1+x}y = \frac{\sin x}{1+x}$$

The IF is

$$\rho(x) = \exp\left(\int \frac{1}{1+x}dx\right) = 1+x$$

Then,

$$
\begin{aligned}
(\rho(x)y)' &= \rho(x)\frac{\sin x}{1+x} \\
((1+x)y)' &= \sin x \\
(1+x)y &= \int \sin x \, dx \\
&= -\cos x + C \\
y &= \frac{-\cos x + C}{1+x}
\end{aligned}
$$

Since $y(0) = 1$,

$$y(0) = \frac{C-1}{1} = C - 1 = 1$$
$$C = 2$$

The PS is

$$y = \frac{2 - \cos x}{1 + x}$$

Problem 1.4.14

Let $x' = \frac{dx}{dy}$

$$\left(\frac{dx}{dy}\right)^{-1} = 2xy^3 \left(\frac{dx}{dy}\right)^{-1} + 2y^4$$

Multiplying the above DE by $x' = \frac{dx}{dy}$ and moving terms around, we get

$$2y^4 x' - 2xy^3 = 1$$

Dividing the above DE by $2y^4$ yields

$$x' - \frac{1}{y}x = \frac{1}{2y^4}$$

a 1st.O linear DE with x as DV and y as IV:

$$x' + P(y)x = Q(y)$$

where

$$P(y) = -\frac{1}{y}, \qquad Q(y) = \frac{1}{2y^4}$$

We get the IF

$$\rho(y) = \exp\left(\int P(y)dy\right) = \exp\left(-\int \frac{1}{y}dy\right) = \frac{1}{y}$$

Using the 1st.O linear DE solution formula, we get

$$x = \frac{1}{\rho(y)}\left(\int (Q(y)\rho(y))dy + C\right)$$
$$= y\left(\int \left(\frac{1}{2y^4}\right)\left(\frac{1}{y}\right)dy + C\right)$$
$$= y\left(-\frac{1}{8}\frac{1}{y^4} + C\right)$$
$$= -\frac{1}{8y^3} + Cy$$

Problem 1.4.15

Substitute $v = \frac{dy}{dx}$

$$v' = y''$$
$$x^2 v' + 3xv = 4x^4$$
$$v' + \frac{3v}{x} = 4x^2$$

$$P(x) = \frac{3}{x}, \qquad Q(x) = 4x^2$$

$$\rho(x) = \exp\left(\int \frac{3}{x} dx\right) = x^3$$

$$\frac{d}{dx}(\rho(x)v) = \rho(x)Q(x)$$

$$x^3 v = \int x^3(4x^2)dx$$

$$v = \frac{4}{6}x^3 + \frac{c}{x^3}$$

Back substituting $v = \frac{dy}{dx}$, we get

$$\frac{dy}{dx} = \frac{4}{6}x^3 + \frac{c}{x^3}$$

$$y = \frac{1}{6}x^4 - \frac{c}{2x^2} + c_1$$

1.5 Substitution Methods

Problem 1.5.1

We can write the DE as

$$\frac{dy}{dx} = \frac{y}{x} + \left(\frac{y}{x}\right)^2$$

Let $u = \frac{y}{x}$. Then, we have $y = ux$ and $\frac{dy}{dx} = u + x\frac{du}{dx}$. Thus,

$$u + x\frac{du}{dx} = u + u^2$$

$$\frac{du}{u^2} = \frac{dx}{x}$$

$$-\frac{1}{u} = \ln x + C$$

Back substituting $u = \frac{y}{x}$, we get

$$y = -\frac{x}{\ln x + C}$$

Problem 1.5.2

Rewrite the DE as

$$\frac{dy}{dx} = \frac{y}{x} + \sqrt{4 + \left(\frac{x}{y}\right)^2}$$

Let $u = \frac{y}{x}$. Then, we have $y = ux$ and $\frac{dy}{dx} = u + x\frac{du}{dx}$. Thus,

$$u + x\frac{du}{dx} = u + \sqrt{4 + \frac{1}{u^2}}$$

$$\frac{u\,du}{\sqrt{1 + (2u)^2}} = \frac{dx}{x}$$

$$\int \frac{u\,du}{\sqrt{1 + (2u)^2}} = \ln|x| + C_1$$

$$\int \frac{1}{8}\frac{d(1 + (2u)^2)}{\sqrt{1 + (2u)^2}} = \ln|x| + C_1$$

$$\frac{1}{8}\left(\frac{(1 + (2u)^2)^{-\frac{1}{2}+1}}{-\frac{1}{2}+1}\right) = \ln|x| + C_1$$

$$\frac{1}{4}\sqrt{1 + (2u)^2} = \ln|x| + C_1$$

$$\sqrt{1 + (2u)^2} = 4\ln|x| + C$$

$$\sqrt{4y^2 + x^2} = 4x \ln|x| + Cx$$

Problem 1.5.3

Let $v = \ln y$. Then, $y = \exp(v)$ and $\frac{dy}{dx} = \exp(v)\frac{dv}{dx}$.

Substituting into the DE, we get

$$\exp(v)\,v' + P(x)\exp(v) = Q(x)v\exp(v)$$

Obviously, $\exp(v) \neq 0$. So, we have

$$v' + P(x) = Q(x)v(x)$$

Problem 1.5.4

$$5y^4 y' - x^2 y' = 2xy$$

$$\left(\frac{dy}{dx}\right)(5y^4 - x^2) = 2xy$$

$$x' = \frac{5y^4 - x^2}{2xy}$$

$$x' = \frac{5y^3}{2x} - \frac{x}{2y}$$

$$x' + \left(\frac{1}{2y}\right)x = \frac{5y^3}{2}x^{-1}$$

$$v = x^2 \Rightarrow v' = 2xx'$$

$$x = v^{\frac{1}{2}} \Rightarrow x' = \left(\frac{1}{2}\right)v^{-\frac{1}{2}}v'$$

$$\left(\frac{1}{2}\right)v^{-\frac{1}{2}}v' + \left(\frac{1}{2y}\right)v^{\frac{1}{2}} = \left(\frac{5}{2}\right)y^3\left(v^{\frac{1}{2}}\right)^{-1}$$

$$v' + \frac{v}{y} = 5y^3$$

$$\rho(y) = \exp\left(\int\left(\frac{1}{y}\right)dy\right) = \exp(\ln y) = y$$

$$(yv)' = 5y^4$$

$$yv = \int 5y^4 dy$$

$$yv = \frac{5y^5}{5} + C$$

$$v = y^4 + \frac{C}{y}$$

$$x^2 = y^4 + \frac{C}{y}$$

$$x = \left(y^4 + \frac{C}{y}\right)^{\frac{1}{2}}$$

Problem 1.5.5

Let $v = \ln y$, we have $y = \exp(v)$ and $y' = \exp(v)\,v'$. Thus,

$$x \exp(v)\, v' - 4x^2 \exp(v) + 2\exp(v)\,v = 0$$

$$v' + \frac{2}{x}v - 4x = 0$$

a linear DE whose IF is

$$\rho(x) = \exp\left(\int \frac{2}{x}dx\right) = x^2$$

Thus,

$$v = x^{-2}\left(\int 4x^3 dx + C\right)$$

$$= x^2 + Cx^{-2}$$

Back substituting, we get

$$y = \exp(x^2 + Cx^{-2})$$

Problem 1.5.6

$$tx' - (m+1)t^m x + 2x \ln x = 0$$

$$\frac{tx'}{x} - (m+1)t^m + 2\ln x = 0$$

Let $y = \ln x$. Then, $y' = \frac{x'}{x}$ and

$$ty' - (m+1)t^m + 2y = 0$$

$$y' + \frac{2}{t}y = (m+1)t^{m-1}$$

Using 1st.O linear DE method, we get

$$\rho(t) = \exp\left(\int \frac{2}{t}dt\right) = t^2$$

$$t^2 y' + 2ty = (m+1)t^{m+1}$$

$$t^2 y = \int (m+1)t^{m+1} dt$$

$$t^2 y = \frac{m+1}{m+2}t^{m+2} + c$$

$$y = \frac{m+1}{m+2}t^m + ct^{-2}$$

$$\ln x = \frac{m+1}{m+2}t^m + ct^{-2}$$

$$x = \exp\left(\frac{m+1}{m+2}t^m + ct^{-2}\right)$$

Problem 1.5.7

Let $v = y^{\frac{1}{3}}$. Then, $y = v^3$ and $y' = 3v^2 v'$. The DE becomes

$$v' - \frac{2}{x} v = 4x^3$$

whose IF is

$$\rho(x) = \exp\left(\int \left(-\frac{2}{x} \right) dx \right) = x^{-2}$$
$$(v\rho)' = 4x$$
$$vx^{-2} = 2x^2 + C$$
$$v = 2x^4 + Cx^2$$
$$y = (2x^4 + Cx^2)^3$$

Problem 1.5.8

Let $v = y'$ and consider v as a function of y. Then,

$$y'' = \frac{dv}{dx} = \frac{dv}{dy}\frac{dy}{dx} = v\frac{dv}{dy}$$

Thus, the DE becomes

$$yv\frac{dv}{dy} = 3v^2$$
$$\frac{dv}{dy} = \frac{3v}{y}$$
$$v = C_1 y^3$$

Thus,

$$\frac{dy}{dx} = C_1 y^3$$
$$-\frac{1}{2} y^{-2} = C_1 x + C_2$$

Finally,

$$(Ax + B)y^2 = 1$$

Problem 1.5.9

Let $v = \frac{dy}{dx} = y'$. Then, $v' = \frac{dv}{dx} = y''$ and the original DE becomes

$$y^3 \frac{dv}{dx} = 3$$
$$y^3 \frac{dv}{dy}\frac{dy}{dx} = 3$$
$$y^3 v \frac{dv}{dy} = 3$$

which is a separable DE in terms of (v, y).

$$\int v dv = \int \frac{3}{y^3} dy$$

$$\frac{v^2}{2} = -\frac{3y^{-2}}{2} + C_1$$

Thus, using $v(x) = \frac{dy}{dx}$, we have

$$\frac{dy}{dx} = \sqrt{-3y^{-2} + C_2}$$

$$\frac{dy}{\sqrt{-\frac{3}{y^2} + C_2}} = dx$$

$$\frac{y dy}{\sqrt{-3 + C_2 y^2}} = dx$$

$$\frac{1}{C_2}\sqrt{C_2 y^2 - 3} = x + C_3$$

Finally, the implicit GS is

$$C_2 y^2 = 3 + (C_2 x + C_4)^2$$

where C_2 and C_4 are constants.

Problem 1.5.10

Method 1: The SOV method. The DE can be written as

$$\frac{y'}{y^3 - y} = x$$

$$\frac{dy}{y^3 - y} = x dx$$

$$\frac{1}{2}\frac{2y dy}{y^2(y^2 - 1)} = x dx$$

$$\frac{1}{2}\left(\frac{1}{y^2 - 1} - \frac{1}{y^2}\right) d(y^2) = x dx$$

$$\frac{1}{2}\left(\ln(y^2 - 1) - (\ln y^2)\right) = \frac{1}{2}x^2 + C_1$$

$$\frac{1}{2}\ln(y^2 - 1) - \ln y = \frac{1}{2}x^2 + C_1$$

$$y^{-2}(y^2 - 1) = C \exp(x^2)$$

$$1 - y^{-2} = C \exp(x^2)$$

Method 2: The Bernoulli DE method.

Let $v = y^{-2}$. Thus, $v' = -2y^{-3}y'$. The DE becomes

$$-\frac{1}{2}v' + xv = x$$

$$v' - 2xv = -2x$$

which is a 1st.O linear DE. Let

$$\rho(x) = \exp\left(\int -2x dx\right)$$
$$= \exp(-x^2)$$

Thus, we know that

$$(\exp(-x^2)\, v)' = -2x \exp(-x^2)$$
$$v = \exp(x^2)\,(\exp(-x^2) + C)$$
$$= 1 + C \exp(x^2)$$

Back substituting $v = y^{-2}$, we have

$$y^{-2} - 1 = C \exp(x^2)$$

Problem 1.5.11

Method 1: The Bernoulli DE method.

$$xy' - y = y^2 \sin x$$
$$y' - \frac{y}{x} = \frac{\sin x}{x} y^2$$

Let $u = y^{1-2} = y^{-1}$. Then,

$$\frac{1}{1-2} u' - \frac{u}{x} = \frac{\sin x}{x}$$
$$u' + \frac{u}{x} = \frac{-\sin x}{x}$$

which is, now, a 1st.O linear DE whose IF is

$$\rho(x) = \exp\left(\int \frac{1}{x} dx\right) = x$$
$$xu' + u = -\sin x$$
$$\int d(xu) = \int (-\sin x) dx + C$$
$$xu = \cos x + C$$

Thus,

$$\frac{x}{y} = \cos x + C$$
$$y = \frac{x}{\cos x + C}$$

Method 2: The S-method.

$$xy' - y = y^2 \sin x$$
$$y' - \frac{y}{x} = \frac{y^2 \sin x}{x}$$
$$y' - \frac{y}{x} = \left(\frac{y}{x}\right) y \sin x$$
$$(u'x + u) - u = u(ux) \sin x$$
$$(u'x) = u^2 x \, \sin x$$

$$x \frac{du}{dx} = u^2 x \sin x$$

$$\int \frac{du}{u^2} = \int \sin x \, dx$$

$$-\frac{1}{u} = -\cos x + C$$

$$\frac{1}{u} = \cos x + C$$

$$\frac{x}{y} = \cos x + C$$

$$y = \frac{x}{\cos x + C}$$

Problem 1.5.12

Let $u = \frac{y}{x}$. Then, $y = ux$ and $\frac{dy}{dx} = u + x \frac{du}{dx}$. Thus, the DE becomes

$$u + x \frac{du}{dx} = -\frac{ux(2x^3 - u^3 x^3)}{x(2u^3 x^3 - x^3)}$$

$$x \frac{du}{dx} = -\frac{u^4 + u}{2u^3 - 1}$$

$$\ln(u + 1) - \ln u + \ln(u^2 - u + 1) = -\ln x + C_1$$

$$\ln \frac{(u^3 - u + 1)(u + 1)}{u} = -\ln x + C_1$$

$$\frac{u^3 + 1}{u} = \frac{C}{x}$$

$$\frac{y^3 + x^3}{xy} = C$$

$$y^3 + x^3 = Cxy$$

Problem 1.5.13

The DE can be written as

$$y' = \frac{y}{x} + \exp\left(\frac{y}{x}\right)$$

Let $u = \frac{y}{x}$. Then, $y = ux$ and $\frac{dy}{dx} = u + x \frac{du}{dx}$. Thus, the DE becomes

$$u + x \frac{du}{dx} = u + \exp(u)$$

$$\exp(-u) \, du = \frac{dx}{x}$$

$$-\exp(-u) = \ln x + C$$

The GS to the DE is

$$-\exp\left(-\frac{y}{x}\right) = \ln x + C$$

Problem 1.5.14

Let $u = x + y$. We have $y' = u' - 1$. Thus,

$$uu' - u = 1$$

$$u' = \frac{1}{u} + 1$$

$$\frac{du}{\frac{1}{u} + 1} = dx$$

$$u - \ln(u + 1) = x + C$$

$$x + y - \ln(x + y + 1) = x + C$$

$$y - \ln(x + y + 1) = C$$

Problem 1.5.15

Let $v = \sin y$. Then, $v' = y' \cos y$. The DE becomes

$$2xvv' = 4x^2 + v^2$$

$$v' - \frac{1}{2x}v = 2xv^{-1}$$

Let $u = v^2$. Then, $u' = 2vv'$

$$u' - \frac{1}{x}u = 4x$$

whose IF is

$$\rho(x) = \exp\left(\int\left(-\frac{1}{x}\right)dx\right) = x^{-1}$$

Then,

$$(x^{-1}u)' = 4$$

$$x^{-1}u = 4x + C$$

$$u = 4x^2 + Cx$$

The GS is

$$\sin^2 y = 4x^2 - Cx$$

Problem 1.5.16

Let $v = y'$. Then, $y'' = v'$ and the DE becomes

$$v' = (x + v)^2$$

Let $u = x + v$. Then, $v' = u' - 1$

$$u' - 1 = u^2$$

$$\frac{du}{u^2 + 1} = dx$$

$$x = \arctan u + C_1$$
$$u = \tan(x - C_1)$$

Thus,

$$y' = v = u - x$$
$$= \tan(x - C_1) - x$$

Finally,

$$y = -\ln(\cos(x - C_1)) - \frac{1}{2}x^2 + C_2$$

Problem 1.5.17

Let $v = \exp(y)$. Then, $y = \ln v$ and we have

$$y' = \frac{1}{v}v'$$

Thus, the DE becomes

$$(x + v)\frac{1}{v}v' = \frac{x}{v} - 1$$
$$(x + v)v' = x - v$$

Let $u = x + v$. Then, $v' = u' - 1$ and

$$uu' - u = -u + 2x$$
$$uu' = 2x$$
$$\frac{1}{2}u^2 = x^2 + C_1$$
$$u^2 = 2x^2 + C$$

Back substituting $u = x + v$ gives

$$v^2 - x^2 + 2xv = C$$

Finally,

$$\exp(2y) - x^2 + 2x\exp(y) = C$$

Problem 1.5.18

We can rewrite the DE as

$$y' = \left(x^{-\frac{3}{2}} - 9x^{\frac{1}{2}}\right)y^2$$

which is a Bernoulli DE. Let $v = y^{-1}$ and we have

$$v' = -y^{-2}y'$$

Thus, the DE becomes

$$v' = 9x^{\frac{1}{2}} - x^{-\frac{3}{2}}$$

It is easy to get

$$v = 6x^{\frac{3}{2}} + 2x^{-\frac{1}{2}} + C$$

Therefore,

$$y = \left(6x^{\frac{3}{2}} + 2x^{-\frac{1}{2}} + C\right)^{-1}$$

Problem 1.5.19

The DE is reformulated as

$$y' + \frac{1}{x}y = \frac{1}{\sqrt{1 + x^4}}y^{-2}$$

a Bernoulli DE. Let $v = y^{1-(-2)} = y^3$, we have

$$v' + \frac{3}{x}v = \frac{3}{\sqrt{1 + x^4}}$$

whose IF is

$$\rho(x) = \exp\left(\int \frac{3}{x}dx\right)$$
$$= x^3$$

Thus,

$$(x^3 v)' = \frac{3x^3}{\sqrt{1 + x^4}}$$
$$v = x^{-3}\int \frac{3x^3 dx}{\sqrt{1 + x^4}}$$
$$= x^{-3}\left(\frac{3}{2}\sqrt{1 + x^4} + C\right)$$

Thus,

$$x^3 y^3 = \left(\frac{3}{2}\sqrt{1 + x^4} + C\right)$$

Problem 1.5.20

Let $v = \exp(y)$, $v' = \exp(y)\, y'$. Thus, the DE becomes

$$v' - \frac{2}{x}v = 2x^2 \exp(2x)$$

whose IF is

$$\rho(x) = \exp\left(\int -\frac{2}{x}dx\right) = x^{-2}$$

Then,

$$(x^{-2}v)' = 2\exp(2x)$$
$$v = x^2 \int 2\exp(2x)\, dx$$
$$= x^2(\exp(2x) + C)$$

Thus,

$$y = 2\ln x + \ln(\exp(2x) + C)$$

Problem 1.5.21

Let $v = \ln y$. Then, $y = \exp(v)$ and $\frac{dy}{dx} = \exp(v)\frac{dv}{dx}$

Using y' and y above, we have

$$\frac{\exp(v)\frac{dv}{dx} - 4xv\exp(v) + 2\frac{\exp(v)}{x}v^n}{\exp(v)} = 0$$

$$\frac{dv}{dx} - 4xv = -\frac{2}{x}v^n$$

which is a Bernoulli DE. Let

$$u = v^{1-n}$$

We have

$$\frac{du}{dx} - 4(1-n)xu = -(1-n)\frac{2}{x}$$

which is a 1st.O DE of u wrt x with IF

$$\rho(x) = \exp\left(\int -4(1-n)x dx\right)$$
$$= \exp(2(n-1)x^2)$$

Thus,

$$u(x) = \exp(-2(n-1)x^2)\left(\int \frac{2(n-1)}{x}\exp(2(n-1)x^2)\,dx + C\right)$$

Back substitution twice gets the GS $y(x)$.

Problem 1.5.22

Let $v = y'$ and consider v as a function of y. Now we have

$$y'' = \frac{dv}{dx} = \frac{dv}{dy}\frac{dy}{dx} = v\frac{dv}{dy}$$

Thus, the DE becomes

$$yv\frac{dv}{dy} + v^2 = yv$$

$$y\frac{dv}{dy} + v = y$$

$$\frac{d}{dy}(yv) = y$$

$$yv = \frac{1}{2}y^2 + C$$

$$v = \frac{1}{2}y + \frac{C}{y}$$

Put $v = y'$ back, we have

$$y' = \frac{1}{2}y + \frac{C}{y}$$

$$\frac{2yy'}{y^2 + C_1} = 1$$

$$\frac{2y}{y^2 + C_1} dy = dx$$

Integrating both sides, we get

$$\ln(y^2 + C_1) = x + C_2$$

Problem 1.5.23

It is easy to observe that $y = 0$ is a solution. For $y \neq 0$, we can rewrite the DE after dividing the original DE by xy:

$$6\frac{y}{x} + 2\left(\frac{y}{x}\right)^2 + \left(9 + 8\frac{y}{x}\right)y' = 0$$

Let $u = \frac{y}{x}$. Then, $y' = u + u'x$ and the DE become

$$6u + 2u^2 + (9 + 8u)(u + u'x) = 0$$

$$\frac{9 + 8u}{3u + 2u^2}u' = -\frac{5}{x}$$

a separable DE whose GS is

$$\left(\frac{3}{u} + \frac{2}{3 + 2u}\right) du = -\frac{5}{x} dx$$

$$3\ln u + \ln(3 + 2u) = -5\ln x + C_1$$

$$u^3(3 + 2u) = Cx^{-5}$$

Back substituting $y = ux$, we get

$$3x^2 y^3 + 2xy^4 = C$$

Problem 1.5.24

(1) From $y = x^{-1} + u$, we have

$$\frac{dy}{dx} = -\frac{1}{x^2} + \frac{du}{dx}$$

Substituting y and y' in the original DE,

$$-\frac{1}{x^2} + \frac{du}{dx} + 7\left(\frac{1}{x} + u\right)\frac{1}{x} - 3\left(\frac{1}{x} + u\right)^2 = \frac{3}{x^2}$$

$$-\frac{1}{x^2} + \frac{du}{dx} + \frac{7}{x^2} + \frac{7u}{x} - 3\left(\frac{1}{x^2} + \frac{2u}{x} + u^2\right) = \frac{3}{x^2}$$

$$\frac{du}{dx} + \frac{u}{x} - 3u^2 = 0$$

$$u' + x^{-1}u = 3u^2$$

which is a Bernoulli DE of u wrt x where $P(x) = x^{-1}, Q(x) = 3, n = 2$.

(2) Let $v = u^{1-n} = u^{1-2} = u^{-1}$, we have

$$\frac{dv}{dx} + (1-n)P(x)v = (1-n)Q(x)$$

$$v' - \frac{v}{x} = -3$$

which is a 1st.O linear DE whose IF is

$$\rho(x) = \exp\left(\int -\frac{1}{x}dx\right) = \frac{1}{x}$$

Then,

$$v = \frac{1}{\rho}\left(\int \rho(x)(-3)dx + C\right) = x\left(-\int \frac{3}{x}dx + C\right)$$
$$= x(-3\ln x + C)$$
$$= -3x\ln x + Cx$$

Substituting back v with u^{-1} and u with $y - x^{-1}$, we have

$$-3x\ln x + Cx = \frac{1}{u} = \frac{1}{y - x^{-1}}$$

$$y = \frac{1}{-3x\ln x + Cx} + \frac{1}{x}$$

Problem 1.5.25

Let $v = y'$, thus, $y'' = v'$ and the original DE becomes

$$x^2 v' + 3xv = 4$$
$$(x^3 v)' = 4x$$
$$x^3 v = 2x^2 + C_1$$
$$v = \frac{2}{x} + \frac{C_1}{x^3}$$

Back substituting $v = y'$, we have

$$y' = \frac{2}{x} + \frac{C_1}{x^3}$$
$$y = \int \left(\frac{2}{x} + \frac{C_1}{x^3}\right)dx$$
$$= 2\ln x - \frac{C_1}{2x^2} + C_2$$

Problem 1.5.26

Convert the DE to the form of Bernoulli DE

$$x' - 1000t^{999}x = -2x^2 t^{-1}$$

Let $v = x^{-1}, v' = -x^{-2}x'$. Plug into the above DE,

$$-x^2 v' - 1000t^{999}x^2 v = -2x^2 t^{-1}$$
$$v' + 1000t^{999}v = 2t^{-1}$$

The IF is

$$\rho = \exp\left(\int 1000t^{999}\, dt\right) = \exp(t^{1000})$$

Multiplying the IF on both sides, we get

$$(\exp(t^{1000})\, v)' = 2t^{-1}\exp(t^{1000})$$

$$v = \exp(-t^{1000})\int 2t^{-1}\exp(t^{1000})\, dt$$

$$x = \exp(t^{1000})\left(\int 2t^{-1}\exp(t^{1000})\, dt\right)^{-1}$$

Problem 1.5.27

$$\frac{dy}{dx} = \frac{x(x+y)}{y(x+y)} = \frac{x}{y}$$

(when $x + y \neq 0$, we divide the DE by $x + y$). This is a Homo DE which is also separable. Let $v = \frac{y}{x}$. Then, $y' = v + v'x$ and the DE becomes

$$v' + \frac{v}{x} = \frac{1}{xv}$$

which is a Bernoulli DE with $n = -1$.

Let $u = v^{1-n} = v^2$. Then, $v = u^{\frac{1}{2}}$ and $v' = \frac{1}{2}u^{-\frac{1}{2}}u'$. Then,

$$v' + \frac{v}{x} = \frac{1}{xv}, \qquad u' + \frac{2u}{x} = \frac{2}{x}$$

where we identify

$$P(x) = \frac{2}{x}, \qquad Q(x) = \frac{2}{x}$$

Thus,

$$\rho(x) = \exp\left(\int P(x)dx\right) = \exp\left(\int \frac{2}{x}dx\right) = \exp(2\ln x) = x^2$$

$$u = \exp\left(-\int P(x)dx\right)\left(\int Q(x)\exp\left(\int P(x)dx\right)dx + C\right)$$

$$= \frac{1}{x^2}\left(\int \left(\frac{2}{x}x^2\right)dx + C\right)$$

$$= \frac{1}{x^2}(x^2 + C)$$

After back substitution $u = \left(\frac{y}{x}\right)^2$, one gets the GS

$$y^2 = x^2 + C$$

Alternatively, SOV will solve the DE more conveniently. The DE can be converted into

$$ydy = xdx$$

Integrating both sides, we get

$$y^2 = x^2 + C$$

Problem 1.5.28

$$y' + \frac{1}{x}y + \frac{1}{2x^2}y^{-1} = 0$$

Dividing the DE by $2x^2y$, we get

$$y' + \frac{1}{x}y = -\frac{1}{2x^2}y^{-1}$$

Let $v = y^{1-n} = y^2$. Then, $y = v^{\frac{1}{2}}$ and $y' = \frac{1}{2}v^{-\frac{1}{2}}v'$. The DE

$$2x^2yy' + 2xy^2 + 1 = 0$$

becomes

$$x^2v' + 2xv + 1 = 0$$

Dividing the above DE by x^2, we get

$$v' + \frac{2}{x}v = -\frac{1}{x^2}, \qquad P(x) = \frac{2}{x}, \qquad Q(x) = -\frac{1}{x^2}$$

$$\rho(x) = \exp\left(\int P(x)dx\right) = \exp\left(\int \frac{2}{x}dx\right) = \exp(2\ln x) = x^2$$

$$v = \exp\left(-\int P(x)dx\right)\left(\int Q(x)\exp\left(\int P(x)dx\right)dx + C\right)$$

$$= \frac{1}{x^2}\left(\int\left(-\frac{1}{x^2}x^2\right)dx + C\right)$$

$$= \frac{1}{x^2}(-x + C)$$

Therefore, the GS is

$$y^2 = \frac{1}{x^2}(-x + C)$$

Problem 1.5.29

Method 1: The SOV method.

$$(x^2 + 1)\frac{dy}{dx} = 2x(y + 1)$$

$$\int \frac{1}{y+1}dy = \int \frac{2x}{x^2+1}dx$$

$$\ln(y + 1) = \ln(x^2 + 1) + \ln(C)$$

$$y = C(x^2 + 1) - 1$$

Method 2: The 1st.O linear DE method.

$$\frac{dy}{dx} - \frac{2x}{x^2+1}y = \frac{2x}{x^2+1}$$

whose IF

$$\rho(x) = \exp\left(-\int \frac{2x}{x^2+1}dx\right) = \frac{1}{x^2+1}$$

Multiplying the IF and integrating, we get

$$\frac{d}{dx}\left(\frac{y}{x^2+1}\right) = \frac{2x}{(x^2+1)^2}$$

$$\frac{y}{x^2+1} = \int \frac{2x}{(x^2+1)^2}dx$$

$$= -\frac{1}{x^2+1} + C$$

$$y = C(x^2+1) - 1$$

Problem 1.5.30

Using substitution $v = y/x$, we get

$$xv' = -b\sqrt{1+v^2}$$

Solving this DE by the SOV method, we get

$$y = \frac{a}{2}\left(\left(\frac{x}{a}\right)^{1-b} - \left(\frac{x}{a}\right)^{1+b}\right)$$

(1) If $b < 1$, the solution y can be zero (answer to (4)).

(2) If $b = 1$, we have

$$y = \frac{a}{2}\left(1 - \left(\frac{x}{a}\right)^2\right)$$

Thus, at $x = 0$, we have $y = a/2$.

(3) If $b > 1$, y blows up easily.

Problem 1.5.31

Let $x = \frac{1}{t} + u$. Then, $\frac{dx}{dt} = -\frac{1}{t^2} + \frac{du}{dt}$ and the DE becomes

$$-\frac{1}{t^2} + \frac{du}{dt} = 3\left(\frac{1}{t} + u\right)^2 - \frac{8}{t}\left(\frac{1}{t} + u\right) + \frac{4}{t^2}$$

$$= -\frac{1}{t^2} - \frac{2u}{t} + 3u^2$$

$$\frac{du}{dt} = -\frac{2u}{t} + 3u^2$$

a Bernoulli DE with

$$P(t) = \frac{2}{t}, \qquad Q(t) = 3, \qquad n = 2$$

Let $v = u^{-1}$, the DE becomes

$$v' - \frac{2}{t}v = -3$$

whose IF is

$$\rho(t) = \exp\left(\int \left(-\frac{2}{t}\right)dt\right) = t^{-2}$$

$$v(t) = \frac{1}{\rho(t)}\int \rho(t)(-3)dt$$

$$= t^2 \left(\int \left(-\frac{3}{t^2} \right) dt + C \right)$$

$$= 3t + Ct^2$$

Given $x = \frac{1}{t} + u$ and $v(t) = u^{-1}(t)$, we have

$$v(t) = \frac{1}{x(t) - \frac{1}{t}}$$

Therefore,

$$x(t) = \frac{1}{3t + Ct^2} + \frac{1}{t}$$

Problem 1.5.32

Let $y = \beta + z$, the original DE becomes a Bernoulli DE

$$y' = (K(x) + y + \beta)(y - \beta)$$
$$= (K(x) + 2\beta + z)z$$
$$= (K(x) + 2\beta)z + z^2$$

or

$$z' - (K(x) + 2\beta)z = z^2$$

which is Bernoulli when

$$P(x) = -(K(x) + 2\beta)$$
$$Q(x) = 1$$
$$n = 2$$

Problem 1.5.33

$$tx' - x = \beta x' x + \beta t$$
$$tx' - \beta x' x = \beta t + x$$
$$x'(t - \beta x) = \beta t + x$$
$$x' = \frac{\beta t + x}{t - \beta x}$$

$$= \frac{\left(\beta + \frac{x}{t} \right)}{1 - \frac{\beta x}{t}}$$

Let $u = \frac{x}{t}$. Then, $x = ut$ and $\frac{dx}{dt} = t\frac{du}{dt} + u$ and

$$t\frac{du}{dt} + u = \frac{\beta + u}{1 - \beta u}$$

$$t\frac{du}{dt} = \frac{\beta + u}{1 - \beta u} - u$$

$$= \frac{\beta + \beta u^2}{1 - \beta u}$$

353

$$\int \frac{1-\beta u}{\beta + \beta u^2}\,du = \int \frac{dt}{t}$$

$$\frac{1}{\beta}\int \frac{1}{1+u^2}\,du - \int \frac{u}{1+u^2}\,du = \ln t + C$$

$$\frac{1}{\beta}\arctan(u) + \frac{1}{2}\ln|1+u^2| = \ln t + C$$

$$\arctan(u) = \frac{\beta}{2}\ln\left(\frac{t^2}{1+u^2}\right) + C_1$$

$$u = \tan\left(\frac{\beta}{2}\ln\left(\frac{t^2}{1+u^2}\right) + C_1\right)$$

After back substitution, one gets

$$x = t\tan\left(\frac{\beta}{2}\ln\left(\frac{t^2}{1+u^2}\right) + C_1\right)$$

Problem 1.5.34

For $n = 0$:

$$y' = by^2 + c$$

$$\frac{dy}{by^2 + c} = dx$$

$$\int \frac{dy}{b\left(y^2 + \frac{c}{b}\right)} = \int dx$$

If $bc > 0$,

$$\frac{1}{b}\int \frac{dy}{y^2 + \left(\sqrt{\frac{c}{b}}\right)^2} = x$$

$$\left(\frac{1}{b}\right)\left(\frac{1}{\sqrt{\frac{c}{b}}}\right)\tan^{-1}\left(\frac{y}{\sqrt{\frac{c}{b}}}\right) = x + A$$

$$\frac{1}{\sqrt{bc}}\tan^{-1}\sqrt{\frac{b}{c}}\,y = x + A$$

$$\tan^{-1}\sqrt{\frac{b}{c}}\,y = \sqrt{bc}(x + A)$$

$$\sqrt{\frac{b}{c}}\,y = \tan\left(\sqrt{bc}(x + A)\right)$$

$$y = \sqrt{\frac{c}{b}}\tan\left(\sqrt{bc}(x + A)\right)$$

If $bc < 0$,

$$\frac{1}{b}\int \frac{dy}{y^2 - \left(\sqrt{-\frac{c}{b}}\right)^2} = x + D$$

$$\frac{1}{2\sqrt{-bc}}\int \left(\frac{1}{y - \sqrt{-\frac{c}{b}}} - \frac{1}{y + \sqrt{-\frac{c}{b}}}\right) dy = x + D$$

$$\frac{1}{2\sqrt{-bc}}\ln \left|\frac{y - \sqrt{-\frac{c}{b}}}{y + \sqrt{-\frac{c}{b}}}\right| = x + E$$

$$\frac{by - \sqrt{-bc}}{by + \sqrt{-bc}} = F \exp(2\sqrt{-bc}x)$$

For $n = -2$:

$$y' = by^2 + cx^{-2}$$

Let $v = y^{-1}$. Then, $v' = -y^{-2}y'$ and $y' = -v^{-2}v'$. The DE becomes

$$-v^{-2}v' = bv^{-2} + cx^{-2}$$
$$v' = -b - cx^{-2}v^2$$

which is a Homo DE and can be solved by the S-method.

Let $z = \frac{v}{x}$. Then,

$$(xz)' = -b - cz^2$$
$$xz' = -cz^2 - z - b$$
$$\int \frac{1}{-cz^2 - z - b} dz = \int \frac{1}{x} dx$$

Case (1): $1 - 4bc > 0$

Let z_1 and z_2 be the roots of $-cz^2 - z - b = 0$. Then, we have

$$\int \frac{1}{c(z_2 - z_1)}\left(\frac{1}{z - z_1} - \frac{1}{z - z_2}\right) dz = \int \frac{1}{x} dx$$
$$\frac{1}{c(z_2 - z_1)}\ln \left|\frac{z - z_1}{z - z_2}\right| = \ln |x| + G_1$$

where G_1 is a constant.

Case (2): $1 - 4bc < 0$

$$\int \frac{1}{-c\left(z + \frac{1}{2c}\right)^2 + \frac{4bc - 1}{-4c}} dz = \int \frac{1}{x} dx$$

Let

$$A = -c, B = \frac{4bc - 1}{-4c}, t = z + \frac{1}{2c}$$

Then, by the assumption, we have $AB > 0$. So

$$\frac{1}{\sqrt{AB}}\int\frac{1}{\left(\sqrt{\frac{A}{B}}t\right)^2+1}\,d\sqrt{\frac{A}{B}}t=\int\frac{1}{x}dx$$

$$\sqrt{\frac{1}{AB}}\,tan^{-1}\left(\sqrt{\frac{A}{B}}t\right)+G_2=ln|x|$$

$$\sqrt{\frac{1}{AB}}\,\arctan\sqrt{\frac{A}{B}}t+G_2=\ln|x|$$

where G_2 is another constant.

Case (3): $1-4bc=0$

Then,

$$-\frac{1}{c}\int\frac{1}{\left(z+\frac{1}{2c}\right)^2}dz=\int\frac{1}{x}dx$$

$$x=H_3\exp\left(\frac{2xy}{2c+xy}\right)$$

One gets different GS's depending on $1-4bc<0,=0,>0$.

1.6 Riccati DEs

Problem 1.6.1

Since $y = y_1 + \frac{1}{v}$, we have $y' = y_1' - \frac{1}{v^2}v'$. Plugging these into the DE, we get

$$y_1' - \frac{1}{v^2}v' = A(x)\left(y_1^2 + \frac{2y_1}{v} + \frac{1}{v^2}\right) + B(x)\left(y_1 + \frac{1}{v}\right) + C(x)$$

Since y_1 is a PS to the DE, $y_1' = A(x)y_1^2 + B(x)y_1 + C(x)$ must be true. Thus,

$$-\frac{1}{v^2}v' = 2A(x)\frac{y_1}{v} + \frac{A(x)}{v^2} + \frac{B(x)}{v}$$

That is,

$$v' + (2Ay_1 + B)v = -A$$

Problem 1.6.2

Let $y = x + \frac{1}{v}$. Then, $y' = 1 - \frac{1}{v^2}v'$ and the DE becomes

$$1 - \frac{1}{v^2}v' + x^2 + \frac{2x}{v} + \frac{1}{v^2} = 1 + x^2$$

i.e.,

$$v' - 2xv = 1$$

whose IF is

$$\rho(x) = \exp\left(\int -2x\,dx\right)$$
$$= \exp(-x^2)$$

Thus,

$$\rho(x)(v' - 2xv) = \rho(x)$$
$$(\rho(x)v)' = \rho(x)$$
$$v = \exp(x^2)\int \exp(-x^2)\,dx$$
$$= \exp(x^2)\left(\frac{\sqrt{\pi}}{2}\operatorname{erf}(x) + C\right)$$

Finally, we have

$$y = x + \frac{1}{v}$$
$$= x + \exp(-x^2)\left(\frac{\sqrt{\pi}}{2}\operatorname{erf}(x) + C\right)^{-1}$$

Note: The error function is defined as

$$\text{erf}(x) = \frac{2}{\sqrt{\pi}} \int_0^x \exp(-t^2)\, dt$$

Problem 1.6.3

Let $-\frac{1}{v} = x - y$. Then,

$$\frac{dv}{v^2 dx} = 1 - \frac{dy}{dx}$$

$$\frac{dy}{dx} = 1 - \frac{dv}{v^2 dx}$$

$$1 - \frac{dv}{v^2 dx} = 1 + \frac{1}{4}\left(-\frac{1}{v}\right)^2$$

$$-\frac{dv}{v^2 dx} = \frac{1}{4v^2}$$

$$-4\, dv = dx$$

Integrating

$$-4 \int dv = \int dx$$

$$-4v = x + c$$

$$v = -\frac{x + c}{4}$$

Back substituting $\frac{1}{v} = y - x$, i.e., $v = \frac{1}{y-x}$, we get

$$\frac{1}{y - x} = -\frac{x + c}{4}$$

$$y = -\frac{4}{x + c} + x$$

Problem 1.6.4

Let $y = x + \frac{1}{v}$. Then, $y' = 1 - \frac{v'}{v^2}$ and the DE becomes

$$1 - \frac{v'}{v^2} - 13\left(x^2 + \left(x + \frac{1}{v}\right)^2\right) + 26x\left(x + \frac{1}{v}\right) = 1$$

$$\frac{v'}{v^2} + \frac{13}{v^2} = 0, \qquad v' = -13, \qquad v = -13x + C$$

Back substituting, we get the GS to the original DE:

$$y = x + \frac{1}{C - 13x}$$

Alternatively, one may find the GS the easier way. The original DE can be written as

$$y' = 1 + 13(x - y)^2$$

Let $u = x - y$. Then, $\frac{du}{dx} = 1 - \frac{dy}{dx}$ and the DE becomes

$$u' + 1 = 1 + 13u^2$$
$$u' = 13u^2$$

Integrating,

$$-\frac{1}{u} = 13x + C$$

Back substituting, we get the same GS

$$\frac{1}{y - x} = 13x + C$$

Problem 1.6.5

Let $y = a + \frac{1}{z}$. Then, $\frac{dy}{dx} = -\frac{1}{z^2}\frac{dz}{dx}$ and plugging them into the original DE yields

$$\frac{dz}{dx} + (f(x) + 2a)z + 1 = 0$$

which is a 1st.O linear DE whose IF is

$$\rho(x) = \exp\left(\int (f(x) + 2a)dx\right)$$

Multiplying the IF on both sides of the z function, we have

$$\rho(x)\frac{dz}{dx} + (f(x) + 2a)\rho(x)z + \rho(x) = 0$$

Using the product rule,

$$d(uv) = udv + vdu$$

we get

$$\frac{d}{dx}(\rho(x)z) = -\rho(x)$$

whose GS is

$$z = -\frac{1}{\rho(x)}\int \rho(x)dx + C$$

Finally, the GS for the original DE is

$$y = a + \frac{1}{z} = a - \frac{\rho(x)}{\int \rho(x)dx - C\rho(x)}$$

where

$$\rho(x) = \exp\left(\int (f(x) + 2a)dx\right)$$

Problem 1.6.6

$$y' = -(a^2 + 4ax^3) + 4x^3y + y^2$$

This DE is Riccati with one given PS $y_1(x) = a$. Making the Riccati substitution

$$y = a + \frac{1}{v}$$

we have

$$-\frac{1}{v^2}\frac{dv}{dx} = \left(a + \frac{1}{v}\right)^2 + 4x^3\left(a + \frac{1}{v}\right) - (a^2 + 4ax^3)$$

Simplifying it, we have

$$v' + (2a + 4x^3)v = -1 \qquad\qquad (A.1)$$

This is a 1st.O linear DE, where

$$P(x) = 2a + 4x^3, Q(x) = -1$$
$$\rho(x) = \exp\left(\int (2a + 4x^3)dx\right) = \exp(2ax + x^4)$$

Multiplying both sides of Eq. (A.1) by $\rho(x)$, we have

$$(\rho(x)v)' = -\rho(x)$$
$$\frac{1}{v} = -\frac{\exp(2ax + x^4)}{\int_c^x \exp(2at + t^4)\,dt}$$

The GS for the original DE is

$$y = a + \frac{1}{v}$$
$$= a - \frac{\exp(2ax + x^4)}{\int_c^x \exp(2at + t^4)\,dt}$$

Problem 1.6.7

The DE is Ricatti with a given solution $y_1(x) = \sin x$ and an appropriate substitution is

$$y = y_1(x) + \frac{1}{v}$$
$$= \sin x + \frac{1}{v}$$
$$y' = \cos x - \frac{v'}{v^2}$$

Substituting back into the original DE gives

$$\cos x - \frac{v'}{v^2} = \frac{2\cos^2 x - \sin^2 x + \left(\sin x + \frac{1}{v}\right)^2}{2\cos x}$$
$$= \frac{2\cos^2 x - \sin^2 x + \sin^2 x + \frac{2\sin x}{v} + \frac{1}{v^2}}{2\cos x}$$
$$= \cos x + \frac{\tan x}{v} + \frac{\frac{1}{v^2}}{2\cos x}$$

The DE can be rewritten as

$$v' + \tan x \, v = -\frac{1}{2}\sec x$$

Multiplying both sides by the IF

$$\rho(x) = \exp\left(\int \tan x \, dx\right) = \exp(\ln|\sec x|) = \sec x$$

we get

$$\sec x \, v' + \sec x \tan x \, v = -\frac{1}{2}\sec^2 x$$

$$\int \frac{d}{dx}(\sec x \, v)dx = -\frac{1}{2}\int \sec^2 x \, dx$$

$$\sec x \, v = -\frac{1}{2}\tan x + C$$

$$v = -\frac{1}{2}\sin x + C\cos x$$

Back substituting $v = (y - \sin x)^{-1}$ gives

$$\frac{1}{y - \sin x} = -\frac{1}{2}\sin x + C\cos x$$

$$y = \frac{2}{C\cos x - \sin x} + \sin x$$

Problem 1.6.8

The DE is Riccati with a known PS $y_1 = x^2$. Introducing the substitution

$$y = y_1 + \frac{1}{v}$$

$$= x^2 + \frac{1}{v}$$

$$y' = 2x - \frac{1}{v^2}\frac{dv}{dx}$$

we convert the original DE into

$$2x - \frac{1}{v^2}\frac{dv}{dx} = \left(x^2 + \frac{1}{v}\right)^2 + \alpha(x)\left(x^2 + \frac{1}{v} - x^2\right) + 2x - x^4$$

$$\frac{dv}{dx} + (\alpha(x) + 2x^2)v = -1 \qquad (A.2)$$

The IF is

$$\rho(x) = \exp(\int (\alpha(x) + 2x^2)dx)$$

$$= \exp\left(\frac{2}{3}x^3 + \int_0^x \alpha(\tau)d\tau\right)$$

where we selected an arbitrary constant 0 as the lower bound for the integration of $\alpha(\tau)$.

Multiplying both sides by $\rho(x)$, we have

$$\frac{d}{dx}\left(\exp\left(\frac{2}{3}x^3 + \int_0^x \alpha(\tau)d\tau\right)v\right) = -\exp\left(\frac{2}{3}x^3 + \int_0^x \alpha(\tau)d\tau\right)$$

$$\exp\left(\frac{2}{3}x^3 + \int_0^x \alpha(\tau)d\tau\right)v = -\int \exp\left(\frac{2}{3}x^3 + \int_0^x \alpha(\tau)d\tau\right)dx$$

$$\frac{1}{v} = -\frac{\exp\left(\frac{2}{3}x^3 + \int_0^x \alpha(\tau)d\tau\right)}{\int_c^x \exp\left(\frac{2}{3}t^3 + \int_0^t \alpha(\tau)d\tau\right)dt}$$

Substituting back, we get the GS

$$y = x^2 + \frac{1}{v}$$

$$= x^2 - \frac{\exp\left(\frac{2}{3}x^3 + \int_0^x \alpha(\tau)d\tau\right)}{\int_c^x \exp\left(\frac{2}{3}t^3 + \int_0^t \alpha(\tau)d\tau\right)dt}$$

For a special case $\alpha(x) = -2x^2$, Eq. (A.2), after the Riccati substitution, becomes

$$\frac{dv}{dx} = -1$$

whose GS is

$$v = -x + c$$

and the GS for the original DE is

$$y = x^2 - \frac{1}{x - c}$$

If we use the above formula for GS for the special case $\alpha(x) = -2x^2$, we have

$$\exp\left(\frac{2}{3}x^3 + \int_0^x \alpha(\tau)d\tau\right) = \exp\left(\frac{2}{3}x^3 + \int_0^x (-2\tau^2)d\tau\right)$$

$$= \exp\left(\frac{2}{3}x^3 - \frac{2}{3}x^3\right)$$

$$= 1$$

$$\int_c^x \exp\left(\frac{2}{3}t^3 + \int_0^t \alpha(\tau)d\tau\right)dt = \int_c^x 1\,dt = x - c$$

Thus, the GS is

$$y = x^2 - \frac{1}{x - c}$$

Problem 1.6.9

The DE is Riccati with a known PS $y_1 = x^2$. Using the Riccati substitution

$$y = y_1 + \frac{1}{v} = x^2 + \frac{1}{v}$$

$$y' = 2x - \frac{v'}{v^2}$$

we convert the DE into

$$x^3 \left(2x - \frac{v'}{v^2} \right) + x^2 \left(x^2 + \frac{1}{v} \right) - \left(x^2 + \frac{1}{v} \right)^2 = 2x^4$$

$$2x^4 - \frac{x^3}{v^2} v' + x^4 + \frac{x^2}{v} - x^4 - \frac{2x^2}{v} - \frac{1}{v^2} = 2x^4$$

$$-\frac{x^3}{v^2} v' - \frac{x^2}{v} = \frac{1}{v^2}$$

$$v' + \frac{v}{x} = -\frac{1}{x^3}$$

Multiplying both sides by the IF

$$\rho(x) = \exp\left(\int \frac{1}{x} dx \right) = x$$

gives

$$xv' + v = -\frac{1}{x^2}$$

$$\int \frac{d}{dx}(xv) dx = -\int \frac{1}{x^2} dx$$

$$xv = \frac{1}{x} + C$$

$$v = \frac{1}{x^2} + \frac{C}{x} = \frac{1 + Cx}{x^2}$$

Back substituting $v = (y - x^2)^{-1}$, we get

$$\frac{1}{y - x^2} = \frac{1 + Cx}{x^2}$$

$$y = x^2 + \frac{x^2}{1 + Cx}$$

1.7 The Exact DEs

Problem 1.7.1

From the DE, we have

$$M(x, y) = 1 + \ln(xy), \quad N(x, y) = \frac{x}{y}$$

We found $M_y(x, y) = \frac{1}{y}$ and $N_x(x, y) = \frac{1}{y}$. Thus, $M_y = N_x$, indicating the DE is exact. Now, the DE can be solved by the exact DE method.

We know

$$M(x, y) = \frac{\partial F(x, y)}{\partial x} = 1 + \ln(xy),$$

$$N(x, y) = \frac{\partial F(x, y)}{\partial y} = \frac{x}{y}$$

Thus,

$$F(x, y) = \int N(x, y) dy$$

$$= \int \frac{x}{y} dy$$

$$= x \ln y + g(x)$$

and

$$\frac{\partial F}{\partial x} = M(x, y)$$

$$\ln y + g'(x) = 1 + \ln(xy)$$

$$g'(x) = 1 + \ln x$$

$$g(x) = x + x \ln x - x + c_2$$

$$= x \ln x + c_2$$

Thus,

$$F(x, y) = x \ln y + x \ln x + c_2$$

Finally, the GS to the DE is

$$x \ln y + x \ln x = C$$

Problem 1.7.2

(1) Multiplying the IF on both sides of the original DE

$$\exp\left(\int P(x)dx\right) A(x, y)dx + \exp\left(\int P(x)dx\right) B(x, y)dy = 0$$

$$M(x, y) = \exp\left(\int P(x)dx\right) A(x, y)$$

$$N(x, y) = \exp\left(\int P(x)dx\right) B(x, y)$$

$$\frac{\partial M}{\partial y} - \frac{\partial B}{\partial x} = \frac{\partial A}{\partial y} \exp\left(\int P dx\right) - \frac{\partial B}{\partial y} \exp\left(\int P dx\right) - BP \exp\left(\int P dx\right)$$

$$= \left(\frac{\partial A}{\partial y} - \frac{\partial B}{\partial y} - BP\right) \exp\left(\int P dx\right)$$

$$= 0$$

The new DE is exact.

(2) The IF is

$$\rho(x) = \exp\left(\int P(x) dx\right) = \exp\left(\int \left(-\frac{3}{x}\right) dx\right) = x^{-3}$$

where we used

$$P(x) = \frac{\left(\dfrac{\partial A(x,y)}{\partial y} - \dfrac{\partial B(x,y)}{\partial x}\right)}{B(x,y)}$$

$$= \frac{-2y - y}{xy}$$

$$= -\frac{3}{x}$$

The new exact DE is

$$x^{-3}(2x - y^2)dx + x^{-3}xy dy = 0$$

Thus,

$$M(x,y) = 2x^{-2} - x^{-3}y^2$$
$$N(x,y) = x^{-2}y$$

Thus,

$$F(x,y) = \int M dx + g(y)$$

$$= -2x^{-1} + \frac{1}{2}x^{-2}y^2 + g(y)$$

With $\partial F/\partial y = N$, we have

$$\frac{\partial F}{\partial y} = x^{-2}y + g'(y) = N = x^{-2}y$$

$$g'(y) = 0$$
$$g(y) = C$$

The GS to the DE is

$$2x^{-1} - \frac{1}{2}x^{-2}y^2 = C_1$$

i.e.,

$$y^2 = C_2 x^2 + 4x$$

Problem 1.7.3

(1) $\qquad \alpha(x)dy + \big(\beta(x)y + \gamma(x)\big)dx = 0$

$$M = \beta(x)y + \gamma(x), \quad N = \alpha(x)$$

If $M_y = N_x$, i.e., the DE is exact, the solution is

$$F(x,y) = \int M dx + \int N dy - \int \left(\frac{d}{dy} \left(\int M dx \right) \right) dy = C$$

If $M_y \neq N_x$, check if

$$\frac{M_y - N_x}{N} = f(x)$$

is a function of purely x, or

$$\frac{M_y - N_x}{M} = g(y)$$

is a function of purely y.

Two IF's are possible

$$\rho(x) = \exp\left(\int f(x) dx \right)$$

$$\rho(y) = \exp\left(-\int g(y) dy \right)$$

and the easiest is usually chosen for further manipulation.

Let $\bar{M} = \rho(x)M$ and $\bar{N} = \rho(x)N$ and replace the M and N in the solution above with $\bar{M}$ and $\bar{N}$.

(2)

$$y' + \frac{\beta(x)}{\alpha(x)} y = -\frac{\gamma(x)}{\alpha(x)}$$

and let

$$P(x) = \frac{\beta(x)}{\alpha(x)}$$

$$Q(x) = -\frac{\gamma(x)}{\alpha(x)}$$

The GS is

$$y = \exp\left(-\int P(x) dx \right) \left(\int Q(x) \exp\left(\int P(x) dx \right) dx + C \right)$$

Problem 1.7.4

(1) Check if this DE is exact.

$$M(x,y) = (P(x)y - Q(x)), \quad N(x,y) = 1$$

$$P(x) = \frac{\partial M(x,y)}{\partial y} \neq \frac{\partial N(x,y)}{\partial x} = 0$$

So, this DE is not exact.

(2) If not, convert it to an exact DE.

For converting a given DE into Exact, we need to calculate two ratios.

$$f(x) = \frac{\dfrac{\partial M(x,y)}{\partial y} - \dfrac{\partial N(x,y)}{\partial x}}{N(x,y)} = P(x)$$

Because $f(x)$ is a function of x, we have the IF

$$\rho(x) = \exp\left(\int f(x)dx\right) = \exp\left(\int P(x)dx\right)$$

By multiplying the IF with the given DE, we obtain an exact DE.

$$\rho(x)\big(P(x)y - Q(x)\big)dx + \rho(x)dy = 0$$

$$\exp\left(\int P(x)dx\right)\big(P(x)y - Q(x)\big)dx + \exp\left(\int P(x)dx\right)dy = 0$$

So, the new $M(x,y)$ and $N(x,y)$ are

$$M(x,y) = \exp\left(\int P(x)dx\right)\big(P(x)y - Q(x)\big)$$

$$N(x,y) = \exp\left(\int P(x)dx\right)$$

Thus,

$$\frac{\partial M(x,y)}{\partial y} = \exp\left(\int P(x)dx\right)$$

$$\frac{\partial N(x,y)}{\partial x} = \exp\left(\int P(x)dx\right)$$

Thus,

$$\frac{\partial M(x,y)}{\partial y} = \frac{\partial N(x,y)}{\partial x}$$

(3) Solve the DE using the exact DE method and the solution may be expressed in terms of the functions $P(x)$ and $Q(x)$.

$$\frac{\partial F(x,y)}{\partial x} = M(x,y) = \exp\left(\int P(x)dx\right)\big(P(x)y - Q(x)\big)$$

$$\frac{\partial F(x,y)}{\partial y} = N(x,y) = \exp\left(\int P(x)dx\right)$$

Integrating the first DE, we get

$$F(x,y) = \int M(x,y)dx + g(y)$$

$$= y\int \exp\left(\int P(x)dx\right)P(x)dx - \int Q(x)\exp\left(\int P(x)dx\right)dx + g(y)$$

Integrating the second DE, we get

$$F(x,y) = \int N(x,y)dy + f(x)$$

$$= y\exp\left(\int P(x)dx\right) + f(x)$$

By comparing the two results, we get the GS

$$y \exp\left(\int P(x)dx\right) - \int Q(x) \exp\left(\int P(x)dx\right)\left(\int P(x)dx\right)dx = C$$

(4) If $P(x) = \frac{1}{x}$ and $Q(x) = \frac{\cos x}{x}$, the corresponding GS can be found via,

$$y \exp\left(\int P(x)dx\right) - \int Q(x) \exp\left(\int P(x)dx\right)dx = C$$

$$y \exp\left(\int \frac{1}{x}dx\right) - \int \frac{\cos x}{x} \exp\left(\int \frac{1}{x}dx\right)dx = C$$

Working out the integrations above, one gets the GS

$$xy - \sin x = C$$

Problem 1.7.5

First, let's check if the DE is exact. We have

$$M(x,y) = 2x - y^2$$
$$N(x,y) = xy$$

Thus, $M_y = -2y$ and $N_x = y$. Obviously, $M_y \neq N_x$, the DE is not exact. But, since

$$M_y - N_x = -3y$$

we have

$$\frac{M_y - N_x}{N} = -\frac{3}{x}$$

which is a function of a single-variable x. Let

$$I(x) = \exp\left(\int \left(-\frac{3}{x}\right)dx\right)$$
$$= x^{-3}$$

we get

$$\left(\frac{2x - y^2}{x^3}\right)dx + x^{-2}ydy = 0$$

which is exact.

$$F(x,y) = \int \frac{y}{x^2}dy$$
$$= \frac{y^2}{2x^2} + g(x)$$

and

$$\frac{\partial F}{\partial x} = M$$

yields

$$\frac{2x - y^2}{x^3} = -y^2x^{-3} + g'(x)$$
$$g(x) = -\frac{2}{x}$$

Therefore, the GS to the DE is

$$\frac{y^2}{2x^2} - \frac{2}{x} = C$$

Problem 1.7.6

Method 1: The S-method.

The DE can be written as

$$y' = -\frac{3x}{4y} - \frac{y}{2x}$$

Let $u = \frac{y}{x}$. Then, $y' = xu' + u$.

The DE becomes

$$xu' + u = -\frac{3}{4u} - \frac{u}{2}$$

$$\frac{4uu'}{2u^2 + 1} = -\frac{3}{x}$$

$$\ln(2u^2 + 1) = -3\ln x + C_1$$

$$2u^2 + 1 = Cx^{-3}$$

Back substituting $u = \frac{y}{x}$ back, we have

$$2\left(\frac{y}{x}\right)^2 + 1 = Cx^{-3}$$

$$2y^2x + x^3 = C$$

Method 2: The exact DE method.

The DE can be written as

$$(3x^2 + 2y^2)dx + 4xydy = 0$$

Thus,

$$M(x, y) = 3x^2 + 2y^2$$

$$N(x, y) = 4xy$$

Since $M_y = N_x = 4y$, this DE is exact.

Thus,

$$F(x, y) = \int N dy$$

$$= 2xy^2 + g(x)$$

On the other hand,

$$\frac{\partial F}{\partial x} = M$$

We have

$$2y^2 + g'(x) = 3x^2 + 2y^2$$

$$g'(x) = 3x^2$$

$$g(x) = x^3$$

Thus, the GS to the DE is

$$2xy^2 + x^3 = C$$

Problem 1.7.7

Method 1: The 1st.O linear DE method.

$$\rho(x) = \exp\left(\int \left(-\frac{x}{x^2+1}\right) dx\right)$$

$$= \exp\left(-\frac{1}{2}\int \frac{2x}{x^2+1} dx\right)$$

$$= \exp\left(-\frac{1}{2}\ln(x^2+1)\right)$$

$$= (x^2+1)^{-\frac{1}{2}}$$

$$y = (x^2+1)^{\frac{1}{2}}\left(\int 2x(x^2+1)(x^2+1)^{-\frac{1}{2}} dx + C\right)$$

$$= (x^2+1)^{\frac{1}{2}}\left(\int 2x(x^2+1)^{\frac{1}{2}} dx + C\right)$$

$$= (x^2+1)^{\frac{1}{2}}\left(\frac{2}{3}(x^2+1)^{\frac{3}{2}} + C\right)$$

$$= \frac{2}{3}(x^2+1)^2 + C(x^2+1)^{\frac{1}{2}}$$

Method 2: The exact DE method.

$$(x^2+1)y' - xy = 2x(x^2+1)^2$$

$$(2x(x^2+1)^2 + xy)dx - (x^2+1)dy = 0$$

$$\frac{\partial M}{\partial y} = x, \qquad \frac{\partial N}{\partial x} = -2x$$

which is not exact. We can compute

$$\frac{M_y - N_x}{N} = -\frac{3x}{x^2+1}$$

$$\rho(x) = \exp\left(\int -\frac{3x}{x^2+1} dx\right)$$

$$= \exp\left(\ln(x^2+1)^{-\frac{3}{2}}\right)$$

$$= (x^2+1)^{-\frac{3}{2}}$$

The new DE is exact.

$$\left(2x(x^2+1)^{-\frac{1}{2}} + x(x^2+1)^{-\frac{3}{2}}y\right) dx - (x^2+1)^{-\frac{1}{2}} dy = 0$$

Thus,

$$F(x,y) = \int N dy = -(x^2+1)^{-\frac{1}{2}}y + g(x)$$

On the other hand,

$$\frac{\partial F}{\partial x} = M$$

$$x(x^2 + 1)^{-\frac{3}{2}}y + g'(x) = 2x(x^2 + 1)^{-\frac{1}{2}} + x(x^2 + 1)^{-\frac{3}{2}}y$$

$$g'(x) = 2x(x^2 + 1)^{\frac{1}{2}}$$

$$g(x) = \frac{2}{3}(x^2 + 1)^{\frac{3}{2}}$$

Therefore, the GS is

$$-(x^2 + 1)^{-\frac{1}{2}}y + \frac{2}{3}(x^2 + 1)^{\frac{3}{2}} = C$$

i.e.,

$$y = \frac{2}{3}(x^2 + 1)^2 + C(x^2 + 1)^{\frac{1}{2}}$$

Problem 1.7.8

Method 1: The exact DE method.

We can write the DE as

$$(x + 3y)dx + (3x - y)dy = 0$$

where $M = x + 3y$ and $N = 3x - y$. Thus, we have

$$M_y = N_x = 3$$

Thus, the DE is exact. Now, we get

$$F(x, y) = \int M(x, y)dx$$

$$= \frac{1}{2}x^2 + 3xy + g(y)$$

Applying it to $N = \frac{\partial F}{\partial y}$, we have

$$3x + g'(y) = 3x - y$$

$$g'(y) = -y$$

$$g(y) = -\frac{1}{2}y^2$$

Thus, GS is

$$\frac{1}{2}x^2 + 3xy - \frac{1}{2}y^2 = C$$

Method 2: The S-method.

We can write the DE as

$$y' = \frac{1 + 3\frac{y}{x}}{\frac{y}{x} - 3}$$

Let $v = \frac{y}{x}$. Then, $y = vx$ and $y' = v + xv'$. This gives

$$v + xv' = \frac{1 + 3v}{v - 3}$$

$$\frac{v-3}{1+6v-v^2}v' = \frac{1}{x}$$

$$-\frac{1}{2}\ln(1+6v-v^2) = \ln x + C_1$$

$$1+6v-v^2 = Cx^{-2}$$

Back substituting $v = y/x$, we have

$$1+6\frac{y}{x}-\left(\frac{y}{x}\right)^2 = Cx^{-2}$$

$$x^2 + 6xy - y^2 = C$$

Problem 1.7.9

We have

$$M(x,y) = y$$
$$N(x,y) = 2x + y^4$$

By checking

$$M_y = 1, \qquad N_x = 2$$

we know the DE is not exact. Since

$$M_y - N_x = -1$$

we can find

$$f(y) = \frac{M_y - N_x}{M} = -\frac{1}{y}$$

which is a single-variable function of y. Let

$$I(y) = \exp\left(-\int\left(-\frac{1}{y}\right)dy\right) = y$$

Now, we have an exact DE

$$y^2 dx + (2xy + y^5)dy = 0$$

To solve this, we have

$$F(x,y) = \int y^2 dx$$

$$= y^2 x + g(y)$$

Since

$$\frac{\partial F}{\partial y} = N$$

we get

$$2xy + g'(y) = 2xy + y^5$$

$$g(y) = \frac{1}{6}y^6$$

Thus, the GS to the DE is

$$y^2 x + \frac{y^6}{6} = C$$

Problem 1.7.10

Rearranging the original DE yields
$$(2x - y^2)dx + \left(-(2xy + 1)\right)dy = 0$$
Then, the two two-variable functions are:
$$M(x,y) = (2x - y^2)$$
$$N(x,y) = -2xy - 1$$
The difference of their partial derivatives is
$$M_y - N_x = (-2y) - (-2y) = 0$$
Thus, the DE is exact.

Expanding the original DE and using $d(uv) = u\,dv + v\,du$, we get,
$$2x\,dx - y^2\,dx - 2xy\,dy - dy = 0$$
One more step,
$$dx^2 - y^2\,dx - x\,dy^2 - dy = 0$$
i.e.,
$$d(x^2 - xy^2 - y) = 0$$
i.e.,
$$x^2 - xy^2 - y = C$$

Problem 1.7.11

Multiplying the DE by dx yields
$$y^3\,dx + (xy^2 - 1)dy = 0$$
The DE can be made exact by multiplying both sides by the IF
$$\rho(y) = \exp\left(-\int \frac{3y^2 - y^2}{y^3}\,dy\right) = y^{-2}.$$
$$y\,dx + (x - y^{-2})dy = 0$$
$$M(x,y) = y, \qquad N(x,y) = x - y^{-2}$$
$$F(x,y) = \int y\,dx = xy + g(y)$$
$$\frac{\partial F}{\partial y} = x + g'(y) = x - y^{-2}$$
$$g'(y) = -y^{-2} \Rightarrow g(y) = y^{-1}$$
The GS is
$$F(x,y) = xy + \frac{1}{y} = C$$

Problem 1.7.12

$$M(x,y) = \cos x + \ln y$$

$$N(x,y) = \frac{x}{y} + \exp(y)$$

Thus,

$$\frac{\partial M}{\partial y} = \frac{1}{y}, \qquad \frac{\partial N}{\partial x} = \frac{1}{y}$$

The DE is exact because $\frac{\partial M}{\partial y} - \frac{\partial N}{\partial x} = 0$.

$$F(x,y) = \int M(x,y)dx + g(y)$$

$$= \int (\cos x + \ln y)dx + g(y)$$

$$= \sin x + x \ln y + g(y)$$

$$\frac{\partial F}{\partial y} = \frac{x}{y} + g' = \frac{x}{y} + \exp(y)$$

$$g' = \exp(y)$$

$$g(y) = \exp(y)$$

Thus, the solution is $F(x,y) = \sin x + x \ln y + \exp(y) = C$.

Problem 1.7.13

Method 1: The exact DE method.

$$(\exp(y) + y \cos x)dx + (x \exp(y) + \sin x)dy = 0$$

$$M = \exp(y) + y \cos x$$

$$N = x \exp(y) + \sin x$$

$$M_y = \exp(y) + \cos x$$

$$N_x = \exp(y) + \cos x$$

$$M = \frac{\partial F}{\partial x} = \exp(y) + y \cos x$$

$$F(x,y) = \int M\, dx = x \exp(y) + y \sin x + A(y)$$

$$\frac{\partial F}{\partial y} = x \exp(y) + \sin x + A'(y) = N = x \exp(y) + \sin x$$

$$x \exp(y) + \sin x + A'(y) = x \exp(y) + \sin x$$

$$A'(y) = 0$$

$$A(y) = C$$

$$F(x,y) = x \exp(y) + y \sin x + C = C_2$$

$$x \exp(y) + y \sin x = C_1$$

Method 2: The S-method.

$$\exp(y) + y \cos x + (x \exp(y) + \sin x)y' = 0$$

$$v = x \exp(y) + y \sin x$$

$$v' = \exp(y) + x \exp(y) y' + y' \sin x + y \cos x$$

$$= (x \exp(y) + \sin x)y' + \exp(y) + y \cos x$$

$$v' = 0$$
$$v = C$$

Thus, $x \exp(y) + y \sin x = C$

Chapter 2 Mathematical Models

2.1 Newton's Law of Cooling

Problem 2.1.1

Newton's law of cooling for constant ambient temperature A,

$$\begin{cases} \dfrac{dT(t)}{dt} = k(A - T) \\ T(t = 0) = T_0 \end{cases}$$

gives us the following PS:

$$T(t) = A + (T_0 - A) \exp(-kt)$$

or

$$t = \frac{1}{k} \ln \frac{T_0 - A}{T - A}$$

We assumed the person died at time $t = 0$ with a body temperature $T_0 = 99$ and the ambient temperature $A = 60$. For two later moments:

(1) t_1 minutes after death, the temperature is $T(t_1) = 85$
$$t_1 = \frac{1}{k} \ln \frac{99 - 60}{85 - 60} = \frac{1}{k} \ln \frac{39}{25}$$

(2) $t_1 + 90$ minutes after death, the temperature is $T(t_1 + 90) = 79$:
$$t_1 + 90 = \frac{1}{k} \ln \frac{99 - 60}{79 - 60} = \frac{1}{k} \ln \frac{39}{19}$$
Dividing the above two equations, we get
$$\frac{t_1 + 90}{t_1} = \frac{\ln \frac{39}{19}}{\ln \frac{39}{25}} = \frac{0.719123}{0.444686} = 1.61715$$

Thus, $t_1 = 146$. The person died 146th minute before noon, *i.e.*, at 9:34 am.

2.2 Torricelli's Law for Draining

Problem 2.2.1

(1) Establish the DE for the draining process:

$$\begin{cases} \dfrac{A(y)dy}{dt} = -k\sqrt{y} \\ y(t=0) = y_0 \end{cases}$$

$$A(y) = \pi r^2 = \pi(R^2 - (R-y)^2) = \pi y(2R - y)$$

$$\dfrac{A(y)dy}{dt} = \dfrac{\pi y(2R-y)dy}{dt}$$

$$= -k\sqrt{y}$$

$$\pi\sqrt{y}(2R-y)dy = -kdt$$

$$\pi \int_{2R}^{0} \sqrt{y}(2R-y)dy = -\int_{0}^{t} kdt$$

$$\pi \left(\dfrac{4}{3}Ry^{\frac{3}{2}} - \dfrac{2}{5}y^{\frac{5}{2}}\right)\Big|_{2R}^{0} = -kt$$

$$\dfrac{4}{3}\pi R\left(0 - (2R)^{\frac{3}{2}}\right) - \dfrac{2}{5}\pi\left(0 - (2R)^{\frac{5}{2}}\right) = -kt$$

Thus, the time taken to drain a full tank is

$$T_1 = \dfrac{16\sqrt{2}}{15}\dfrac{\pi}{k}R^{\frac{5}{2}}$$

(2) If we double the radius of the container while keeping other parameters unchanged, the time to drain is

$$T_2 = \dfrac{16\sqrt{2}}{15}\dfrac{\pi}{k}(2R)^{\frac{5}{2}}$$

$$= \dfrac{128}{15}\dfrac{\pi}{k}(R)^{\frac{5}{2}}$$

Alternatively, one may also start from scratch to compute,

$$A(y) = \pi(4R^2 - (2R-y)^2) = \pi y(4R - y)$$

$$\dfrac{A(y)dy}{dt} = \dfrac{\pi y(4R-y)dy}{dt} = -k\sqrt{y}$$

$$\pi \int_{4R}^{0} \sqrt{y}(4R-y)dy = -\int_{0}^{T} kdt$$

$$\dfrac{8}{3}\pi R\left(0 - (4R)^{\frac{3}{2}}\right) - \dfrac{2}{5}\pi\left(0 - (4R)^{\frac{5}{2}}\right) = -kT$$

$$T_2 = \dfrac{128}{15}\dfrac{\pi}{k}R^{\frac{5}{2}}$$

Thus,

$$\dfrac{T_2}{T_1} = 2^{\frac{5}{2}} = 4\sqrt{2} \approx 5.657$$

The ratio of the total amount of liquid, measured in volume, is

$$\frac{V_2}{V_1} = 2^3 = 8$$

Notice, the time ratio (5.657) is less than the volume ratio (8).

Problem 2.2.2

For draining the upper half:

$$\frac{dV}{dt} = -k\sqrt{y} = \frac{A(y)dy}{dt}$$

$$r^2 + (y - R)^2 = R^2$$

$$A(y) = \pi r^2 = \pi(R^2 - (y - R)^2) = \pi y(2R - y)$$

Thus,

$$\frac{\pi y(2R - y)dy}{dt} = -k\sqrt{y}$$

$$\pi\sqrt{y}(2R - y)dy = -kdt$$

$$\pi \int_{2R}^{R} \left(2Ry^{\frac{1}{2}} - y^{\frac{3}{2}}\right)dy = -k\int_0^{T_1} dt$$

$$\pi\left(\frac{2Ry^{\frac{3}{2}}}{\frac{3}{2}} - \frac{y^{\frac{5}{2}}}{\frac{5}{2}}\right)\Bigg|_{2R}^{R} = -kT_1$$

$$-\frac{\pi}{k}\left(\frac{4}{3}Ry^{\frac{3}{2}} - \frac{2}{5}y^{\frac{5}{2}}\right)\Bigg|_{2R}^{R} = T_1$$

$$-\frac{\pi}{k}\left(\frac{4}{3}R\left(R^{\frac{3}{2}}\right) - \frac{2}{5}R^{\frac{5}{2}} - \left(\frac{4}{3}R\,(2R)^{\frac{3}{2}} - \frac{2}{5}(2R)^{\frac{5}{2}}\right)\right) = T_1$$

$$T_1 = \frac{16\sqrt{2} - 14}{15}\left(\frac{\pi}{k}R^{\frac{5}{2}}\right)$$

For draining the lower half:

$$r^2 + (R - y)^2 = R^2$$

$$A(y) = \pi r^2 = \pi(R^2 - (R - y)^2) = \pi y(2R - y)$$

$$\pi\sqrt{y}(2R - y)dy = -kdt$$

$$\pi \int_{R}^{0} \left(2Ry^{\frac{1}{2}} - y^{\frac{3}{2}}\right)dy = -k\int_0^{T_2} dt$$

$$\pi\left(\frac{2Ry^{\frac{3}{2}}}{\frac{3}{2}} - \frac{y^{\frac{5}{2}}}{\frac{5}{2}}\right)\Bigg|_{R}^{0} = -kT_2$$

$$-\frac{\pi}{k}\left(\frac{4}{3}Ry^{\frac{3}{2}} - \frac{2}{5}y^{\frac{5}{2}}\right)\Bigg|_{R}^{0} = T_2$$

$$-\frac{\pi}{k}\left(-\left(\frac{4}{3}R\left(R^{\frac{3}{2}}\right) - \frac{2}{5}R^{\frac{5}{2}}\right)\right) = T_2$$

$$\frac{\pi}{k}\left(\frac{4}{3}R^{\frac{5}{2}} - \frac{2}{5}R^{\frac{5}{2}}\right) = T_2$$

$$T_2 = \frac{14}{15}\left(\frac{\pi}{k}R^{\frac{5}{2}}\right)$$

Therefore,

$$\frac{T_1}{T_2} = \frac{16\sqrt{2} - 14}{14} \approx 0.616 < 1$$

Thus,

$$T_2 > T_1$$

Thus, it takes 38.4% more time to drain the lower half than the upper half regardless of the container size and the draining constant.

Problem 2.2.3

Let y be the vertical liquid surface height from the leaking hole. The liquid surface disk radius is r and the cup bottom radius is αR.

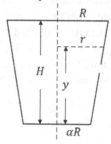

Thus, we have

$$\frac{R - \alpha R}{H} = \frac{r - \alpha R}{y}$$

Thus,

$$r = R\left(\alpha + (1 - \alpha)\frac{y}{H}\right)$$

We can make a few remarks for the above formula: (1) if $\alpha = 1, r = R$ and it's a cylinder; (2) $\forall\, \alpha < 1$ and for $y = H$ or $r = R$, at the top of the cup; (3) $\forall\, \alpha < 1$ and for $y = 0$ or $r = \alpha R$, at the bottom of the cup; additionally, $r \to 0$ if $\alpha \to 0$, a case of a cone.

Applying the law for draining, we have

$$\pi R^2 \left(\alpha + (1 - \alpha)\frac{y}{H}\right)^2 \frac{dy}{dt} = -ky^{\frac{1}{2}}$$

The time to leak from the top $(y = H)$ to height $(y = h)$ is $T(h)$

$$\int_H^h y^{-\frac{1}{2}}\left(\alpha + (1 - \alpha)\frac{y}{H}\right)^2 dy = \int_0^{T(h)}\left(-\frac{k}{\pi R^2}\right)dt$$

Thus, we get

$$T(h) = \frac{\pi R^2}{k} \int_h^H \frac{\left(\alpha + (1-\alpha)\frac{y}{H}\right)^2 dy}{\sqrt{y}}$$

$$= \frac{\pi R^2}{k} \int_h^H \left(\alpha^2 y^{-\frac{1}{2}} + \frac{2\alpha(1-\alpha)}{H} y^{\frac{1}{2}} + \frac{(1-\alpha)^2}{H^2} y^{\frac{3}{2}}\right) dy$$

$$= \frac{\pi R^2}{k} \left(2\alpha^2 y^{\frac{1}{2}} + \frac{4\alpha(1-\alpha)}{3H} y^{\frac{3}{2}} + \frac{2(1-\alpha)^2}{5H^2} y^{\frac{5}{2}}\right)\Big|_h^H$$

$$= \frac{\pi R^2}{k} \left(\left(2\alpha^2 + \frac{4\alpha(1-\alpha)}{3} + \frac{2(1-\alpha)^2}{5}\right) H^{\frac{1}{2}}\right.$$
$$\left. - \left(2\alpha^2 h^{\frac{1}{2}} + \frac{4\alpha(1-\alpha)}{3H} h^{\frac{3}{2}} + \frac{2(1-\alpha)^2}{5H^2} h^{\frac{5}{2}}\right)\right)$$

The above is the general formula for the time it takes for a cup to leak from its rim to any height h. Let's examine a few special cases:

(1) If $h = H$, leaking just got started and the time is
$$T(h = H) = 0$$

(2) If $h = 0$, the liquid surface has reached the bottom of the cup (the leaking hole is now set at the bottom instead of the mid-height) and the time is
$$T(h = 0) = \frac{\pi R^2}{k} \left(2\alpha^2 + \frac{4\alpha(1-\alpha)}{3} + \frac{2(1-\alpha)^2}{5}\right) H^{\frac{1}{2}}$$

(3) If $h = \frac{H}{2}$, the liquid surface has reached the mid-height of the cup and the time is
$$T(h = \frac{H}{2}) = \frac{\pi R^2}{k} H^{\frac{1}{2}} \left(\left(2\alpha^2 + \frac{4}{3}\alpha(1-\alpha) + \frac{2}{5}(1-\alpha)^2\right)\right.$$
$$\left. - \left(2\alpha^2 + \frac{2}{3}\alpha(1-\alpha) + \frac{1}{10}(1-\alpha)^2\right)\frac{1}{\sqrt{2}}\right)$$
$$= \frac{\pi R^2}{k} H^{\frac{1}{2}} \left(\frac{2}{15}(8\alpha^2 + 4\alpha + 3) - \frac{1}{30\sqrt{2}}(43\alpha^2 + 14\alpha + 3)\right)$$

For our problem, $\alpha = 0.9$, the leaking time is
$$T_1 = \frac{\pi R^2}{k} H^{\frac{1}{2}} \left(\frac{2}{15}\left(8\left(\frac{9}{10}\right)^2 + 4\left(\frac{9}{10}\right) + 3\right)\right.$$
$$\left. - \frac{1}{30\sqrt{2}}\left(43\left(\frac{9}{10}\right)^2 + 14\left(\frac{9}{10}\right) + 3\right)\right)$$

$$= \frac{\pi R^2}{k} v \left(\frac{1744 - \frac{1841}{\sqrt{2}}}{1000} \right)$$

$$\approx (0.4422) \frac{\pi R^2}{k} H^{\frac{1}{2}}$$

Next, we turn the cup upside down,

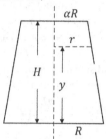

Thus, we have

$$\frac{R - \alpha R}{H} = \frac{R - r}{y}$$

Thus,

$$r = R \left(1 - (1 - \alpha) \frac{y}{H} \right)$$

which makes good sense because, just as the previous case, (1) if $\alpha = 1$, $r = R$ and it's a cylinder; (2) $\forall \, \alpha < 1$, if $y = H, r = \alpha R$, at the top of the cup; (3) $\forall \, \alpha < 1$, if $y = 0, r = R$, at the bottom of the cup.

Similarly, applying the law for draining, we have

$$\pi R^2 \left(1 - (1 - \alpha) \frac{y}{H} \right)^2 \frac{dy}{dt} = -k y^{\frac{1}{2}}$$

$$\int_H^h \frac{\left(1 - (1 - \alpha) \frac{y}{H} \right)^2}{\sqrt{y}} \, dy = \int_0^{T(h)} \left(-\frac{k}{\pi R^2} \right) dt$$

Thus, we get

$$T(h) = \frac{\pi R^2}{k} \int_h^H \frac{\left(1 - (1 - \alpha) \frac{y}{H} \right)^2 dy}{\sqrt{y}}$$

$$= \frac{\pi R^2}{k} \int_h^H \left(y^{-\frac{1}{2}} - \frac{2(1 - \alpha)}{H} y^{\frac{1}{2}} + \frac{(1 - \alpha)^2}{H^2} y^{\frac{3}{2}} \right) dx$$

$$= \frac{\pi R^2}{k} \left(2y^{\frac{1}{2}} - \frac{4(1 - \alpha)}{3H} y^{\frac{3}{2}} + \frac{2(1 - \alpha)^2}{5H^2} y^{\frac{5}{2}} \right) \Bigg|_h^H$$

$$= \frac{\pi R^2}{k} \left(\left(2 \left(H^{\frac{1}{2}} - h^{\frac{1}{2}} \right) - \frac{4(1-\alpha)}{3H} \left(H^{\frac{3}{2}} - h^{\frac{3}{2}} \right) \right. \right.$$

$$\left. \left. + \frac{2(1-\alpha)^2}{5H^2} \left(H^{\frac{5}{2}} - h^{\frac{5}{2}} \right) \right) \right)$$

The above is the general formula, for the second case, for the time the cup leaks from its rim to any height h. Let's examine a few special cases:

(1) If $h = H$, leaking just got started and the time is
$$T(h = H) = 0$$

(2) If $h = 0$, liquid surface reached the bottom of the cup (the leaking hole is at the bottom instead of the mid-height) and the time is
$$T(h = 0) = \frac{\pi R^2}{k} \left(2 - \frac{4(1-\alpha)}{3} + \frac{2(1-\alpha)^2}{5} \right) H^{\frac{1}{2}}$$

(3) If $h = \frac{H}{2}$, liquid surface reached the mid-height of the cup and the time is

$$T \left(h = \frac{H}{2} \right) = \frac{\pi R^2}{k} \left(\left(2 - \frac{4(1-\alpha)}{3} + \frac{2(1-\alpha)^2}{5} \right) H^{\frac{1}{2}} \right.$$

$$\left. - \left(2h^{\frac{1}{2}} - \frac{4(1-\alpha)}{3H} h^{\frac{3}{2}} + \frac{2(1-\alpha)^2}{5H^2} h^{\frac{5}{2}} \right) \right)$$

$$= \frac{\pi R^2}{k} H^{\frac{1}{2}} \left(\left(2 \left(1 - 2^{-\frac{1}{2}} \right) - \frac{4(1-\alpha)}{3} \left(1 - 2^{-\frac{3}{2}} \right) \right. \right.$$

$$\left. \left. + \frac{2(1-\alpha)^2}{5} \left(1 - 2^{-\frac{5}{2}} \right) \right) \right)$$

Applying $\alpha = 0.9$, we get
$$T_2 = \frac{\pi R^2}{k} H^{\frac{1}{2}} \left(\left(2 \left(1 - 2^{-\frac{1}{2}} \right) - \frac{0.4}{3} \left(1 - 2^{-\frac{3}{2}} \right) + \frac{0.02}{5} \left(1 - 2^{-\frac{5}{2}} \right) \right) \right)$$

$$\approx (0.5029) \frac{\pi R^2}{k} H^{\frac{1}{2}}$$

It turns out the second case takes slightly longer to leak.
$$\frac{T_2}{T_1} = \frac{0.5029}{0.4422} = 1.137$$

Thus, the second case take 13.7% longer to leak. Go figure it out!

Problem 2.2.4

Referring to Problem 2.2.2, we have the time for draining the upper half with draining constant k

$$t_1 = \frac{16\sqrt{2} - 14}{15}\left(\frac{\pi}{k}R^{\frac{5}{2}}\right)$$

and the time for draining the lower half with draining constant k_1

$$t_2 = \frac{14}{15}\left(\frac{\pi}{k_1}R^{\frac{5}{2}}\right)$$

To make $t_1 = t_2$, we have

$$\frac{16\sqrt{2} - 14}{15}\frac{\pi}{k}R^{\frac{5}{2}} = \frac{14}{15}\frac{\pi}{k_1}R^{\frac{5}{2}}$$

$$\frac{16\sqrt{2} - 14}{k} = \frac{14}{k_1}$$

$$k_1 = \frac{14}{16\sqrt{2} - 14}k = \frac{7}{8\sqrt{2} - 7}k \approx 1.623k$$

Thus, if the draining constant is 6.23% bigger while draining the lower half, the time will be the same.

Problem 2.2.5

Torricelli's law tells us that

$$\frac{dV}{dt} = -k\sqrt{y} = \frac{A(y)dy}{dt}$$

$$A(y) = \pi r^2 = \pi(R^2 - (R - y)^2) = \pi y(2R - y)$$

$$\frac{\pi y(2R - y)dy}{dt} = -k\sqrt{y}$$

$$\pi\sqrt{y}(2R - y)dy = -kdt$$

$$-\frac{\pi}{k}\int_{2R}^{y_m}\left(2Ry^{\frac{1}{2}} - y^{\frac{3}{2}}\right)dy = \int_0^{t_m}dt = t_m$$

We seek the y_m such that the time it takes to drain the container from height $2R$ to height y_m is equal to the time it takes to drain the container from height y_m to height 0, *i.e.*,

$$-\frac{\pi}{k}\int_{2R}^{y_m}\left(2Ry^{\frac{1}{2}} - y^{\frac{3}{2}}\right)dy = -\frac{\pi}{k}\int_{y_m}^0\left(2Ry^{\frac{1}{2}} - y^{\frac{3}{2}}\right)dy = \frac{T}{2}$$

where T is the time to drain the container fully. Thus, we may find T by

$$T = \int_0^T dt$$

$$= -\frac{\pi}{k}\int_{2R}^0\left(2Ry^{\frac{1}{2}} - y^{\frac{3}{2}}\right)dy = -\frac{\pi}{k}\left(\frac{4}{3}Ry^{\frac{3}{2}} - \frac{2}{5}y^{\frac{5}{2}}\right)\Big|_{2R}^0$$

$$= \frac{16\sqrt{2}}{15} \frac{\pi}{k} R^{\frac{5}{2}}$$

We want y_m such that

$$-\frac{\pi}{k} \int_{y_m}^{0} \left(2Ry^{\frac{1}{2}} - y^{\frac{3}{2}} \right) dy = \frac{T}{2} = \frac{8\sqrt{2}}{15} \frac{\pi}{k} R^{\frac{5}{2}}$$

$$-\frac{\pi}{k} \left(\frac{4}{3} Ry^{\frac{3}{2}} - \frac{2}{5} y^{\frac{5}{2}} \right) \Big|_{y_m}^{0} = \frac{8\sqrt{2}}{15} \frac{\pi}{k} R^{\frac{5}{2}}$$

$$\frac{4}{3} Ry_m^{\frac{3}{2}} - \frac{2}{5} y_m^{\frac{5}{2}} = \frac{8\sqrt{2}}{15} R^{\frac{5}{2}}$$

Then, the volume of the leftover liquid is given by

$$\int_{0}^{y_m} A(y) dy = \int_{0}^{y_m} \pi(2Ry - y^2) dy$$

2.3 Population Model

Problem 2.3.1

From the population model, we have

$$\frac{dP}{dt} = (\beta - \delta)P$$
$$= (k_1 P - k_2 P)P$$
$$= (k_1 - k_2)P^2$$

where k_1 and k_2 are both constants. Let $k = k_1 - k_2 =$ another constant.

(1) From the problem, we know

$$\beta = \frac{k_1}{\sqrt{P}}, \qquad \delta = \frac{k_2}{\sqrt{P}}$$

Thus,

$$\frac{dP}{dt} = (\beta - \delta)P$$
$$= \left(\frac{k_1}{\sqrt{P}} - \frac{k_2}{\sqrt{P}}\right)P$$
$$= (k_1 - k_2)\sqrt{P}$$

Using $k = k_1 - k_2$ defined earlier yields

$$\frac{dP}{\sqrt{P}} = kdt$$
$$2\sqrt{P} = kt + C_1$$
$$P(t) = \left(\frac{1}{2}kt + C\right)^2$$

Plugging $P(0) = P_0$ into it, we get $C^2 = P_0$. Thus,

$$P(t) = \left(\frac{1}{2}kt + \sqrt{P_0}\right)^2$$

(2) Given $P_0 = 100$ and $P(t = 6) = 169$, we get

$$169 = \left(\frac{1}{2} \times 6k + \sqrt{100}\right)^2$$

This gives $k = 1$. Thus, it is straightforward to evaluate $P(12) = 16^2 = 256$.

Problem 2.3.2

(1) From the problem, we get

$$\beta = k_1 P, \qquad \delta = k_2 P$$

$$\frac{dP}{dt} = kP^2$$

Solving this separable DE, we get

$$\frac{1}{P} = -kt + C$$

$$P(t) = \frac{1}{C - kt}$$

Applying the IC $P(0) = P_0$, we get $C = \frac{1}{P_0}$ and the population $P(t) = \frac{1}{\frac{1}{P_0} - kt} = \frac{P_0}{1 - kP_0 t}$.

(2) When $t \to 1/(kP_0)$, or, $kP_0 t \to 1$, the population approaches infinity, i.e.,

$$\lim_{t \to 1/(kP_0)} P(t) = \lim_{kP_0 t \to 1} \frac{P_0}{1 - kP_0 t} \to \infty$$

(3) Given $P_0 = 6$ and $P(10) = 9$, we get

$$9 = \frac{6}{1 - k \cdot 6 \cdot 10}$$

$$k = \frac{1}{180}$$

Thus, the doomsday occurs at

$$t = \frac{1}{kP_0} = 30 \text{ (Months)}$$

(4) Suppose $\beta < \delta$, we know that $k < 0$. Thus, we know that

$$1 - kP_0 t > 0, \qquad t > 0$$

and

$$P(t) \to 0, \qquad t \to \infty$$

Problem 2.3.3

We can always write the logistic DE in the following form

$$\frac{dP}{dt} = kP(M - P)$$

where

$$b = k, \qquad M = \frac{a}{b}$$

Since, for $k = b > 0$, the limiting population will be M, we can easily calculate that

$$M = \frac{a}{b} = \frac{B_0/P_0}{D_0/P_0^2} = \frac{B_0 P_0}{D_0}$$

Problem 2.3.4

(1) This is a 1st.O separable DE.

$$\int \frac{dP}{P(\ln M - \ln P)} = \int k\,dt$$

Let $\ln P = Z$. Then, $P\,dZ = dP$.

Substituting P by Z and dZ, we have

$$\int \frac{P\,dZ}{P(\ln M - Z)} = \int \frac{dZ}{\ln M - Z} = \int k\,dt$$

Thus,

$$-\ln(\ln M - Z) = kt + C$$
$$\ln(\ln M - \ln P) = -kt - C$$
$$\ln M - \ln P = K\exp(-kt)$$
$$P(t) = \exp(\ln M - K\exp(-kt))$$

Applying the IC $P(t = 0) = P_0$, we get

$$K = \ln M - \ln P_0$$

Thus,

$$P(t) = \exp(\ln M - (\ln M - \ln P_0)\exp(-kt))$$

(2) If $P_0 > M$,

$$\ln M - \ln P_0 < 0$$

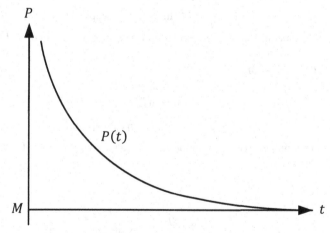

Figure A.1 If $P_0 > M$, the population will decrease and approach value M as $t \to \infty$.

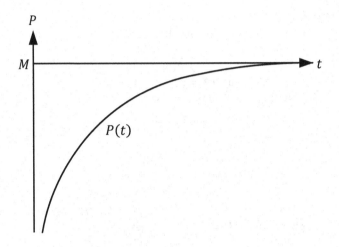

Figure A.2 If $P_0 < M$, the population will increase and approach value M as $t \to \infty$.

$$(\ln M - \ln P_0) \exp(-kt) < 0$$
$$\frac{dP}{dt} = kP(\ln M - \ln P_0) \exp(-kt) < 0$$

This means that $P(t)$ is a monotonically decreasing function of t. When $t \to \infty$, $P(t) \to M$.

(3) If $P_0 < M$,

$$\ln M - \ln P_0 > 0$$
$$(\ln M - \ln P_0) \exp(-kt) > 0$$
$$\frac{dP}{dt} = kP(\ln M - \ln P_0) \exp(-kt) > 0$$

This means $P(t)$ is a monotonically increasing function of t. When $t \to \infty$, $P(t) \to M$.

Problem 2.3.5

The DE is

$$\frac{dP}{dt} = \beta_0 \exp(-\alpha t) P$$

which is a separable DE. We have

$$\frac{dP}{P} = \beta_0 \exp(-\alpha t) \, dt$$
$$\ln P = -\frac{\beta_0}{\alpha} \exp(-\alpha t) + C$$

Since $P(0) = P_0$, we get

$$C = \ln P_0 + \frac{\beta_0}{\alpha}$$

Thus,

$$\ln P = \ln P_0 + \frac{\beta_0}{\alpha}(1 - \exp(-\alpha t))$$

$$P(t) = P_0 \exp\left(\frac{\beta_0}{\alpha}(1 - \exp(-\alpha t))\right)$$

Now, it is clear that for $t \to \infty$, $\exp(-\alpha t) \to 0$ and, thus,

$$\lim_{t \to \infty} P(t) = P_0 \exp\left(\frac{\beta_0}{\alpha}\right)$$

Problem 2.3.6

Since $k > 0$, $M > 0$ and $P > 0$, for the DE

$$\frac{dP}{dt} = kP(M - P)$$

we consider the following two cases: $M > P$ and $M < P$.

(1) If $M > P$, we have

$$\frac{dP}{dt} = kP(M - P) > 0$$

Thus, the function $P(t)$ is monotonically increasing wrt t.

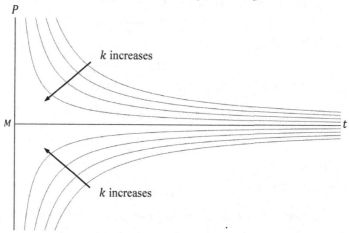

Figure A.3 Function $P(t)$ is close to value M as k increases.

Since dP/dt is the slope of the graph, as k increases, dP/dt increases and the curve approaches closer to M.

(2) If $M < P$,

$$\frac{dP}{dt} = kP(M - P) < 0$$

We know that $P(t)$ is a monotonically decreasing function in t. As k increases, the $|dP/dt|$ increases, the curve approaches M.

From above, we can conclude, as k increases, the curve gets closer to the limiting population M.

Problem 2.3.7

Birth rate is $\beta = 0$ and death rate is $\delta = k/\sqrt{p}$ where k is a constant.

By general population function, we have

$$\frac{dP}{dt} = (\beta - \delta)P$$

$$= \left(0 - \frac{k}{\sqrt{P}}\right)P$$

$$= -k\sqrt{P}$$

with IC $P(0) = P_0$. Solving this 1st.O separable DE, we have

$$\frac{dP}{\sqrt{P}} = -kdt$$

$$2\sqrt{P} = -kt + C$$

Applying the IC $P(0) = P_0$, we get

$$C = 2\sqrt{P_0}$$

Thus,

$$P(t) = \left(-\frac{kt}{2} + \sqrt{P_0}\right)^2$$

Using the IC $P(0) = P_0 = 900$ and $P(6) = 441$, we get

$$441 = \left(-\frac{6k}{2} + \sqrt{900}\right)^2$$

we found

$$k = 3$$

Therefore,

$$P(t) = \left(-\frac{3t}{2} + 30\right)^2$$

The condition that all fish are dead means $P(t) = 0$ for which we have

$$\left(-\frac{3t}{2} + 30\right)^2 = 0 \Rightarrow t = 20 \text{ (Weeks)}$$

Problem 2.3.8

From the IC, we have

$$a = \frac{B(0)}{P(0)} = \frac{8}{120}$$

$$b = \frac{D(0)}{P^2(0)} = \frac{6}{120^2}$$

To solve this DE, we should write the DE in the standard logistic DE form

$$\frac{dP}{dt} = kP(M - P)$$

where

$$k = b, \qquad M = \frac{a}{b}$$

Since we know the solution to the logistic DE

$$P(t) = \frac{MP_0}{P_0 + (M - P_0)\exp(-kMt)}$$

Thus, for $P(t) = 0.95M$, we can solve for t

$$0.95M = \frac{MP_0}{P_0 + (M - P_0)\exp(-kMt)}$$

gives $t = 27.69$.

Problem 2.3.9

With $\alpha = k_1 P, \beta = k_2 P$ and $k = k_2 - k_1$, the population model is given by

$$\frac{dP}{dt} = (\beta - \alpha)P = (k_2 - k_1)P^2 = kP^2$$

where we used $\beta - \alpha = kP > 0$.
Using SOV, we get

$$-\frac{1}{P} = kt + C$$

(1) Using the IC, we get the PS

$$P(t) = \left(\frac{1}{P_0} - kt\right)^{-1}$$

(2) Derive the doomsday formula.
When the denominator is zero, the doomsday arrives. Thus,

$$T = \frac{1}{P_0 k}$$

(3) Calculate the exact time of the doomsday.
First, calculate the rate difference birth and death.

$$P(12) = \left(\frac{1}{P_0} - (\beta - \alpha)t\right)^{-1}$$

$$= \left(\frac{1}{2011} - 12k\right)^{-1}$$

$$= 4027$$

Thus,

$$k \approx 2.0745 \cdot 10^{-5}$$

Second, calculate the time

$$T = \frac{1}{P_0 k} = \frac{1}{2011(2.0745 \cdot 10^{-5})} \approx 23.970$$

(4) Calculate the population.

Because $(\alpha - \beta) = -kP > 0$, we have $(-k) > 0$. Thus,

$$\lim_{t \to +\infty} P(t) = \lim_{t \to +\infty} \left(\frac{1}{P_0} - kt\right)^{-1} = 0$$

As time goes to infinity, the population vanishes.

Problem 2.3.10

The population modeling is given by

$$\frac{dP}{dt} = (\beta - \alpha)P = \left(0 - \frac{k}{\sqrt{P}}\right)P = -kP^{\frac{1}{2}}$$

whose GS is

$$2\sqrt{P} = -kt + C$$

Using the IC, we get

$$C = 2\sqrt{4000}$$
$$2\sqrt{P} = -kt + 2\sqrt{4000}$$

When we plug in the population value of 2014 wrt time 11, the unknown coefficient is

$$k = \frac{2\sqrt{4000} - 2\sqrt{2014}}{11} \approx 3.3396$$

So, the time for all of the fish in the lake to die, *i.e.*, $P(t) = 0$, is

$$T = \frac{C}{k} = 37.876$$

In order for the population not to change with time, we must set $k = 0$. To do so, we need 4000 rather than 2014.

2.4 Acceleration-Velocity Model

Problem 2.4.1

From the given condition, the acceleration, $a(t) = -k$, where k is a positive constant, *i.e.,*

$$\frac{dv}{dt} = -k$$

$$v(t) = \int (-k)dt$$

$$= -kt + C$$

Plugging in the IC $v(0) = 88$, we get $C = 88$ and the velocity

$$v(t) = -kt + 88$$

The car will stop skidding once the velocity is zero, *i.e.,*

$$v(t) = -kt + 88 = 0$$

resulting in the stopping time

$$t_S = \frac{88}{k}$$

The car will stop skidding when $t_S = 88/k$. On the other hand,

$$v(t) = \frac{dx}{dt}$$

$$x(t) = \int v \, dt$$

$$= -\frac{k}{2}t^2 + 88t + C$$

Considering $x(0) = 0$, we have

$$C = 0$$

Thus, the displacement is

$$x(t) = -\frac{k}{2}t^2 + 88t$$

The displacement at $t_S = 88/k$ is 176, *i.e.,*

$$-\frac{k}{2}\left(\frac{88}{k}\right)^2 + 88\frac{88}{k} = 176$$

Solving this equation, we get $k = 22$ ft/s and $t_S = 4$ seconds.
Thus, the car provides a constant deceleration of 22 ft/s² and the car continues to skid for 4 seconds.

Problem 2.4.2

$$\frac{dv}{dt} = a(t) = 50 \sin 5t$$

$$v(t) = \int 50 \sin 5t \, dt$$

$$= -10 \cos 5t + C$$

$$v(0) = -10 \Rightarrow -10 + C = -10 \Rightarrow C = 0$$

$$v(x) = -10 \cos 5t$$

$$\frac{dx}{dt} = v(t)$$

$$x(t) = \int v(t) dt$$

$$= \int -10 \cos 5t \, dt$$

$$= -2 \sin 5t + C$$

$$x(0) = 8 \Rightarrow C = 8$$

$$x(t) = 8 - 2 \sin 5t$$

Problem 2.4.3

$$v(t) = \int a(t) dt$$

$$= \int \frac{1}{(t+1)^n} dt$$

$$= -\frac{1}{n-1} \cdot \frac{1}{(t+1)^{n-1}} + C$$

IC $v(0) = 0$ gives

$$C = \frac{1}{n-1}$$

Thus,

$$v(t) = \frac{1}{n-1}\left(1 - \frac{1}{(t+1)^{n-1}}\right)$$

and

$$x(t) = \int v(t) dt$$

$$= \int \frac{1}{n-1}\left(1 - \frac{1}{(t+1)^{n-1}}\right) dt$$

$$= \frac{1}{n-1} t + \frac{1}{(n-1)(n-2)} \cdot \frac{1}{(t+1)^{n-2}} + C$$

IC $x(0) = 0$ gives

$$C = -\frac{1}{(n-1)(n-2)}$$

Thus,

$$x(t) = \frac{1}{n-1}t + \frac{1}{(n-1)(n-2)}\frac{1}{(t+1)^{n-2}} - \frac{1}{(n-1)(n-2)}$$

Problem 2.4.4

(1) This is a separable DE which can be written as

$$\frac{dv}{v} = -kdt$$
$$\ln v = -kt + C$$
$$v = C\exp(-kt)$$

From the IC $v(0) = v_0$, we get $C = v_0$ and the velocity

$$v(t) = v_0\exp(-kt)$$

Also, we know

$$dx = v(t)dt$$
$$dx = v_0\exp(-kt)\,dt$$
$$x = -\frac{v_0}{k}\exp(-kt) + C$$

Applying the IC $x(0) = x_0$, we have

$$C = x_0 + \frac{v_0}{k}$$

Thus,

$$x(t) = x_0 + \frac{v_0}{k}(1 - \exp(-kt))$$

(2) As $t \to \infty$, $\exp(-kt) \to 0$. Thus,

$$\lim_{t\to\infty} x(t) = x_0 + \frac{v_0}{k}$$

Problem 2.4.5

$$\frac{dv}{dt} = -kv^{\frac{3}{2}}$$
$$\int_{v_0}^{v}\frac{dv}{v^{3/2}} = \int_{0}^{t} -kdt$$
$$-2v^{-\frac{1}{2}} + 2v_0^{-\frac{1}{2}} = -kt$$
$$2v^{-\frac{1}{2}} = \frac{\sqrt{v_0}kt + 2}{\sqrt{v_0}}$$
$$v(t) = \frac{4v_0}{\left(\sqrt{v_0}kt + 2\right)^2}$$

For the displacement, we have

$$dx = v(t)dt$$

$$dx = \frac{4v_0}{\left(\sqrt{v_0}kt + 2\right)^2}dt$$

$$x(t) = -\frac{4\sqrt{v_0}}{k\left(2 + kt\sqrt{v_0}\right)} + C$$

Using the IC $x(0) = x_0$, we get

$$C = x_0 + \frac{2\sqrt{v_0}}{k}$$

Thus,

$$x(t) = x_0 + \frac{2\sqrt{v_0}}{k}\left(1 - \frac{2}{kt\sqrt{v_0} + 2}\right)$$

As $t \to \infty, \frac{2}{kt\sqrt{v_0}+2} \to 0$. Thus,

$$\lim_{t \to \infty} x(t) = x_0 + \frac{2\sqrt{v_0}}{k}$$

a finite displacement as the object moves infinitely.

Note: If the exponent is greater than $3/2$, the variable t will be found in the numerator of function $x(t)$ to enble $x(t) \to \infty$ as $t \to \infty$.

Problem 2.4.6

In this type of problems, only two of the three variables (x, v, t) are independent. One always selects the two variables that are relevant to the underlying problem to formulate, and to solve, the DE.

$$\frac{dv}{dt} = \frac{dv}{dx}\frac{dx}{dt} = \frac{dv}{dx}v$$

Case 1:

$$\begin{cases} m\dfrac{dv}{dx}v = -k_1v \\ v(x = 0) = v_0 \\ v(x_M) = 0 \end{cases}$$

$$-\int_{v_0}^{0} \frac{m}{k_1}dv = \int_{0}^{x_M} dx$$

$$x_M = \frac{m}{k_1}v_0$$

Case 2:

$$\begin{cases} m\dfrac{dv}{dx}v = -k_2 v^{\frac{3}{2}} \\ v(x = 0) = v_0 \\ v(x_M) = 0 \end{cases}$$

$$-\int_{v_0}^{0} \frac{m}{k_2} v^{1-\frac{3}{2}} dv = \int_{0}^{x_M} dx$$

$$x_M = \frac{m}{k_2} 2\sqrt{v_0}$$

Case 3:

$$\begin{cases} m\dfrac{dv}{dx}v = -k_3 v^2 \\ v(x = 0) = v_0 \\ v(x_M) = 0 \end{cases}$$

$$-\int_{v_0}^{0} \frac{m}{k_3} \frac{dv}{v} = \int_{0}^{x_M} dx$$

$$x_M = -\frac{m}{k_3} \ln v \Big|_{v_0}^{0} \to \infty$$

Problem 2.4.7

Let

$$v = \frac{dr}{dt}$$

and convert the acceleration, using the chain rule, from a formula involving v and t to another one involving v and r:

$$\frac{dv}{dt} = \frac{dv}{dr}\frac{dr}{dt} = v\frac{dv}{dr}$$

Thus, the DE becomes

$$v\frac{dv}{dr} = -\frac{GM_e}{r^2} + \frac{GM_m}{(S-r)^2}$$

Using the SOV method, we get

$$\int v\,dv = \int \left(-\frac{GM_e}{r^2} + \frac{GM_m}{(S-r)^2} \right) dr$$

$$\frac{v^2}{2} = \frac{GM_e}{r} + \frac{GM_m}{S-r} + C$$

The above relationship can also be obtained by conservation of total energy. Using the IC's $r'(t = 0) = v(t = 0) = v_0$ and $r(t = 0) = 0$, we know $v(r = 0) = v_0$. This gives

$$C = \frac{v_0^2}{2} - \frac{GM_e}{R} - \frac{GM_m}{S-R}$$

Thus, the velocity function is

$$v = \sqrt{2GM_e\left(\frac{1}{r}-\frac{1}{R}\right) + 2GM_m\left(\frac{1}{S-r}-\frac{1}{S-R}\right) + v_0^2}$$

The attractions to the object from the Earth and the Moon balance out at a peculiar point r_0, i.e.,

$$-\frac{GM_e}{r_0^2} + \frac{GM_m}{(S-r_0)^2} = 0$$

Solving this equation for r_0, we get

$$r_0 = \frac{\sqrt{M_e}S}{\sqrt{M_e}-\sqrt{M_m}}$$

Plugging the values of $M_m = 7.35 \times 10^{22}$ (kg), $S = 3.844 \times 10^8$ (m), $M_e = 5.975 \times 10^{24}$ (kg) into the above formula, we get the value of r_0. Using $R = 6.378 \times 10^6$ (m) and setting $v = 0$ if the object will rest after reaching the peculiar balancing point, we found $v_0 = 11,109$ (m/s).

This result is interesting. For comparison, the well-known escape speed from the Earth (without consideration of the attractions from other objects and the air and other resistances) is 11,180 (m/s) while the speed to reach the Earth-Moon influence-free zone, as shown above, is $v_0 = 11,109$ (m/s), a tiny bit smaller by 71 (m/s) or 0.6%. In other words, the pulling force of the Moon on the "escaping object" is truly insignificant. Of course, the speed of 71 (m/s) is approximately 159 (mph), which is higher than most planes' take off speeds of 140 (mph).

Problem 2.4.8

(1) The EoM is

$$m\frac{dv}{dt} = -mg - kv$$

Considering $v = \frac{dy}{dt}$, we get, as usual,

$$\frac{dv}{dt} = \frac{dv}{dy}\frac{dy}{dt} = v\frac{dv}{dy}$$

Thus,

$$mv\frac{dv}{dy} = -mg - kv$$

$$v\frac{dv}{dy} = -g - \frac{k}{m}v$$

$$\frac{vdv}{-g - \frac{k}{m}v} = dy$$

$$\int \frac{vdv}{-g - \frac{k}{m}v} = y + C$$

This is a separable DE whose solution can be obtained conveniently via the following steps.

Let

$$w = -g - \frac{k}{m}v$$

$$v = -\frac{mw + mg}{k}$$

$$dv = -\frac{m}{k}dw$$

$$\int \frac{mw + mg}{kw}\frac{m}{k}dw = \left(\frac{m}{k}\right)^2 (w + g\,ln|w|)$$

Thus,

$$\left(\frac{m}{k}\right)^2 \left(-g - \frac{k}{m}v + g\,ln\left|-g - \frac{k}{m}v\right|\right) = y + C$$

Applying the IC's

$$\begin{cases} y(0) = 0 \\ v(0) = v_0 \end{cases}$$

we get

$$C = \left(\frac{m}{k}\right)^2 \left(-g - \frac{k}{m}v_0 + g\,ln\left|-g - \frac{k}{m}v_0\right|\right)$$

Thus,

$$y(t) = \left(\frac{m}{k}\right)^2 \left(-\frac{k}{m}(v - v_0) + g\,ln\left|\frac{mg + kv}{mg + kv_0}\right|\right)$$

At the highest point y_{max} we have $v = 0$

$$y_{max} = \left(\frac{m}{k}\right)^2 \left(\frac{k}{m}v_0 + g\,ln\left|\frac{mg}{mg + kv_0}\right|\right)$$

(2)

$$m\frac{dv}{dt} = -mg - kv \Rightarrow \frac{dv}{dt} = -g - \frac{kv}{m}$$

which is a separable DE. Thus,

$$\int \frac{dv}{-g - \frac{k}{m}v} = \int dt$$

$$t + C = -\frac{m}{k}ln\left|-g - \frac{k}{m}v\right| = -\frac{m}{k}ln\left|\frac{-mg - kv}{m}\right|$$

$$v(0) = v_0 \Rightarrow C = -\frac{m}{k} \ln \left| \frac{-mg - kv_0}{m} \right|$$

$$y(t) = -\frac{m}{k} \ln \left| \frac{-mg - kv}{m} \right| + \frac{m}{k} \ln \left| \frac{-mg - kv_0}{m} \right|$$

$$= \frac{m}{k} \ln \left| \frac{mg + kv_0}{mg + kv} \right|$$

At the highest point, we have $v = 0$

$$t_{up} = \frac{m}{k} \ln \left| \frac{mg + kv_0}{mg} \right| = \frac{m}{k} \ln \left| 1 + \frac{kv_0}{mg} \right|$$

Problem 2.4.9

(1) The DE is

$$\frac{dv}{dt} = -kv$$

$$\ln v = -kt + C$$

$$v = C \exp(-kt)$$

Using $v(0) = 40$ and $v(10) = 20$, we find

$$C = 40, \quad k = \frac{\ln 2}{10}$$

i.e.,

$$v = 40 \exp\left(-\frac{\ln 2}{10} t \right) = 40 \exp\left(-\frac{t}{10} \ln 2 \right) = 40 \left(2^{-\frac{t}{10}} \right)$$

Similarly,

$$\frac{dx}{dt} = v$$

gives

$$\int_0^x dx = \int_0^t 40 \exp(-kt) \, dt$$

$$x = \frac{40}{k} (1 - \exp(-kt))$$

when $t \to \infty$, we have

$$x = \lim_{t \to \infty} \left(\frac{40}{k} (1 - \exp(-kt)) \right) = \frac{400}{\ln 2} \approx 577 \ (\text{ft})$$

(2) The DE becomes

$$\frac{dv}{dt} = -kv^2$$

This gives

$$\frac{1}{v} = kt + C$$

which is

$$v = \frac{1}{kt + C}$$

Using $v(0) = 40$ and $v(10) = 20$, we find

$$C = \frac{1}{40}, \qquad k = \frac{1}{400}$$

Thus,

$$v = \frac{40}{1 + 40kt} = \frac{40}{1 + \frac{t}{10}}$$

Finally,

$$\frac{dx}{dt} = v(t) = \frac{40}{1 + \frac{t}{10}}$$

We get

$$x(t) = 400 \ln\left(1 + \frac{t}{10}\right) + C$$

Using $x(0) = 0$, we get $C = 0$. Thus

$$x(t) = 400 \ln\left(1 + \frac{t}{10}\right)$$

Thus, at $t = 60$,

$$x(t = 60) = 400 \ln\left(1 + \frac{60}{10}\right) \approx 778$$

Problem 2.4.10

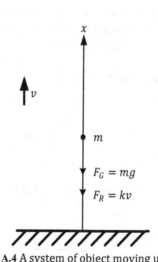

Figure A.4 A system of object moving upward.

(1) The total force on this object is $F = -(F_R + F_G)$. Using Newton's law $F = ma = m\frac{dv}{dt}$, we have

$$F = -mg - kv = m\frac{dv}{dt}$$

which is a 1st.O separable DE.

Solving this DE with IC $v(0) = v_0$, we get

$$v(t) = \left(v_0 + \frac{mg}{k}\right)\exp\left(-\frac{k}{m}t\right) - \frac{mg}{k}$$

(2) At the point it reverses, the velocity of the object vanishes at $v(t) = 0$, thus,

$$\left(v_0 + \frac{mg}{k}\right)\exp\left(-\frac{k}{m}t\right) - \frac{mg}{k} = 0$$

$$t = -\frac{m}{k}\ln\frac{mg}{mg + kv_0}$$

It depends on the initial velocity v_0, i.e., the bigger it is, the longer it takes to reverse the direction.

(3) When the object keeps a constant speed, the acceleration vanishes. Thus, $a = -mg - kv = 0 \Rightarrow v = -\frac{mg}{k}$, which is independent of the initial velocity v_0.

Problem 2.4.11

Let t_1 be the time you accelerate and t_2 be the time you decelerate. $t_1 + t_2 = t$, which is the time you travel from one stop sign to the other.

The driver accelerates at a_1 to reach the highest speed $a_1 t_1$ from which the driver can decelerate to stop at the other end. Thus,

$$a_1 t_1 = a_2 t_2$$

The total distance has to satisfy

$$\frac{1}{2}a_1 t_1^2 + \frac{1}{2}a_2 t_2^2 = L$$

Plugging

$$t_2 = \frac{a_1}{a_2}t_1$$

into the above, we have

$$\frac{1}{2}a_1 t_1^2 + \frac{1}{2}\frac{a_1^2}{a_2^2}t_1^2 = L$$

which gives

$$t_1 = \sqrt{\frac{2a_2 L}{a_1 a_2 + a_1^2}}$$

$$t_2 = \sqrt{\frac{2a_1L}{a_1a_2 + a_2^2}}$$

If $v_m \geq a_1t_1 (= a_2t_2)$, the result stays the same.

If $v_m < a_1t_1$, you should accelerate to v_m (need time t_1). Then, travel at a constant velocity v_m for time t_3. Finally, it decelerates from v_m to velocity 0 (need time t_2). The total time is $t = t_1 + t_2 + t_3$.

Here

$$t_1 = \frac{v_m}{a_1} \Rightarrow L_1 = \frac{v_m}{2a_1^2}$$

$$t_2 = \frac{v_m}{a_2} \Rightarrow L_2 = \frac{v_m}{2a_2^2}$$

and

$$L_3 = L - L_1 - L_2$$

$$t_3 = \frac{1}{v_m}\left(L - \frac{v_m}{2a_1^2} - \frac{v_m}{2a_2^2}\right)$$

Finally,

$$t = \frac{v_m}{a_1} + \frac{v_m}{a_2} + \frac{1}{v_m}\left(L - \frac{v_m}{2a_1^2} - \frac{v_m}{2a_2^2}\right)$$

Problem 2.4.12

We start with the usual force equation

$$m\frac{dv}{dt} = m\frac{dv}{dt}\frac{dy}{dy} = m\frac{dy}{dt}\frac{dv}{dy} = mv\frac{dv}{dy} = -\frac{GMm}{(y+R)^2}$$

where M is the mass of the Earth, m is the mass of the projectile, y is the height from the Earth's surface and $r = y + R$.

Thus, we will separate this equation and integrate

$$\int_{v_0}^{v} v\,dv = \int_0^y -\frac{GM}{(y+R)^2}\,dy$$

to obtain

$$v^2 = v_0^2 - 2GM\left(\frac{1}{R} - \frac{1}{R+y}\right)$$

We found the velocity to be

$$v(y) = \sqrt{v_0^2 - 2GM\left(\frac{1}{R} - \frac{1}{R+y}\right)} = \sqrt{v_0^2 - 2GM\left(\frac{1}{R} - \frac{1}{r}\right)}$$

The velocity will become imaginary when

$$v_0{}^2 < 2GM \left(\frac{1}{R} - \frac{1}{r} \right)$$

Problem 2.4.13

Consider v to be a function of y. Thus,

$$y'' = \frac{dv}{dt} = \frac{dv}{dy}\frac{dy}{dt} = v\frac{dv}{dy}$$

and the original DE can be written as

$$v\frac{dv}{dy} = -\frac{GM}{(y+R)^2} - \beta \exp(-y)$$

$$\int v\,dv = \int \left(-\frac{GM}{(y+R)^2} - \beta \exp(-y) \right) dy$$

$$\frac{v^2}{2} = \frac{GM}{y+R} + \beta \exp(-y) + C$$

Plugging the IC $y(0) = 0$ and $y'(0) = v_0$, we have

$$\frac{v_0^2}{2} = \frac{GM}{R} + \beta + C$$

$$C = \frac{v_0^2}{2} - \frac{GM}{R} - \beta$$

$$v^2 = v_0^2 - \frac{2GMy}{(y+R)R} + 2\beta(\exp(-y) - 1)$$

At $v = 0$, the object reaches the max-height y_m which is given implicitly by

$$\frac{2GMy_m}{(y_m + R)R} - 2\beta \exp(-y_m) = v_0^2 - 2\beta$$

For the total time T, we should consider

$$v = \frac{dy}{dt}$$

and it gives

$$\sqrt{v_0^2 - \frac{2GMy}{(y+R)R} + 2\beta(\exp(-y) - 1)} = \frac{dy}{dt}$$

$$T = \int_0^{y_m} \left(v_0^2 - \frac{2GMy}{(y+R)R} + 2\beta(\exp(-y) - 1) \right)^{-\frac{1}{2}} dy$$

Problem 2.4.14

According to Newton's law, we have

$$m\frac{dv}{dt} = \alpha v + \beta v^2$$

Since

$$v = \frac{dx}{dt} = \frac{dx}{dv}\frac{dv}{dt}$$

then

$$\frac{dv}{dt} = v\frac{dv}{dx}$$

Plugging this back

$$mv\frac{dv}{dx} = \alpha v + \beta v^2$$

$$m\frac{dv}{dx} = \alpha + \beta v$$

$$\frac{m}{\alpha + \beta v}dv = dx$$

$$\frac{m}{\beta}\ln(\alpha + \beta v) = x + C$$

At $x = 0$, it gives $v = v_0$. So,

$$C = \frac{m}{\beta}\ln(\alpha + \beta v_0)$$

Therefore,

$$x(v) = \frac{m}{\beta}\ln\left(\frac{\alpha + \beta v}{\alpha + \beta v_0}\right)$$

The max-distance is

$$x_m(v = 0) = \frac{m}{\beta}\ln\left(\frac{\alpha}{\alpha + \beta v_0}\right)$$

If we double the initial speed,

$$x(v) = \frac{m}{\beta}\ln\left(\frac{\alpha + \beta v}{\alpha + 2\beta v_0}\right)$$

$$x_m(v = 0) = \frac{m}{\beta}\ln\left(\frac{\alpha}{\alpha + 2\beta v_0}\right)$$

Problem 2.4.15

Assuming the following:

v_0 and v_1 are entering and exiting Medium-1, respectively.
v_1 and v_2 are entering and exiting Medium-2, respectively.
v_2 and v_3 are entering and exiting Medium-3, respectively.

Using the following relationship for all three cases,

$$\frac{dv}{dt} = \frac{dv}{dt}\frac{dx}{dx} = v\frac{dv}{dx}$$

The EoM for Medium-1:

$$mv\frac{dv}{dx} = -kv$$

Integrating both sides, we get:

$$\int_{v_0}^{v_1} dv = \int_0^L \left(-\frac{k}{m}\right) dx$$

Thus,

$$v_1 = v_0 - \frac{k}{m}L$$

The EoM for Medium-2:

$$mv\frac{dv}{dx} = -kv^2$$

Integrating both sides, we get

$$\int_{v_1}^{v_2} \left(\frac{1}{v}\right) dv = \int_0^L \left(-\frac{k}{m}\right) dx$$

Thus,

$$v_2 = v_1 \exp\left(-\frac{k}{m}L\right)$$
$$= \left(v_0 - \frac{k}{m}L\right)\exp\left(-\frac{k}{m}L\right)$$

For Medium-3, using results from Medium-1, we get

$$v_3 = v_2 - \frac{k}{m}L$$

Forcing $v_3 = 0$, we have

$$v_3 = v_2 - \frac{k}{m}L = 0$$

or

$$\left(v_0 - \frac{k}{m}L\right)\exp\left(-\frac{k}{m}L\right) - \frac{k}{m}L = 0$$

Solving the above, we get

$$v_0 = \frac{k}{m}L\left(1 + \exp\left(\frac{k}{m}L\right)\right)$$

which is the special initial speed.

Problem 2.4.16

Method 1: Working in (x, v) system.
For all three cases, we have the following general DE:

$$\begin{cases} m\dfrac{dv}{dt} = m\dfrac{dv}{dx}\dfrac{dx}{dt} = m\dfrac{dv}{dx}v = -kv^{\alpha} \\ v(x = 0) = v_0 \end{cases}$$

We can easily get the following,

$$\int_{v_0}^{0} v^{1-\alpha}dv = -\int_{0}^{x_M}\frac{k}{m}dx$$

The bullet stops at the max distance x_M.

$$x_M = \frac{m}{k}\int_{0}^{v_0} v^{1-\alpha}dv = \frac{m}{k}\frac{v_0^{2-\alpha}}{2-\alpha}$$

For Case 1, $\alpha = 1$,

$$x_M = \frac{m}{k}\frac{v_0^1}{1} = \frac{mv_0}{k}$$

For Case 2, $\alpha = 3/2$,

$$x_M = \frac{m}{k}\frac{v_0^{1/2}}{1/2} = \frac{2mv_0^{1/2}}{k}$$

For Case 3, $\alpha = 2$,

$$x_M \to \infty$$

Method 2: Working in the (v,t) system and, then, the (x,t) system.
(i)

$$m\frac{dv}{dt} = -k_1 v$$

$$\frac{dv}{v} = -\left(\frac{k_1}{m}\right)dt$$

$$\ln v = -\left(\frac{k_1}{m}\right)t + C$$

$$v = \exp\left(-\frac{k_1}{m}t + C\right)$$

$$v = C_1\exp\left(-\frac{k_1}{m}t\right)$$

$$v(0) = v_0 = C_1$$

So, we have

$$v(t) = v_0\exp\left(-\frac{k_1}{m}t\right)$$

$$\frac{dx}{dt} = v_0\exp\left(-\frac{k_1}{m}t\right)$$

$$dx = v_0\exp\left(-\frac{k_1}{m}t\right)dt$$

$$x = \frac{v_0\exp\left(-\frac{k_1}{m}t\right)}{-\dfrac{k_1}{m}} + C_2$$

$$x = -\frac{mv_0}{k_1}\exp\left(-\frac{k_1}{m}t\right) + C_2$$

$$x(0) = x_0 = -\frac{mv_0}{k_1} + C_2$$

$$C_2 = x_0 + \frac{mv_0}{k_1}$$

So, we have

$$x(t) = x_0 + \frac{mv_0}{k_1}\left(1 - \exp\left(-\frac{k_1}{m}t\right)\right)$$

As t goes to infinity, $\exp\left(-\frac{k_1}{m}t\right)$ goes to 0. Thus, x goes to

$$x_0 + \frac{mv_0}{k_1}$$

So, the total distance traveled is x at infinity minus x_0, or

$$\frac{mv_0}{k_1}$$

(ii)

$$m\frac{dv}{dt} = -k_2 v^{\frac{3}{2}}$$

$$\frac{dv}{dt} = -\frac{k_2}{m}v^{\frac{3}{2}}$$

$$v^{-\frac{3}{2}}dv = -\frac{k_2}{m}dt$$

$$\frac{v^{-\frac{1}{2}}}{-\frac{1}{2}} = -\frac{k_2}{m}t + C$$

$$v^{-\frac{1}{2}} = \frac{k_2}{2m}t + \frac{-C}{2}$$

$$v = \left(C_1 + \frac{k_2}{2m}t\right)^{-2}$$

$$v(0) = v_0 = C_1^{-2}$$

$$C_1 = \frac{1}{\sqrt{v_0}}$$

$$v = \left(\frac{1}{\sqrt{v_0}} + \frac{k_2}{2m}t\right)^{-2}$$

$$x = \int v\, dt$$

$$u = \frac{1}{\sqrt{v_0}} + \frac{k_2}{2m}t$$

$$du = \frac{k_2}{2m}dt$$

$$\frac{2m}{k_2}du = dt$$

So, we have

$$x = \frac{2m}{k_2}\int u^{-2}du$$

$$x = -\frac{2m}{k_2}\left(\frac{1}{\sqrt{v_0}} + \frac{k_2}{2m}t\right)^{-1} + C_3$$

Applying the IC $x(0) = x_0$, we have

$$x_0 = -\frac{2m}{k_2}\sqrt{v_0} + C_3$$

$$x(t) = -\frac{2m}{k_2}\left(\frac{1}{\sqrt{v_0}} + \frac{k_2}{2m}t\right)^{-1} + x_0 + \frac{2m}{k_2}\sqrt{v_0}$$

$$= -\frac{2m}{k_2}\left(\frac{2m}{2m\sqrt{v_0}} + \frac{k_2 t\sqrt{v_0}}{2m\sqrt{v_0}}\right)^{-1} + x_0 + \frac{2m}{k_2}\sqrt{v_0}$$

$$= -\frac{2m}{k_2}\left(\frac{2m\sqrt{v_0}}{2m + k_2 t\sqrt{v_0}}\right) + x_0 + \frac{2m}{k_2}\sqrt{v_0}$$

So,

$$\lim_{t\to\infty} x(t) = x_0 + \frac{2m}{k_2}\sqrt{v_0}$$

Thus, the total distance is

$$\lim_{t\to\infty} x(t) - x_0 = \frac{2m}{k_2}\sqrt{v_0}$$

(iii)

$$m\frac{dv}{dt} = -k_3 v^2$$

$$v^{-2}dv = -\frac{k_3}{m}dt$$

$$-v^{-1} = -\frac{k_3}{m}t + C$$

$$v = \left(\frac{k_3}{m}t + C_2\right)^{-1}$$

$$v(0) = v_0 = \frac{1}{C_2}$$

$$v(t) = \left(\frac{k_3}{m}t + \frac{1}{v_0}\right)^{-1}$$

$$x = \int v\,dt$$

$$x = \frac{m}{k_3}\ln\left(\frac{k_3}{m}t + \frac{1}{v_0}\right) + C_3$$

$$x(0) = x_0 = \frac{m}{k_3}\ln\frac{1}{v_0} + C_3 = -\frac{m}{k_3}\ln v_0 + C_3$$

$$x(t) = \frac{m}{k_3}\ln\left(\frac{k_3}{m}t + \frac{1}{v_0}\right) + x_0 + \frac{m}{k_3}\ln v_0$$

As t goes to infinity, $x(t)$ goes to infinity and, in this situation, the bullet will fly to infinity *and beyond.*

Problem 2.4.17

The EoM is

$$\frac{dv}{dt} = -\frac{\alpha(\beta + v^2)}{m}$$

(1) The position function, given by X, is a function of t only.
Rearranging the EoM and using $\frac{dX}{dt} = v$, we have

$$\frac{dv}{dt}\frac{dX}{dX} = -\frac{\alpha(\beta + v^2)}{m}$$

$$v\frac{dv}{dX} = -\frac{\alpha(\beta + v^2)}{m}$$

Solving the above by the SOV method, we get

$$\ln(v^2 + \beta) = -\frac{2\alpha}{m}X + C$$

Applying the IC, we found

$$\ln(v_0^2 + \beta) = C$$

Thus, the velocity and position are related by

$$\ln(v^2 + \beta) = -\frac{2\alpha}{m}X + \ln(v_0^2 + \beta)$$

When $v \to 0$, the bullet reaches the farthest, *i.e.*,

$$\ln\beta = -\frac{2\alpha}{m}X_{max} + \ln(v_0^2 + \beta)$$

The max-distance is

$$X_{max} = \frac{m}{2\alpha}\ln\frac{(v_0^2 + \beta)}{\beta}$$

(2) Applying the IC and using the original EoM, we compose

$$\begin{cases} \dfrac{dv}{dt} = -\dfrac{\alpha(\beta + v^2)}{m} \\ v(t = 0) = v_0 \end{cases}$$

Looking up the table of integrals or integrate manually, we obtain

$$\frac{1}{\sqrt{\beta}}\tan^{-1}\frac{v}{\sqrt{\beta}} = C - \frac{\alpha}{m}t$$

The explicit expression of the velocity is

$$v = \sqrt{\beta}\tan\left(\sqrt{\beta}\left(C - \frac{\alpha}{m}t\right)\right)$$

Applying the IC, we get

$$v(t = 0) = v_0 = \sqrt{\beta}\tan(\sqrt{\beta}C)$$

The explicit expression of C is

$$C = \frac{\tan^{-1}\left(\frac{v_0}{\sqrt{\beta}}\right)}{\sqrt{\beta}}$$

When the velocity becomes zero, the time reaches its maximum.

$$\frac{\alpha}{m}T = \frac{\tan^{-1}\left(\frac{v_0}{\sqrt{\beta}}\right)}{\sqrt{\beta}}$$

Finally,

$$T = \frac{m \cdot \tan^{-1}\left(\frac{v_0}{\sqrt{\beta}}\right)}{\alpha\sqrt{\beta}}$$

As long as the time is in the interval $t \in [0, T]$, the bullet is in motion.

Problem 2.4.18

The net force

$$F = mg - \frac{mg}{\rho} - \mu v$$

(positive sign means the force is downwards).
So

$$\frac{dv}{dt} = \frac{F}{m} = \frac{mg\left(1 - \frac{1}{\rho}\right) - \mu v}{m} = \frac{\mu\left(\frac{mg}{\mu} - \frac{mg}{\mu\rho} - v\right)}{m}$$

$$\int \frac{m}{\mu\left(\frac{mg}{\mu} - \frac{mg}{\mu\rho} - v\right)} dv = \int dt$$

$$v(t) = \exp\left(-\frac{\mu t}{m}\right)\left(\frac{mg}{\mu\rho} - \frac{mg}{\mu}\right) + \left(\frac{mg}{\mu} - \frac{mg}{\mu\rho}\right)$$

Thus,

$$H = \int_0^T v(t)\, dt$$

$$= \exp\left(-\frac{\mu}{m}T\right)\left(\frac{mg}{\mu\rho} - \frac{mg}{\mu}\right)\left(-\frac{m}{\mu}\right) + \left(\frac{mg}{\mu} - \frac{mg}{\mu\rho}\right)T - \left(\frac{mg}{\mu\rho} - \frac{mg}{\mu}\right)\left(-\frac{m}{\mu}\right)$$

Once we find the solution for T, the time that the egg hits the bottom, we can just plug it into the velocity $v(t)$ to get the impact speed, at the time T.

Alternative Method:

Recognizing

$$\frac{dv}{dt} = \frac{dv}{dy}\frac{dy}{dt} = \frac{dv}{dy}v$$

we have

$$m\frac{dv}{dt} = m\frac{dv}{dy}v$$

$$= mg - \frac{mg}{\rho} - \mu v$$

resulting in

$$\frac{dv}{dy}v = g\left(1 - \frac{1}{\rho}\right) - \frac{\mu}{m}v$$

Let

$$\alpha = g\left(1 - \frac{1}{\rho}\right), \qquad \beta = -\frac{\mu}{m}$$

we get the following separable DE

$$\frac{dv}{dy}v = \alpha + \beta v$$

which can be solved using two definite integrals

$$\int_0^{v_f} \frac{vdv}{\alpha + \beta v} = \int_0^H dy$$

where v_f is the speed of impact. So

$$H = \left(\frac{v}{\beta} - \frac{\alpha\ln(\alpha + \beta v)}{\beta^2}\right)\Bigg|_0^{v_f}$$

$$= \frac{v_f}{\beta} - \frac{\alpha\ln(\alpha + \beta v_f)}{\beta^2} + \frac{\alpha\ln(\alpha)}{\beta^2}$$

One can solve for v_f now.

2.5 Windy Day Plane Landing

Problem 2.5.1

$$v_x = -v_0 \cos \alpha = -v_0 \frac{x}{\sqrt{x^2 + y^2}}$$

$$v_y = -v_0 \sin \alpha - w = -v_0 \frac{y}{\sqrt{x^2 + y^2}} - w$$

Thus,

$$\frac{dy}{dx} = \frac{\left(\frac{dy}{dt}\right)}{\left(\frac{dx}{dt}\right)} = \frac{v_y}{v_x} = \frac{-v_0 \sin \alpha - w}{-v_0 \cos \alpha} = \frac{\sin \alpha}{\cos \alpha} + \frac{w}{v_0 \cos \alpha}$$

Therefore,

$$\frac{dy}{dx} = \frac{y}{x} + \frac{w}{v_0}\left(\frac{\sqrt{x^2 + y^2}}{x}\right) = \frac{y}{x} + \frac{w}{v_0}\sqrt{1 + \frac{y^2}{x^2}}$$

Let $u = \frac{y}{x}$. Then, $y = ux$ and $\frac{dy}{dx} = u + \frac{du}{dx}x$. Plugging into the above DE, we get

$$u + \frac{du}{dx}x = u + \frac{w}{v_0}\sqrt{1 + u^2}$$

$$\frac{du}{dx}x = \frac{w}{v_0}\sqrt{1 + u^2}$$

$$\ln\left(u + \sqrt{1 + u}\right) = \frac{w}{v_0}\ln(x) + C$$

Applying the IC $y(x = M) = 0$, we get

$$y(x) = \frac{M}{2}\left(\left(\frac{x}{M}\right)^{1-\frac{w}{v_0}} - \left(\frac{x}{M}\right)^{1+\frac{w}{v_0}}\right)$$

If $w > v_0$,

$$\lim_{x \to 0} \left(\frac{x}{M}\right)^{1-\frac{w}{v_0}} \to \infty$$

and the path length is infinite.

If $w < v_0$,

$$y(x) = \frac{M}{2}\left(\left(\frac{x}{M}\right)^{1-\frac{w}{v_0}} - \left(\frac{x}{M}\right)^{1+\frac{w}{v_0}}\right)$$

$$L = \int_0^M \sqrt{dx^2 + dy^2} = \int_0^M \sqrt{1 + \left(\frac{dy}{dx}\right)^2}\, dx$$

If $w = 0$, the trajectory is a straight line:

$$y(x) = \frac{M}{2}\left(\left(\frac{x}{M}\right)^{1-\frac{0}{v_0}} - \left(\frac{x}{M}\right)^{1+\frac{0}{v_0}}\right) = 0$$

and the path length is

$$L = M$$

Problem 2.5.2

From the problem, we know that with closed parachute, we have

$$m\frac{dv}{dt} = mg - kv$$

This is a linear DE and we have

$$v = \frac{mg}{k} + C_1 \exp\left(-\frac{kt}{m}\right)$$

Given $v(0) = 0$, we know that

$$C_1 = -\frac{mg}{k}$$

Hence, we have

$$v = \frac{mg}{k}\left(1 - \exp\left(-\frac{kt}{m}\right)\right)$$

and

$$x = \int v dt$$

$$= \frac{mg}{k}t + \frac{m^2 g}{k^2}\exp\left(-\frac{kt}{m}\right) + C_2$$

Since $x(0) = 0$, we have

$$C_2 = -\frac{m^2 g}{k^2}$$

Thus,

$$x(t) = \frac{mg}{k}\left(t + \frac{m}{k}\exp\left(-\frac{kt}{m}\right) - \frac{m}{k}\right)$$

Suppose he opens the parachute at $t = t_1$, we know that

$$x_1 = \frac{mg}{k}\left(t_1 + \frac{m}{k}\exp\left(-\frac{kt_1}{m}\right) - \frac{m}{k}\right)$$

and

$$v_1 = \frac{mg}{k}\left(1 - \exp\left(-\frac{kt_1}{m}\right)\right)$$

Starting from t_1, the parachute will be open and the DE becomes

$$mv' = mg - nkv$$

and we get

$$v = \frac{mg}{nk} + C_3 \exp\left(-\frac{nkt}{m}\right)$$

where from $v(t_1) = v_1$, we have

$$C_3 = \frac{n-1}{n}\frac{mg}{k}\exp\left(\frac{nkt_1}{m}\right) - \frac{mg}{k}\exp\left(\frac{(n-1)kt_1}{m}\right)$$

For x, we have

$$x = \int v dt$$

$$= \frac{mg}{nk}t - \frac{C_3 m}{nk}\exp\left(-\frac{nkt}{m}\right) + C_4$$

where C_4 satisfies $x(t_1) = x_1$. After finding C_4, we have

$$x_2 = H - x_1$$

Solving for t_2 such that $x(t_2) = x_2$ and make $v(t_2) = v_0$, we can find the optimal t_1.

Problem 2.5.3

Before opening the parachute:

In vertical (y-) direction,

$$m\frac{dv_y}{dt} = mg$$

$$\int_0^{v_y} dv_y = \int_0^t g dt, \qquad v_y = gt$$

$$\frac{dy}{dt} = gt, \qquad \int_0^y dy = \int_0^t gt dt$$

$$y = \frac{1}{2}gt^2 \Rightarrow \frac{H}{2} = \frac{1}{2}gt_1^2$$

$$t_1 = \sqrt{\frac{H}{g}}$$

$$v_{y_1} = gt_1 = \sqrt{gH}$$

In horizontal (x-) direction, zero force and constant velocity v_0

$$x = v_0 t \Rightarrow t = \frac{x}{v_0}$$

Thus,

$$y = \frac{1}{2}gt^2 = \frac{1}{2}g\left(\frac{x}{v_0}\right)^2 \Rightarrow \frac{dy}{dx} = \frac{g}{v_0^2}x$$

415

Thus,

$$x_1 = v_0 t_1 = v_0 \sqrt{\frac{H}{g}}$$

So, the length

$$L_1 = \int_0^{x_1} \sqrt{1 + \left(\frac{dy}{dx}\right)^2}\, dx = \int_0^{v_0\sqrt{\frac{H}{g}}} \sqrt{1 + \left(\frac{g}{v_0^2}x\right)^2}\, dx$$

$$= \frac{1}{2}\sqrt{\frac{v_0^2 H}{g} + H^2} + \frac{v_0^2}{2g}\ln\left(\frac{\sqrt{gH}}{v_0} + \sqrt{\frac{gH}{v_0^2} + 1}\right)$$

After opening the parachute:

In vertical (y-) direction:

$$m\frac{dv_y}{dt} = mg - \alpha v_y$$

$$\int_{v_{y1}}^{v_y} \frac{m\, dv_y}{mg - \alpha v_y} = \int_0^t dt, v_y$$

$$= \frac{mg}{\alpha} + \left(\sqrt{gH} - \frac{mg}{\alpha}\right)\exp\left(-\frac{\alpha t}{m}\right)$$

$$\frac{dy}{dt} = \frac{mg}{\alpha} + \left(\sqrt{gH} - \frac{mg}{\alpha}\right)\exp\left(-\frac{\alpha t}{m}\right), \int_0^y dy$$

$$= \int_0^t \left(\frac{mg}{\alpha} + \left(\sqrt{gH} - \frac{mg}{\alpha}\right)\exp\left(-\frac{\alpha t}{m}\right)\right) dt$$

$$y = \frac{mg}{\alpha}t - \frac{m}{\alpha}\left(\sqrt{gH} - \frac{mg}{\alpha}\right)\exp\left(-\frac{\alpha t}{m}\right) + \frac{m}{\alpha}\left(\sqrt{gH} - \frac{mg}{\alpha}\right)$$

$$\frac{H}{2} = \frac{mg}{\alpha}t_2 - \frac{m}{\alpha}\left(\sqrt{gH} - \frac{mg}{\alpha}\right)\exp\left(-\frac{\alpha t_2}{m}\right) + \frac{m}{\alpha}\left(\sqrt{gH} - \frac{mg}{\alpha}\right)$$

We could get t_2 implicitly.

In horizontal (x-) direction:

$$m\frac{dv_x}{dt} = -\alpha v_x$$

$$-\int_{v_0}^{v_x} \frac{m}{\alpha v_x}\, dv_x = \int_0^t dt, v_x = v_0\exp\left(-\frac{\alpha t}{m}\right)$$

$$\frac{dx}{dt} = v_0\exp\left(-\frac{\alpha t}{m}\right), \quad \int_0^x dx = \int_0^t v_0\exp\left(-\frac{\alpha t}{m}\right) dt$$

$$x = -\frac{mv_0}{\alpha}\exp\left(-\frac{\alpha t}{m}\right) + \frac{mv_0}{\alpha} \Rightarrow t = -\frac{m}{\alpha}\ln(1 - \frac{\alpha}{mv_0}x)$$

$$y = \frac{mg}{\alpha}t - \frac{m}{\alpha}\left(\sqrt{gH} - \frac{mg}{\alpha}\right)\exp\left(-\frac{\alpha}{m}t\right) + \frac{m}{\alpha}\left(\sqrt{gH} - \frac{mg}{\alpha}\right)$$

$$= -\frac{m^2 g}{\alpha^2} \ln\left(1 - \frac{\alpha}{mv_0}x\right) - \frac{m}{\alpha}\left(\sqrt{gH} - \frac{mg}{\alpha}\right)(1 - \frac{\alpha}{mv_0}x) + \frac{m}{\alpha}\left(\sqrt{gH} - \frac{mg}{\alpha}\right)$$

$$\frac{dy}{dx} = \frac{mg}{\alpha v_0 - \frac{\alpha^2}{m}x} + \frac{1}{v_0}\left(\sqrt{gH} - \frac{mg}{\alpha}\right)$$

$$x_2 = -\frac{mv_0}{\alpha}\exp\left(-\frac{\alpha t_2}{m}\right) + \frac{mv_0}{\alpha}$$

So, the length

$$L_2 = \int_0^{x_2} \sqrt{1 + \left(\frac{dy}{dx}\right)^2}\, dx$$

$$= \int_0^{x_2} \left(1 + \left(\frac{mg}{\alpha v_0 - \frac{\alpha^2}{m}x} + \frac{1}{v_0}\left(\sqrt{gH} - \frac{mg}{\alpha}\right)\right)^2\right)^{\frac{1}{2}} dx$$

The length of the trajectory is $L = L_1 + L_2$.

Problem 2.5.4

From the problem, we know that the resistance in vector form is

$$\vec{R} = -\alpha|\vec{v}|\frac{\vec{v}}{|\vec{v}|}$$

$$= -\alpha\vec{v}$$

Now, it is clear to see the resistance in x and y directions are independent from each other. Thus, we only need to consider the y axis. Before the parachute opens, we have

$$\frac{dv_y}{dt} = -g$$

which gives

$$v_y = -gt + C_1$$

given $v_y(0) = 0, C_1 = 0$ we have

$$v_y = -gt$$

$$y = -\frac{1}{2}gt^2 + C_2$$

Plugging in $y(0) = H, C_2 = H$

$$y = H - \frac{1}{2}gt^2$$

If the man opens the parachute at $y = \frac{1}{2}H$, the time it takes from start to open the parachute is

$$\frac{1}{2}H = H - \frac{1}{2}gt_1^2$$

$$t_1 = \sqrt{\frac{H}{g}}$$

and at that time,

$$v_y = -\sqrt{gH}$$

After opening the parachute, the EoM becomes

$$\frac{dv_y}{dt} = -g - \frac{\alpha}{m}v_y$$

which is a separable DE. The GS is

$$-\frac{m}{\alpha}\ln\left(\frac{\alpha}{m}v_y + g\right) = t + C_3$$

Given $v_y(0) = -\sqrt{gH}$, we have

$$C_3 = -\frac{m}{\alpha}\ln\left(g - \frac{\alpha}{m}\sqrt{gH}\right)$$

Thus,

$$\frac{m}{\alpha}\ln\left(\frac{g - \frac{\alpha}{m}\sqrt{gH}}{g + \frac{\alpha}{m}v_y}\right) = t$$

$$(mg - \alpha\sqrt{gH})\exp\left(-\frac{\alpha}{m}t\right) - mg = \alpha v_y$$

$$v_y = \left(\frac{mg}{\alpha} - \sqrt{gH}\right)\exp\left(-\frac{\alpha}{m}t\right) - \frac{mg}{\alpha}$$

and

$$y = -\frac{m}{\alpha}\left(\frac{mg}{\alpha} - \sqrt{gH}\right)\exp\left(-\frac{\alpha}{m}t\right) - \frac{mg}{\alpha}t + C_4$$

Given $y(0) = 0$, we have

$$C_4 = \frac{m^2 g}{\alpha^2} - \frac{m\sqrt{gH}}{\alpha}$$

and solving for $y(t_2) = \frac{1}{2}H$ gives an implicit equation:

$$\frac{1}{2}\frac{\alpha H}{mg} = \left(\sqrt{\frac{H}{g}} - \frac{m}{\alpha}\right)\left(\exp\left(-\frac{\alpha}{m}t\right) - 1\right) - t_2$$

The total time is $t = t_1 + t_2$.

In the case he opens the parachute at $y = \frac{1}{4}H$, we have

$$t_1 = \sqrt{\frac{3H}{2g}}$$

and at that time

$$v_y(t_1) = -\sqrt{\frac{3gH}{2}}$$

The equation after opening the parachute becomes

$$v_y = \left(\frac{mg}{\alpha} - \sqrt{\frac{3}{2}gH}\right)\exp\left(-\frac{\alpha}{m}t\right) - \frac{mg}{\alpha}$$

and the implicit expression for t_2 is

$$\frac{3\,\alpha H}{4\,mg} = \left(\sqrt{\frac{3H}{2g}} - \frac{m}{\alpha}\right)\left(\exp\left(-\frac{\alpha}{m}t_2\right) - 1\right) - t_2$$

Problem 2.5.5

This is a problem of Newtonian mechanics $F_g + F_d = ma$. Total weight of the man and his parachute m with acceleration a, gravitational force F_g and a drag force F_d. Let x be the distance above the Earth's surface, with positive direction downward, $= \frac{dv}{dt}$, where $v = \frac{dx}{dt}$ is the velocity and $F_g = -mg$.

Since the drag force is assumed to be proportional to velocity, $F_d = -kv$ (with closed paracute) and $F_d = -nkv$ (with open parachute). The deployment occurs at time t_0.

During the initial fall with parachute closed, we have

$$m\frac{dv}{dt} = mg - kv, v(0) = 0$$

an IVP. The DE is separable,

$$\frac{dv}{dt} = g - \frac{kv}{m}$$

Thus,

$$\frac{dv}{g - \dfrac{kv}{m}} = dt$$

$$-\frac{m}{k}\frac{d\left(g - \dfrac{kv}{m}\right)}{\left(g - \dfrac{kv}{m}\right)} = dt$$

$$-\frac{m}{k}\ln\left(g - \frac{kv}{m}\right) = t + C$$

$$g - \frac{kv}{m} = C_1 \exp\left(-\frac{kt}{m}\right) \Rightarrow v(t) = \frac{m}{k}\left(g - C_1 \exp\left(-\frac{kt}{m}\right)\right)$$

$$t = 0, v(0) = 0$$

$$t = 0, v(0) = 0$$

$$C_1 = g \frac{dx}{dt} = v = \frac{m}{k} g \left(1 - \exp\left(-\frac{kt}{m} \right) \right)$$

$$x = \frac{m}{k} g \left(t + \frac{m}{k} \exp\left(-\frac{kt}{m} \right) \right) + C_2$$

$$x(0) = 0 \Rightarrow C_2 = -\frac{m^2}{k^2} g$$

So, the distance of closed parachute fall is

$$x_1 = \frac{m}{k} g \left(t_0 + \frac{m}{k} \exp\left(-\frac{kt_0}{m} \right) \right) - \frac{m^2}{k^2} g$$

At t_0,

$$V(t_0) = \frac{m}{k} \left(1 - \exp\left(-\frac{kt_0}{m} \right) \right)$$

Starting with the opening parachute t_0 time, the DE becomes

$$m \frac{dv}{dt} = mg - nkv$$

$$m \frac{dv}{dx} \frac{dx}{dt} = mg - nkv$$

$$m \frac{dv}{dx} v = mg - nkv$$

whose solution is

$$v(t) = \frac{m}{nk} \left(g - C_3 \exp\left(-\frac{nkt}{m} \right) \right)$$

Since

$$v(t_0) = \frac{m}{k} g \left(1 - \exp\left(-\frac{kt_0}{m} \right) \right)$$

$$C_3 = \exp\left(\frac{kt_0}{m} \right) ng \left(\frac{1}{n} - 1 + \exp\left(-\frac{kt_0}{m} \right) \right)$$

$$\frac{dx}{dt} = \frac{m}{nk} \left(g - C_3 \exp\left(-\frac{nkt}{m} \right) \right)$$

$$x_2 = \frac{m}{nk} g \left(t + C_3 \frac{m}{nk} \exp\left(-\frac{kt}{m} \right) \right) + C_4$$

$$x_2(t_0) = x_1$$

which means

$$\frac{m}{nk} g \left(t_0 + C_3 \frac{m}{nk} \exp\left(-\frac{kt_0}{m} \right) \right) + C_4 = \frac{m}{k} g \left(t_0 + \frac{m}{k} \exp\left(-\frac{kt_0}{m} \right) \right) - \frac{m^2}{k^2} g$$

One can easily find the constant C_4

$$H = x_1 + x_2$$

$$x_2 = \frac{m}{nk} g \left(t_1 + C_3 \frac{m}{nk} \exp\left(-\frac{kt_1}{m} \right) \right) + C_4$$

$$= H - \frac{m}{k} g \left(t_0 + \frac{m}{k} \exp\left(-\frac{kt_0}{m}\right) \right) - \frac{m^2}{k^2} g$$

The falling time t_1 is in principle solvable from the above equation, but it is tedious and may require a numerical solution technique. Therefore the total falling time is $t_1 + t_0$, the speed he hits the ground is

$$v(t_1) = \frac{m}{nk} \left(g - C_3 \exp\left(-\frac{nkt}{m}\right) \right)$$

We can adjust the t_0 to enable the quickest fall and lightest hit on the ground.

2.6 A Swimmer's Problem

Problem 2.6.1

(1) The total mass for a boat with n boaters is given by $M + nm$. The total force propelling the boat is nf_0. Water resistance is $\mu_0 v$, which acts in the opposite direction. Thus, we have the following DE

$$(M + nm)\frac{dv}{dt} = nf_0 - \mu_0 v$$

(2) This is a separable equation. Thus, we have

$$\int (M + nm)\frac{dv}{nf_0 - \mu_0 v} = \int dt$$

$$-\frac{M + nm}{\mu_0}\ln(nf_0 - \mu_0 v) = t + c$$

With $v(t = 0) = 0$, one can get

$$c = -\frac{M + nm}{\mu_0}\ln(nf_0)$$

Plugging this value of c into the solution, we get

$$v(t) = \frac{nf_0}{\mu_0}\left(1 - \exp\left(-\frac{\mu_0 t}{M + nm}\right)\right)$$

(3) The total travel time T follows the following equation,

$$L = \int_0^T v(t)\, dt = \int_0^T \frac{nf_0}{\mu_0}\left(1 - \exp\left(-\frac{\mu_0 t}{M + nm}\right)\right) dt$$

$$= \frac{nf_0}{\mu_0}\left(T + \frac{M + nm}{\mu_0}\exp\left(-\frac{\mu_0 T}{M + nm}\right) - \frac{M + nm}{\mu_0}\right)$$

Differentiating T wrt n (assuming real variable instead of integer for convenience) leads to $\frac{dT}{dn} < 0$ and, thus, the more boaters, the shorter the time. The answer is "To win, use as many boaters as practical."

An alternate argument: one can compute the boat speed wrt n.

$$\frac{dv}{dn} = \frac{f_0}{\mu_0}\left(1 - \exp\left(-\frac{\mu_0 t}{M + nm}\right)\right) + \frac{nmf_0 t}{(M + nm)^2}\exp\left(-\frac{\mu_0 t}{M + nm}\right)$$

Now

$$\frac{f_0}{\mu_0}\left(1 - \exp\left(-\frac{\mu_0 t}{M + nm}\right)\right) \geq 0 \ \forall\, t \geq 0$$

and

$$\frac{nmf_0\mu_0 t}{\mu_0(M + nm)^2}\exp\left(-\frac{\mu_0 t}{M + nm}\right) > 0$$

Thus,

$$\frac{dv}{dn} > 0 \ \forall \ t \geq 0$$

In other words, v is an increasing function of n. So, at any instant, the more boaters, the faster the boat travels. "The added weight of the boater has not slowed down the boat, as long as the added boater does not pedal backward."

2.7 River Ferryboat-Docking Problem

Problem 2.7.1

Assume the jet's speed is v_0 and the jet's heading angle is α. Without wind, the jet's velocity is $[v_0 \cos \alpha, \ v_0 \sin \alpha]$. With wind, the jet's velocity can be $[v_0 \cos \alpha, \ v_0 \sin \alpha + v_w]$.

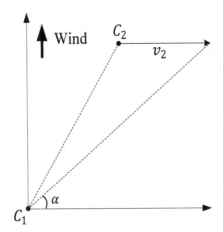

Figure A.5 The jet landing model.

$$v_x = \frac{dx}{dt} = v_0 \cos \alpha = v_0 \frac{x}{\sqrt{x^2 + y^2}} \qquad (A.3)$$

$$v_y = \frac{dy}{dt} = v_0 \sin(\alpha) + v_w = v_0 \frac{y}{\sqrt{x^2 + y^2}} + v_w \qquad (A.4)$$

$$\frac{(A.4)}{(A.3)} \Rightarrow \frac{dy}{dx} = \frac{v_0 \dfrac{y}{\sqrt{x^2 + y^2}} + v_w}{v_0 \dfrac{x}{\sqrt{x^2 + y^2}}} \Rightarrow \frac{dy}{dx} = \frac{y}{x} + \frac{v_w}{v_0} \sqrt{1 + \left(\frac{y}{x}\right)^2}$$

Problem 2.7.2

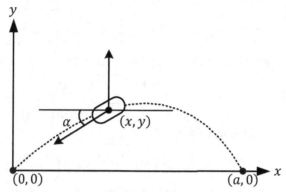

Figure A.6 The ferryboat model.

(1) Establish the DEs for the boat's trajectory,

$$\begin{cases} v_x = \dfrac{dx}{dt} = -V_0 \cos \alpha & \text{(A.5)} \\[2mm] v_y = \dfrac{dy}{dt} = W(x) - V_0 \sin \alpha & \text{(A.6)} \end{cases}$$

We know that

$$\begin{cases} \sin \alpha = \dfrac{y}{\sqrt{x^2 + y^2}} \\[3mm] \cos \alpha = \dfrac{x}{\sqrt{x^2 + y^2}} \end{cases}$$

Dividing Eq. (A.6) by Eq. (A.5), we get

$$\frac{dy}{dx} = \frac{W(x) - V_0 \sin \alpha}{-V_0 \cos \alpha} = \frac{W(x)}{-V_0 \cos \alpha} + \tan \alpha$$

$$= -\frac{\dfrac{w_0 x(a-x)}{a^2}}{V_0 \cos \alpha} + \frac{y}{x} = -\frac{w_0 x(a-x)}{V_0 a^2}\sqrt{1 + \left(\frac{y}{x}\right)^2} + \frac{y}{x}$$

$$= -\frac{x(a-x)}{b^2}\sqrt{1 + \left(\frac{y}{x}\right)^2} + \frac{y}{x}$$

where

$$\frac{1}{b^2} = \frac{w_0}{V_0 a^2}$$

(2) By substituting $v = y/x$, we transform the DE into

$$v + xv' = -\frac{x(a-x)}{b^2}\sqrt{1 + v^2} + v$$

$$v' = -\frac{a-x}{b^2}\sqrt{1 + v^2}$$

By the SOV method, we have

$$\int \frac{dv}{\sqrt{1+v^2}} = -\int \frac{a-x}{b^2} dx$$

$$\ln\left(v + \sqrt{1+v^2}\right) = \frac{x^2 - 2ax}{2b^2} + C$$

With the IC $y(\alpha) = 0$, we find $C = a^2/(2b^2)$. Finally, we have

$$\ln\left(v + \sqrt{1+v^2}\right) = \frac{1}{2}\left(\frac{x-a}{b}\right)^2$$

$$v + \sqrt{1+v^2} = \exp\left(\frac{(x-a)^2}{2b^2}\right)$$

Substituting back v with y/x,

$$\frac{y}{x} + \sqrt{1 + \left(\frac{y}{x}\right)^2} = \exp\left(\frac{(x-a)^2}{2b^2}\right)$$

(3) If the river flows slowly relative to the boat, $b \to \infty$, the result becomes $y \to 0$, which is a straight line.

2.8 Equation for Compound Interest

Problem 2.8.1

(1) Establish the DE

$$dZ = Zrdt - Q_0 t dt$$

which gives

$$\frac{dZ}{dt} = Zr - Q_0 t$$

and IC $Z(0) = Z_0$

(2) We rewrite the DE

$$Z' - Zr = -Q_0 t$$

which is a linear DE. Thus,

$$(Z \exp(-rt))' = -Q_0 t \exp(-rt)$$

$$Z \exp(-rt) = -Q_0 \left(-\frac{\exp(-rt)}{r^2} - \frac{t \exp(-rt)}{r} \right) + C$$

$$Z = \frac{Q_0}{r^2} + \frac{Q_0 t}{r} + C \exp(-rt)$$

Plug in the IC $Z(0) = Z_0$, we have

$$Z_0 = \frac{Q_0}{r^2} + C$$

$$C = Z_0 - \frac{Q_0}{r^2}$$

Finally,

$$Z(t) = \frac{Q_0}{r} t + \frac{Q_0}{r^2} + \left(Z_0 - \frac{Q_0}{r^2} \right) \exp(-rt)$$

Problem 2.8.2

(1) The change of loan dx after time interval dt

$$dx = xrdt - Ddt$$

Thus, we have the equation

$$\frac{dx}{dt} = xr - D$$

with $x(0) = Z$.

(2) The above DE can be solved easily by the SOV as

$$\int \frac{dx}{xr - D} = \int dt$$

$$\frac{1}{r} \ln(xr - D) = t + C$$

$$x(t) = \frac{1}{r}\left(D + \exp(r(t + C))\right)\Bigg\} \Rightarrow C = \frac{1}{r}\ln(rZ - D)$$
$$x(0) = Z$$

$$x(t) = \frac{1}{r}(D + \exp(rt + \ln(rZ - D)))$$

$$= \frac{1}{r}(D + (rZ - D)\exp(rN))$$

(3) Obviously, if $rZ - D > 0$, the loan will grow with time. O.W., it will decrease. Thus, the critical point is $rZ - D = 0$, *i.e.*, $D = rZ$ (The Young's only pays off the interest).

(4) Pay off in N days,

$$x(N) = \frac{1}{r}(D + (rZ - D)\exp(rN)) = 0$$

$$N = -\frac{1}{r}\ln\left(1 - \frac{rZ}{D}\right)$$

(5) Pay off in $N/2$ days,

$$\frac{N}{2} = -\frac{1}{r}\ln\left(1 - \frac{rZ}{E}\right)$$

where E is the daily payment to pay off the loan in half time. We also know

$$N = -\frac{1}{r}\ln\left(1 - \frac{rZ}{E}\right)$$

$$\frac{N}{2} = -\frac{1}{2r}\ln\left(1 - \frac{rZ}{E}\right) = -\frac{1}{r}\ln\sqrt{1 - \frac{rZ}{E}}$$

Comparing it with the previous equation, we have

$$\sqrt{1 - \frac{rZ}{D}} = 1 - \frac{rZ}{E}$$

Let $Y = rZ/D$ and $rZ = YD$. Then,

$$\sqrt{1 - Y} = 1 - \frac{YD}{E}$$

$$E = \frac{YD}{1 - \sqrt{1 - Y}}$$

Problem 2.8.3

(1) The DE can be written as $dZ = Zrdt - Wdt$ which can be solved easily by the SOV method

$$\int_{Z_0}^{Z} \frac{dZ}{Z - \frac{W}{r}} = \int_{0}^{t} rdt$$

$$Z(t) = \frac{W}{r} + \left(Z_0 - \frac{W}{r}\right)\exp(rt)$$

Now, the pay-off time is when

$$Z(T) = \frac{W}{r} + \left(Z_0 - \frac{W}{r}\right)\exp(rt) = 0$$

Solving the above produces

$$T = -\frac{1}{r}\ln\left(1 - \frac{rZ_0}{W}\right)$$

(2) According to the result above,

$$T_1 = T\left(Z_0, W, r \to 0\right)$$

$$= \lim_{r \to 0}\left(-\frac{1}{r}\ln\left(1 - \frac{rZ_0}{W}\right)\right) = -\frac{1}{r}\left(-\frac{rZ_0}{W}\right) = \frac{Z_0}{W}$$

$$T_2 = T\left(Z_0, W, r = \frac{W}{2Z_0}\right) = \frac{Z_0}{W}(2\ln 2).$$

(3) With equation in (1) we have

$$T = -\frac{1}{r}\ln\left(1 - \frac{rZ_0}{W}\right)$$

Now, with payment W_2, the loan is paid off in half time, *i.e.*,

$$\frac{T}{2} = -\frac{1}{r}\ln\left(1 - \frac{rZ_0}{W_2}\right)$$

Solving these two AEs, we get new payment,

$$W_2 = W + \sqrt{W(W - rZ_0)}$$

Problem 2.8.4

Borrower: Mr. Wyze

$z_0 = $ The initial loan amount.

$w_0 = $ Periodic payment per unit time.

$r_w = r = $ Interest rate.

$z_w(t)= $ Amount owed to the bank at time t.

$dz_w= $ Decrement of the loan due to the payments made

$$dz_w = z_w(t)rdt - w_0 dt$$

$$z'_w = z_w r - w_0$$

$$\int_{z_0}^{z_w} \frac{z'_w}{z_w - \frac{w_0}{r}} = \int_{t=0}^{t} r\,dt$$

$$z_w(t) = \frac{w_0}{r} + \left(z_0 - \frac{w_0}{r}\right)\exp(rt)$$

Borrower: Mr. Fulesch

$z_0 = $ The initial loan amount.

$w_0 = $ Periodic payment per unit time.

$r_f = \frac{1}{5}r(1 + t) = $ Interest rate.

$z_f(t)$ = Amount owed to the bank at time t.

dz_f = Decrement of the loan due to the payments made

$$dz_f = z_f(t)r_f dt - w_0 dt$$

$$z_f' = z_f \frac{r}{5}(1+t) - w_0 \tag{A.7}$$

which is a 1st.O linear DE.

$$y' + P(x)y = Q(x)$$

$$P(t) = -\frac{r}{5}(1+t)$$

$$\rho(t) = \exp\left(-\frac{r}{5}\int_0^t (1+t)\,dt\right) = C\exp\left(-\frac{r}{5}\left(t + \frac{t^2}{2}\right)\right)$$

$$Q(t) = -w_0$$

$$\frac{d}{dt}\left(z_f(t)\rho(t)\right) = Q(t)\rho(t)$$

$$z_f(t) = z_0 - w_0\rho(t)\int_0^t \rho(t)\,dt \tag{A.8}$$

Solve for t with Eq. (A.7) and Eq. (A.8).

Problem 2.8.5

With the new periodic payment $(1+\alpha)W_0$,

$$dz = zr\,dt - (1+\alpha)W_0\,dt$$

$$\frac{dz}{z - \dfrac{(1+\alpha)W_0}{r}} = r\,dt$$

$$\int_{z_0}^0 \frac{dz}{z - \dfrac{(1+\alpha)W_0}{r}} = \int_0^{T_\alpha} r\,dt$$

$$\ln\left(\frac{-\dfrac{(1+\alpha)W_0}{r}}{z_0 - \dfrac{(1+\alpha)W_0}{r}}\right) = rT_\alpha$$

$$T_\alpha = \frac{1}{r}\ln\left(\frac{(1+\alpha)W_0}{(1+\alpha)W_0 - z_0 r}\right)$$

Similarly, we find the payoff time with periodic payment W_0

$$T = \frac{1}{r}\ln\left(\frac{W_0}{W_0 - z_0 r}\right)$$

With periodic payment W_0, the total interests paid to the bank is

$$z_0 rT = z_0 \ln\left(\frac{W_0}{W_0 - z_0 r}\right)$$

With periodic payment $(1+\alpha)W_0$, the total interests paid to the bank is

$$Z_0 r T_\alpha = Z_0 \ln\left(\frac{(1+\alpha)W_0}{(1+\alpha)W_0 - Z_0 r}\right)$$

Problem 2.8.6

(1) With double periodic payment, we have

$$dz = zrdt - 2Wdt$$

$$\frac{dz}{z - \frac{2W}{r}} = rdt$$

$$\int_{Z_0}^{0} \frac{dz}{z - \frac{2W}{r}} = \int_{0}^{T_{2w}} rdt$$

$$\ln\left(\frac{-\frac{2W}{r}}{Z_0 - \frac{2W}{r}}\right) = rT_{2w}$$

$$T_{2w} = \frac{1}{r}\ln\left(\frac{2W}{2W - Z_0 r}\right)$$

(2) Doubling rate, we have

$$dz = 2zrdt - W_{2r}dt$$

$$\frac{dz}{z - \frac{W_{2r}}{2r}} = 2rdt$$

$$\int_{Z_0}^{0} \frac{dz}{z - \frac{W_{2r}}{2r}} = \int_{0}^{T_1} 2rdt$$

$$\ln\left(\frac{-\frac{W_{2r}}{2r}}{Z_0 - \frac{W_{2r}}{2r}}\right) = 2rT_1$$

$$W_{2r} = \frac{2Z_0 r}{1 - \exp(-2rT_1)}$$

Problem 2.8.7

(1) With periodic payment nW, we have

$$dz = zrdt - nWdt$$

$$\frac{dz}{z - \frac{nW}{r}} = rdt$$

$$\int_{Z_0}^{0} \frac{dz}{z - \frac{nW}{r}} = \int_{0}^{T_w} rdt$$

$$\ln\left(\frac{-\frac{nW}{r}}{Z_0 - \frac{nW}{r}}\right) = rT_w$$

$$T_w = \frac{1}{r}\ln\left(\frac{nW}{nW - Z_0 r}\right)$$

(2) With interest rate br, we have

$$dz = bzrdt - W_r dt$$

$$\frac{dz}{z - \frac{W_r}{br}} = brdt$$

$$\int_{Z_0}^{0} \frac{dz}{z - \frac{W_r}{br}} = \int_{0}^{T} brdt$$

$$\ln\left(\frac{-\frac{W_r}{br}}{Z_0 - \frac{W_r}{br}}\right) = brT$$

$$W_r = \frac{brZ_0}{1 - \exp(-brT)}$$

Problem 2.8.8

$$dW = Wrdt - W_0 dt$$

and

$$\int_{Z_0}^{0} \frac{dW}{Wr - W_0} = \int_{0}^{T_0} dt = T_0$$

or

$$T_0 = -\frac{1}{r}\ln\left(1 - \frac{rZ_0}{W_0}\right)$$

If payment is ar

$$dW = Wardt - W_0 dt$$

and

$$\int_{Z_0}^{0} \frac{dW}{War - W_0} = \int_{0}^{T_1} dt = T_1$$

or

$$T_1 = -\frac{1}{ar}\ln\left(1 - \frac{arZ_0}{aW_0}\right)$$

Let

$$T(r) = -\frac{1}{r}\ln\left(1 - \frac{rZ_0}{W_0}\right) = f(r)g(r)$$

where

$$f(r) = \frac{1}{r}$$

and

$$g(r) = -\ln\left|1 - \frac{rZ_0}{W_0}\right|$$

T is always greater than zero and $T = f(r)g(r) > 0$

$$f(r) = \frac{1}{r} > 0 \Rightarrow g(r) = -\ln\left|1 - \frac{rZ_0}{W_0}\right| > 0 \Rightarrow \left|1 - \frac{rZ_0}{W_0}\right| < 1$$

$$T_0 = T(r) = f(r)g(r)$$
$$T_1 = T(\alpha r)g(\alpha r)$$

Let

$$A = \frac{Zr}{W_0}$$

Then,

$$\begin{cases} T_0 = -\dfrac{\ln|1 - A|}{r} \\ T_1 = -\dfrac{\ln|1 - \alpha A|}{\alpha r} \end{cases}$$

(1) If $a > 1$,

$$\alpha r > r \Rightarrow \frac{1}{\alpha r} < \frac{1}{r} \Rightarrow f(\alpha r) < f(r) \tag{I}$$

Case 1:

$$g(r) = -\ln(1 - A) \text{ and } g(\alpha r) = -\ln(1 - \alpha A)$$
$$1 > \alpha A > A > 0 \Rightarrow 0 < 1 - \alpha A < 1 - A < 1$$
$$-\ln(1 - \alpha A) > -\ln(1 - A) > 0$$
$$g(\alpha r) > g(r) \tag{II}$$

Case 2:

$$g(r) = -\ln(A - 1) \text{ and } g(\alpha r) = -\ln(\alpha A - 1)$$

Then,

$$1 > \alpha A - 1 > A - 1 > 0$$
$$0 < -\ln(\alpha A - 1) < -\ln(A - 1)$$
$$\Rightarrow g(\alpha r) < g(r) \tag{III}$$

If (I) and (II) are satisfied, it is difficult to get a result.
If (I) and (III) are satisfied, $T_1 < T_0$.
So, the function should be

$$T_0 = -\frac{1}{r}\ln\left(\frac{Zr}{W_0} - 1\right)$$

$$T_1 = -\frac{1}{\alpha r}\ln\left(\frac{Z\alpha r}{W_0} - 1\right)$$

This means $Zr > W_0$.

(2) If $0 < \alpha < 1$,

With a similar method, we have $T_1 > T_0$

(3) If $\alpha = 0$, the DE becomes

$$dW = -W_0 dt$$

coupling with the IC $W(0) = Z_0$, an IVP. Solving it, we get

$$T_1 = \frac{Z_0}{W_0}$$

Chapter 3 Linear DEs of Higher Order

3.1 Classification of Linear DEs

Problem 3.1.1

$$\text{LHS} = y' + y^2 = -\frac{1}{x^2} + \left(\frac{1}{x}\right)^2 = 0 = \text{RHS}$$

For $y = \frac{C}{x}$ where $C \neq 1$ and $C \neq 0$, we have

$$\text{LHS} = -\frac{C}{x^2} + \left(\frac{C}{x}\right)^2 = \frac{C(C-1)}{x^2} \neq 0$$

Problem 3.1.2

From $y = x^3$, we know $y' = 3x^2$ and $y'' = 6x$. Thus,

$$\text{LHS} = yy''$$
$$= x^3 \times 6x$$
$$= 6x^4 = \text{RHS}$$

For $y = Cx^3$ where $C \neq 1$, we have

$$\text{LHS} = Cx^3 \times 6Cx$$
$$= 6C^2x^4 \neq 6x^4$$

Problem 3.1.3

For $y_1 = 1$, we know that $y_1' = 0$ and $y_1'' = 0$. Thus,

$$\text{LHS} = y_1 y_1'' + (y_1')^2$$
$$= 0 = \text{RHS}$$

For $y_2 = \sqrt{x}$, we know that

$$y_2' = \frac{1}{2\sqrt{x}}$$

$$y_2'' = -\frac{1}{4x\sqrt{x}}$$

Thus,

$$\text{LHS} = \sqrt{x}\left(-\frac{1}{4x\sqrt{x}}\right) + \left(\frac{1}{2\sqrt{x}}\right)^2$$
$$= 0 = \text{RHS}$$

However, for $y = y_1 + y_2$, we have

$$\text{LHS} = (\sqrt{x} + 1)\left(-\frac{1}{4x\sqrt{x}}\right) + \left(\frac{1}{2\sqrt{x}}\right)^2$$

$$= -\frac{1}{4x\sqrt{x}} \neq 0$$

Problem 3.1.4

(a) y_C is the GS to the Homo DE, thus,
$$y_C'' + py_C' + qy_C = 0$$
y_P is a PS to the InHomo DE. Thus,
$$y_P'' + py_P' + qy_P = f(x)$$
Let $y(x) = y_C + y_P$. Now, we have
$$\begin{aligned} \text{LHS} &= (y_C + y_P)'' + p(y_C + y_P)' + q(y_C + y_P) \\ &= (y_C'' + py_C' + qy_C) + (y_P'' + py_P' + qy_P) \\ &= 0 + f(x) = \text{RHS} \end{aligned}$$

(b) From part (a), we get the GS
$$y = y_C + y_P = 1 + C_1 \cos x + C_2 \sin x$$
Applying the IC's, we have
$$y(0) = -1 = 1 + C_1$$
$$y'(0) = -1 = C_2$$
yield $C_1 = -2$ and $C_2 = -1$. The PS is
$$y(x) = 1 - 2\cos x - \sin x$$

3.2 Linear Independence

Problem 3.2.1

Part 1:

$$
\begin{aligned}
L_1L_2x(t) &= L_1\big((D^2 + \alpha_2 D + \beta_2)x\big)\\
&= L_1(x'' + \alpha_2 x' + \beta_2 x)\\
&= (D^2 + \alpha_1 D + \beta_1)(x'' + \alpha_2 x' + \beta_2 x)\\
&= x^{(4)} + (\alpha_1 + \alpha_2)x''' + (\alpha_1\alpha_2 + \beta_1 + \beta_2)x'' + (\alpha_1\beta_2 + \alpha_2\beta_1)x'\\
&\quad +\beta_1\beta_2 x\\
L_2L_1x(t) &= L_2\big((D^2 + \alpha_1 D + \beta_1)x\big)\\
&= L_2(x'' + \alpha_1 x' + \beta_1 x)\\
&= (D^2 + \alpha_2 D + \beta_2)(x'' + \alpha_1 x' + \beta_1 x)\\
&= x^{(4)} + (\alpha_1 + \alpha_2)x''' + (\alpha_1\alpha_2 + \beta_1 + \beta_2)x'' + (\alpha_1\beta_2 + \alpha_2\beta_1)x'\\
&\quad +\beta_1\beta_2 x
\end{aligned}
$$

$$L_1L_2x(t) = L_2L_1x(t)$$

Part 2:

$$L_1 \equiv D + t, \qquad L_2 \equiv tD + 1$$
$$
\begin{aligned}
L_1L_2x(t) &= L_1\big((tD + 1)x\big)\\
&= L_1(tx' + x)\\
&= (D + t)(tx' + x)\\
&= tx'' + (t^2 + 1)x' + tx\\
L_2L_1x(t) &= L_2\big((D + t)x\big)\\
&= L_2(x' + tx)\\
&= (tD + 1)(x' + tx)\\
&= tx'' + (t^2 + 1)x' + 2tx
\end{aligned}
$$
$$L_1L_2x(t) \neq L_2L_1x(t)$$

Problem 3.2.2

Note that

$$\cosh x = \frac{\exp(x) + \exp(-x)}{2}; \sinh x = \frac{\exp(x) - \exp(-x)}{2}$$

i.e.,

$$\cosh x + \sinh x = \exp(x)$$

Thus, the linear combination of the three functions can be zero and, thus, they are LD.

Problem 3.2.3

Let $c_1 = 1, c_2 = c_3 = 0$. Then, we have
$$c_1 \cdot 0 + c_2 \sin x + c_3 \exp(x) = 0$$
So, these functions are LD.

Note: 0 is always LD with any other function.

Problem 3.2.4

Let $c_1 = 15, c_2 = -16, c_3 = -6$. Then, we have
$$15 \cdot 2x - 16 \cdot 3x^2 - 6(5x - 8x^2) = 0$$
So, these functions are LD.

Problem 3.2.5

Let $c_1 = 1, c_2 = -1, c_3 = 1$. Then, we have
$$x^2 - x^3 + (x^3 - x^2) = 0$$
So, these functions are LD.

Problem 3.2.6

$$f(x) = \pi$$
$$f'(x) = 0$$
$$g(x) = \cos^2 x + \sin^2 x = 1$$
$$g'(x) = 0$$

The Wronskian is

$$W(f, g) = \begin{vmatrix} f & g \\ f' & g' \end{vmatrix}$$
$$= \begin{vmatrix} \pi & 1 \\ 0 & 0 \end{vmatrix} = 0$$

Thus, $f(x)$ and $g(x)$ are not LI.

Problem 3.2.7

(1) Since both y_1 and y_2 are the solutions to the DE
$$A(x)y'' + B(x)y' + C(x)y = 0$$
we get
$$A(x)y_1'' = -B(x)y_1' - C(x)y_1$$
$$A(x)y_2'' = -B(x)y_2' - C(x)y_2$$

The Wronskian of y_1 and y_2 is
$$W(y_1, y_2) = \begin{vmatrix} y_1 & y_2 \\ y_1' & y_2' \end{vmatrix}$$
$$= y_1 y_2' - y_2 y_1'$$

Thus, we have

$$A(x)\frac{dW}{dx} = A(x)(y_1y_2' - y_2y_1')'$$
$$= A(x)(y_1y_2'' - y_2y_1'')$$
$$= y_1(Ay_2'') - y_2(Ay_1'')$$

Substituting Ay_1'' and Ay_2'', we have

$$A(x)\frac{dW}{dx} = y_1(-By_2' - Cy_2) - y_2(-By_1' - Cy_1)$$
$$= -B(y_1y_2' - y_2y_1')$$
$$= -B(x)W(x)$$

(2) This separable DE can be solved as

$$\frac{dW}{W} = -\frac{B}{A}dx$$
$$\ln W = \int -\frac{B(x)}{A(x)}dx + K$$
$$W = K\exp\left(-\int \frac{B(x)}{A(x)}dx\right)$$

Problem 3.2.8

For $f_i(x) = x^i$, we know that

$$f_i^{(i)}(x) = i!, \qquad f_i^{(j)}(x) = 0, \qquad j > i$$

Thus, the Wronskian

$$W(f_0, f_1, \dots, f_n) = \begin{vmatrix} f_0 & f_1 & \cdots & f_n \\ f_0' & f_1' & \cdots & f_n' \\ \vdots & \vdots & \ddots & \vdots \\ f_0^{(n)} & f_1^{(n)} & \cdots & f_n^{(n)} \end{vmatrix}$$

is actually an upper triangular matrix with diagonal elements

$$f_i^{(i)} = i!$$

Thus, we have

$$W(f_0, f_1, \dots, f_n) = \prod_{i=0}^{n} i! \neq 0$$

We know functions $f_0, f_1, \dots, f_n$ are LI.

Problem 3.2.9

Suppose $x_1, \dots, x_n$ are LI solutions to the Homo DE.

$$W(t) = \begin{vmatrix} x_1(t) & x_2(t) & \cdots & x_n(t) \\ x_1'(t) & x_2'(t) & \cdots & x_n'(t) \\ \vdots & \vdots & \ddots & \vdots \\ x_1^{(n-1)}(t) & x_2^{(n-1)}(t) & \cdots & x_n^{(n-1)}(t) \end{vmatrix}$$

Taking the derivative of $W(t)$ we get

$$\frac{W(t)}{dt} = \frac{d}{dt} \begin{vmatrix} x_1(t) & x_2(t) & \cdots & x_n(t) \\ x_1'(t) & x_2'(t) & \cdots & x_n'(t) \\ \vdots & \vdots & \ddots & \vdots \\ x_1^{(n-1)}(t) & x_2^{(n-1)}(t) & \cdots & x_n^{(n-1)}(t) \end{vmatrix}$$

$$= \begin{vmatrix} x_1'(t) & x_2'(t) & \cdots & x'_n(t) \\ x_1'(t) & x_2'(t) & \cdots & x_n'(t) \\ \vdots & \vdots & \ddots & \vdots \\ x_1^{(n-1)}(t) & x_2^{(n-1)}(t) & \cdots & x_n^{(n-1)}(t) \end{vmatrix}$$

$$+ \begin{vmatrix} x_1(t) & x_2(t) & \cdots & x_n(t) \\ x_1''(t) & x_2''(t) & \cdots & x_n''(t) \\ \vdots & \vdots & \ddots & \vdots \\ x_1^{(n-1)}(t) & x_2^{(n-1)}(t) & \cdots & x_n^{(n-1)}(t) \end{vmatrix} + \cdots$$

$$+ \begin{vmatrix} x_1(t) & x_2(t) & \cdots & x_n(t) \\ x_1'(t) & x_2'(t) & \cdots & x_n'(t) \\ \vdots & \vdots & \ddots & \vdots \\ x_1^{(n)}(t) & x_2^{(n)}(t) & \cdots & x_n^{(n)}(t) \end{vmatrix}$$

All terms except the last one are zero. So,

$$\frac{W(t)}{dt} = \begin{vmatrix} x_1(t) & x_2(t) & \cdots & x_n(t) \\ x_1'(t) & x_2'(t) & \cdots & x_n'(t) \\ \vdots & \vdots & \ddots & \vdots \\ x_1^{(n)}(t) & x_2^{(n)}(t) & \cdots & x_n^{(n)}(t) \end{vmatrix}$$

Plugging $x_1, \ldots, x_n$ into the Homo DE, we have

$$x_1^n = P_1(t)x_1^{n-1} + \cdots + P_n(t)x_1$$
$$x_2^n = P_1(t)x_2^{n-1} + \cdots + P_n(t)x_2$$
$$\vdots$$
$$x_n^n = P_1(t)x_n^{n-1} + \cdots + P_n(t)x_n$$

Put this into the DE wrt W

$$\frac{dW(t)}{dt} = \begin{vmatrix} x_1(t) & x_2(t) & \cdots & x_n(t) \\ x_1'(t) & x_2'(t) & \cdots & x_n'(t) \\ \vdots & \vdots & \ddots & \vdots \\ x_1^{(n)}(t) & x_2^{(n)}(t) & \cdots & x_n^{(n)}(t) \end{vmatrix}$$

$$\times \begin{vmatrix} x_1(t) & x_2(t) & \cdots & x_n(t) \\ x_1'(t) & x_2'(t) & \cdots & x_n'(t) \\ \vdots & \vdots & \ddots & \vdots \\ -P_1(t)x_1^{(n-1)}(t) & -P_1(t)x_2^{(n-1)}(t) & \cdots & -P_1(t)x_n^{(n-1)}(t) \end{vmatrix}$$

$$+\begin{vmatrix} x_1(t) & x_2(t) & \cdots & x_n(t) \\ x_1'(t) & x_2'(t) & \cdots & x_n'(t) \\ \vdots & \vdots & \ddots & \vdots \\ -P_2(t)x_1^{(n-2)}(t) & -P_2(t)x_2^{(n-2)}(t) & \cdots & -P_2(t)x_n^{(n-2)}(t) \end{vmatrix}+\dots$$

$$+\begin{vmatrix} x_1(t) & x_2(t) & \cdots & x_n(t) \\ x_1'(t) & x_2'(t) & \cdots & x_n'(t) \\ \vdots & \vdots & \ddots & \vdots \\ -P_n(t)x_1^{(n-1)}(t) & -P_n(t)x_2^{(n-1)}(t) & \cdots & -P_n(t)x_n^{(n-1)}(t) \end{vmatrix}$$

Now everything except the first term cancels out

$$\frac{dW(t)}{dt}=\begin{vmatrix} x_1(t) & x_2(t) & \cdots & x_n(t) \\ x_1'(t) & x_2'(t) & \cdots & x_n'(t) \\ \vdots & \vdots & \ddots & \vdots \\ -P_1(t)x_1^{(n-1)}(t) & -P_1(t)x_2^{(n-1)}(t) & \cdots & -P_1(t)x_n^{(n-1)}(t) \end{vmatrix}$$

$$=-P_1(t)\begin{vmatrix} x_1(t) & x_2(t) & \cdots & x_n(t) \\ x_1'(t) & x_2'(t) & \cdots & x_n'(t) \\ \vdots & \vdots & \ddots & \vdots \\ x_1^{(n-1)}(t) & x_2^{(n-1)}(t) & \cdots & x_n^{(n-1)}(t) \end{vmatrix}$$

$$=-P_1(t)W(t)$$

Thus,

$$\frac{dW(t)}{dt}=-P_1(t)W(t)$$

where $W(t)$ is a function of $P_1(t)$. Easily, we solve the above DE to get

$$W(t)=C\exp\left(-\int P_1(t)\,dt\right)$$

Problem 3.2.10

We know that $\exp\big((r_1+r_2+r_3)x\big)>0\ \forall x, r_1, r_2$ and r_3. Notice that

$$\begin{vmatrix} 1 & 1 & 1 \\ r_1 & r_2 & r_3 \\ r_1^2 & r_2^2 & r_3^2 \end{vmatrix}=(r_2-r_1)(r_3-r_1)(r_3-r_2)$$

is a Vandermonde determinant which, for distinct r_1, r_2 and r_3, is non-zero. Therefore, the Wronskian is

$$W=\exp\big((r_1+r_2+r_3)x\big)\begin{vmatrix} 1 & 1 & 1 \\ r_1 & r_2 & r_3 \\ r_1^2 & r_2^2 & r_3^2 \end{vmatrix}\neq 0$$

Thus, $\exp(r_1x)$, $\exp(r_2x)$, and $\exp(r_3x)$ are LI.

Problem 3.2.11

First, if $\exists\ 1 \le i, j \le n,\ i \ne j$ such that $r_i = r_j$ from the property of determinants, we know that $V = 0$. Now, suppose r_i's are mutually distinct. Some simple calculations suggest

$$V = \prod_{1 \le j < i \le n} (r_i - r_j)$$

We will prove this by induction. It is obviously true for $n = 2$. Suppose it is true for $n = k$. Then, for $n = k + 1$, we have

$$V_{k+1} = \begin{vmatrix} 1 & 1 & \cdots & 1 \\ r_1 & r_2 & \cdots & r_{k+1} \\ \vdots & \vdots & \ddots & \vdots \\ r_1^k & r_2^k & \cdots & r_{k+1}^k \end{vmatrix}$$

$$= \begin{vmatrix} 0 & 0 & \cdots & 0 & 1 \\ r_1 - r_{k+1} & r_2 - r_{k+1} & \cdots & r_k - r_{k+1} & r_{k+1} \\ \vdots & \vdots & \cdots & \ddots & \vdots \\ r_1^k - r_{k+1}^k & r_2^k - r_{k+1}^k & \cdots & r_k^k - r_{k+1}^k & r_{k+1}^k \end{vmatrix}$$

Similarly, without changing V, we can subtract, in order, r_{k+1} times row k from row $k + 1, r_{k+1}$ times row $k - 1$ from row k, etc, until we subtract r_{k+1} times row 2 from row 3.

$$V_{k+1} = \begin{vmatrix} 0 & \cdots & 0 & 1 \\ (r_1 - r_{k+1})r_1^0 & \cdots & (r_k - r_{k+1})\, r_k^0 & 0 \\ \vdots & \ddots & \vdots & \vdots \\ (r_1 - r_{k+1})r_1^{k-1} & \cdots & (r_k - r_{k+1})r_k^{k-1} & 0 \end{vmatrix}$$

$$= (-1)^k V_k \prod_{i=1}^{k} (r_i - r_{k+1})$$

$$= V_k \prod_{i=1}^{k} (r_{k+1} - r_i)$$

$$= \prod_{1 \le j < i \le k+1} (r_i - r_j)$$

Here finishes the proof.

Problem 3.2.12

We get the Wronskian

$$W(f_1, f_2, \ldots, f_n) = \begin{vmatrix} f_1 & f_2 & \cdots & f_n \\ f_1' & f_2' & \cdots & f_n' \\ \vdots & \vdots & \ddots & \vdots \\ f_1^{(n-1)} & f_2^{(n-1)} & \cdots & f_n^{(n-1)} \end{vmatrix}$$

$$= \begin{vmatrix} \exp(r_1 x) & \exp(r_2 x) & \cdots & \exp(r_n x) \\ r_1 \exp(r_1 x) & r_2 \exp(r_2 x) & \cdots & r_n \exp(r_n x) \\ \vdots & \vdots & \ddots & \vdots \\ r_1^{n-1} \exp(r_1 x) & r_2^{n-1} \exp(r_2 x) & \cdots & r_n^{n-1} \exp(r_n x) \end{vmatrix}$$

$$= \exp\big((r_1 + r_2 + \cdots + r_n)x\big) \begin{vmatrix} 1 & 1 & \cdots & 1 \\ r_1 & r_2 & \cdots & r_n \\ \vdots & \vdots & \ddots & \vdots \\ r_1^{n-1} & r_2^{n-1} & \cdots & r_n^{n-1} \end{vmatrix}$$

Since $\exp\big((r_1 + r_2 + \cdots + r_n)x\big) > 0$, and the second part is a Vandermonde determinant, we know that for $r_1, r_2, \ldots, r_n$ that are distinct, the Wronskian is non-zero. Hence, f_i are LI.

3.3 Homogeneous DEs

Problem 3.3.1

(1) The roots

$$x_{1,2} = \frac{1}{2}\left(-i \pm \sqrt{i^2 - 8}\right)$$
$$= i, -2i$$

(2) The roots

$$x_{1,2} = \frac{1}{2}\left(2i \pm \sqrt{(2i)^2 - 12}\right)$$
$$= 3i, -i$$

Problem 3.3.2

(a) We know that any complex number z can be written as $z = a + ib$ where a and b are the real and imaginary parts respectively, and this number can also be written in polar coordinates as

$$z = a + ib$$
$$= r(\cos\theta + i\sin\theta)$$
$$= r\exp(i\theta)$$

where $r = \sqrt{a^2 + b^2}$, $\theta = \tan^{-1}\left(\frac{b}{a}\right)$, and $\exp(i\theta) = \cos\theta + i\sin\theta$ according to the Euler's formula. We can express any complex numbers in any of the three forms, e.g.,

$$4 = 4 + i0 = 4\exp(i0)$$
$$-2 = -2 + i0 = 2\exp(i\pi)$$
$$3i = 0 + 3i = 3\exp\left(i\frac{\pi}{2}\right)$$
$$1 + i = \sqrt{2}\exp\left(i\frac{\pi}{4}\right)$$
$$-1 + i\sqrt{3} = 2\exp\left(i\frac{2\pi}{3}\right)$$

For the last two complex numbers, we offer more details:

$$1 + i \Rightarrow r = \sqrt{2}, \qquad \theta = \frac{\pi}{4} \Rightarrow 1 + i = \sqrt{2}\exp\left(i\frac{\pi}{4}\right)$$
$$-1 + i\sqrt{3} \Rightarrow r = 2, \qquad \theta = \frac{2}{3}\pi \Rightarrow -1 + i\sqrt{3} = 2\exp\left(i\frac{2\pi}{3}\right)$$

(b) We have

$$2 - 2i\sqrt{3} = 4\exp\left(-i\frac{\pi}{3}\right)$$
$$-2 + 2i\sqrt{3} = 4\exp\left(i\frac{2\pi}{3}\right)$$

Thus,

$$2 - 2i\sqrt{3} = \left(\pm 2 \exp\left(-i\frac{\pi}{6}\right)\right)^2$$

$$-2 + 2i\sqrt{3} = \left(\pm 2 \exp\left(i\frac{\pi}{3}\right)\right)^2$$

Problem 3.3.3

Let $t = \exp(u)$. Then, $u = \ln t$

$$tx' = \frac{dx}{du}$$

$$t^2 x'' = \frac{d^2 x}{du^2} - \frac{dx}{du}$$

$$t^3 x''' = \frac{d^3 x}{du^3} - 3\frac{d^2 x}{du^2} + 2\frac{dx}{du}$$

Here $x' = \frac{dx}{du}$

$$x''' - 3x'' + 2x' + 6(x'' - x') + 7x' + x = 0$$

$$x''' + 3x'' + 3x' + x = 0$$

$$r^3 + 3r^2 + 3r + 1 = 0$$

$$(r + 1)^3 = 0$$

$$r_1 = r_2 = r_3 = -1 \qquad .$$

The GS is

$$x(u) = C_1 \exp(-u) + C_2 u \exp(-u) + C_3 u^2 \exp(-u)$$

$$x(t) = C_1 t^{-1} + C_2 t^{-1} \ln t + C_3 t^{-1} \ln^2 t$$

$$x_1 = t^{-1}, \qquad x_2 = t^{-1} \ln t, \qquad x_3 = t^{-1} \ln^2 t$$

$$W(x_1, x_2, x_3) =$$

$$\begin{vmatrix} t^{-1} & -t^{-2} & 2t^{-3} \\ t^{-1}\ln(t) & t^{-2}(1 - \ln(t)) & t^{-3}(2\ln(t) - 3) \\ t^{-1}\ln^2(t) & t^{-2}(2\ln(t) - \ln^2(t)) & t^{-3}(2 - 6\ln(t) + 2\ln^2(t)) \end{vmatrix}$$

$$= t^{-1}t^{-2}t^{-3} \begin{vmatrix} 1 & \ln(t) & \ln^2(t) \\ -1 & 1 - \ln(t) & 2\ln(t) - \ln^2(t) \\ 2 & 2\ln(t) - 3 & 2 - 6\ln(t) + 2\ln^2(t) \end{vmatrix}$$

$$= t^{-6} \begin{vmatrix} 1 & \ln(t) & \ln^2(t) \\ 0 & 1 & 2\ln(t) \\ 0 & -3 & 2 - 6\ln(t) \end{vmatrix} = t^{-6} \begin{vmatrix} 1 & 2\ln(t) \\ -3 & 2 - 6\ln(t) \end{vmatrix}$$

$$= t^{-6}(2 - 6\ln(t) + 6\ln(t)) = 2t^{-6}$$

$$W(x_1, x_2, x_3) = \frac{2}{t^6}$$

Problem 3.3.4

This is a 2nd.O c-coeff DE whose C-Eq is

$$r^2 + 3r + 2 = 0$$
$$(r+1)(r+2) = 0$$

whose roots are

$$r_{1,2} = -1, -2$$

Thus, the GS is

$$y = C_1 \exp(-x) + C_2 \exp(-2x)$$

Given IC $y(0) = 1, y'(0) = 6$, we have

$$\begin{cases} C_1 + C_2 = 1 \\ -C_1 - 2C_2 = 6 \end{cases}$$

This gives

$$\begin{cases} C_1 = 8 \\ C_2 = -7 \end{cases}$$

Therefore,

$$y(x) = 8\exp(-x) - 7\exp(-2x)$$

At the highest point, the velocity vanishes, which means

$$y'(x) = -8\exp(-x) + 14\exp(-2x) = 0$$

Solving this, we get

$$x = \ln\frac{7}{4}$$

and

$$y\left(\ln\frac{7}{4}\right) = \frac{16}{7}$$

Thus, the highest point is

$$\left(\ln\frac{7}{4}, \frac{16}{7}\right)$$

Problem 3.3.5

This is a 2nd.O c-coeff DE whose C-Eq is

$$r^2 + 3r + 2 = 0$$
$$(r+1)(r+2) = 0$$

whose roots are

$$r_{1,2} = -1, -2$$

Thus, the GS is

$$y = C_1 \exp(-x) + C_2 \exp(-2x)$$

From the given IC, we have for y_1

$$\begin{cases} C_1 + C_2 = 3 \\ -C_1 - 2C_2 = 1 \end{cases}$$

This gives

$$\begin{cases} C_1 = 7 \\ C_2 = -4 \end{cases}$$

$$y_1 = 7\exp(-x) - 4\exp(-2x)$$

Similarly, we have for y_2

$$\begin{cases} C_1 + C_2 = 0 \\ -C_1 - 2C_2 = 1 \end{cases}$$

This gives

$$\begin{cases} C_1 = 1 \\ C_2 = -1 \end{cases}$$

$$y_2 = \exp(-x) - \exp(-2x)$$

To find their intersection, let

$$y_1(x) = y_2(x)$$

This gives

$$7\exp(-x) - 4\exp(-2x) = \exp(-x) - \exp(-2x)$$

$$x = -\ln 2$$

Thus,

$$y_{1,2}(-\ln 2) = -2$$

Hence, the intersection is

$$(-\ln 2, -2)$$

Problem 3.3.6

The C-Eq is

$$3r^3 + 2r^2 = 0$$

whose roots are

$$r_{1,2,3} = 0, 0, -\frac{2}{3}$$

Thus, the GS is

$$y = C_1 + C_2 x + C_3 \exp\left(-\frac{2}{3}x\right)$$

Given the IC, we get

$$\begin{cases} C_1 + C_3 = -1 \\ C_2 - \frac{2}{3}C_3 = 0 \\ \frac{4}{9}C_3 = 1 \end{cases}$$

which gives

$$\begin{cases} C_1 = -\dfrac{13}{4} \\[2mm] C_2 = \dfrac{3}{2} \\[2mm] C_3 = \dfrac{9}{4} \end{cases}$$

Therefore, the PS is

$$y = \frac{1}{4}\left(-13 + 6x + 9\exp\left(-\frac{2}{3}x\right)\right)$$

Problem 3.3.7

From the solution, we know that the C-Eq has three repeated roots $-2i$ and three repeated roots $2i$. Thus, the C-Eq should be

$$(r + 2i)^3(r - 2i)^3 = 0$$
$$r^6 + 12r^4 + 48r^2 + 64 = 0$$

and the DE is

$$y^{(6)} + 12y^{(4)} + 48y'' + 64 = 0$$

Problem 3.3.8

The C-Eq is

$$r^3 - 1 = 0$$
$$(r - 1)(r^2 + r + 1) = 0$$

whose roots are

$$r_{1,2,3} = 1, \frac{-1 + \sqrt{3}i}{2}, \frac{-1 - \sqrt{3}i}{2}$$

Thus, the GS is

$$y = C_1 \exp(x) + C_2 \exp\left(\frac{-1 + \sqrt{3}i}{2}x\right) + C_3 \exp\left(\frac{-1 - \sqrt{3}i}{2}x\right)$$

From the IC, we have

$$\begin{cases} C_1 + C_2 + C_3 = 1 \\[2mm] C_1 + \dfrac{-1 + \sqrt{3}i}{2}C_2 + \dfrac{-1 - \sqrt{3}i}{2}C_3 = 0 \\[2mm] C_1 + \dfrac{-1 - \sqrt{3}i}{2}C_2 + \dfrac{-1 + \sqrt{3}i}{2}C_3 = 0 \end{cases}$$

This gives

$$C_1 = C_2 = C_3 = \frac{1}{3}$$

Thus, we have

$$y = \frac{1}{3}\exp(x) + \frac{1}{3}\exp\left(\frac{-1+\sqrt{3}i}{2}x\right) + \frac{1}{3}\exp\left(\frac{-1-\sqrt{3}i}{2}x\right)$$

which can be written as

$$y = \frac{1}{3}\exp(x) + \frac{2}{3}\exp\left(-\frac{1}{2}x\right)\cos\frac{\sqrt{3}}{2}x$$

Problem 3.3.9

For the 4th-order c-coeff DE whose C-Eq is

$$r^4 - r^3 - r^2 - r - 2 = 0$$
$$(r+1)(r-2)(r^2+1) = 0$$

The GS is

$$y = C_1 \exp(-x) + C_2 \exp(2x) + C_3 \cos x + C_4 \sin x$$

From the given IC, we have

$$\begin{cases} C_1 + C_2 + C_3 = 0 \\ -C_1 + 2C_2 + C_4 = 0 \\ C_1 + 4C_2 - C_3 = 0 \\ -C_1 + 8C_2 - C_4 = 30 \end{cases}$$

Solving for the IC we have,

$$C_1 = -5, \qquad C_2 = 2, \qquad C_3 = 3, \qquad C_4 = -9$$

Thus, the PS is

$$y = -5\exp(-x) + 2\exp(2x) + 3\cos x - 9\sin x$$

Problem 3.3.10

If $x > 0$, the DE is $y'' + y = 0$ with the GS

$$y = A\cos x + B\sin x$$

If $x < 0$, the DE becomes $y'' - y = 0$ with the GS

$$y = C\exp(x) + D\exp(-x)$$

Applying the IC's $y_1(0) = 1$ and $y_1'(0) = 0$, we have

$$\begin{cases} A = 1 \\ B = 0 \end{cases}, \qquad \begin{cases} C + D = 1 \\ C - D = 0 \end{cases} \Rightarrow C = D = \frac{1}{2}$$

Thus,

$$y = \begin{cases} \cos x, & x \geq 0 \\ \cosh x, & x < 0 \end{cases}$$

Similarly, one can get y_2,

$$y = \begin{cases} \sin x, & x \geq 0 \\ \sinh x, & x < 0 \end{cases}$$

Problem 3.3.11

This is a higher order c-coeff DE whose corresponding C-Eq is

$$r^3(r-2)(r+3)(r^2+1) = 0$$

whose roots are $r_{1,2,3,4,5,6,7} = 0, 0, 0, 2, -3, i, -i$

So, the solution is

$$y = C_1 + C_2 x + C_3 x^2 + C_4 \exp(2x) + C_5 \exp(-3x) + C_6 \sin x + C_7 \cos x$$

Problem 3.3.12

The C-Eq is

$$r^2 - 2r + 2 = 0$$
$$r_{1,2} = 1 \pm i$$

The GS is

$$y = \exp(x)\,(C_1 \sin x + C_2 \cos x)$$

From the given IC, we have

$$C_{1,2} = 5, 0$$

Thus, we have

$$y = 5 \exp(x) \sin x$$

Problem 3.3.13

The C-Eq is

$$r^3 + 9r = 0$$
$$r_{1,2,3} = 0, 3i, -3i$$

The GS is

$$y = C_1 + C_2 \sin 3x + C_3 \cos 3x$$

From the given IC, we have

$$\begin{cases} C_1 + C_3 = 3 \\ 3C_2 = -1 \\ -9C_3 = 2 \end{cases}$$

This gives

$$\begin{cases} C_1 = \dfrac{29}{9} \\ C_2 = -\dfrac{1}{3} \\ C_3 = -\dfrac{2}{9} \end{cases}$$

Thus, the solution is

$$y = \frac{29}{9} - \frac{1}{3}\sin 3x - \frac{2}{9}\cos 3x$$

Problem 3.3.14

The C-Eq is

$$(r-1)^3(r-2)^2(r-3)(r^2+9) = 0$$

whose roots are

$$r_{1,2,3,4,5,6,7,8} = 1, 1, 1, 2, 2, 3i, -3i$$

Thus, the GS is

$$y = (C_1 + C_2 x + C_3 x^2)\exp(x) + (C_4 + C_5 x)\exp(2x) + C_6 \exp(3x)$$
$$+C_7 \cos 3x + C_8 \sin 3x$$

Problem 3.3.15

This is a 2nd.O c-coeff DE whose C-Eq is

$$r^2 - r - 15 = 0$$

with roots

$$r_{1,2} = \frac{1 \pm \sqrt{61}}{2}$$

Thus, the GS is

$$y = c_1 \exp\left(\frac{1 + \sqrt{61}}{2}x\right) + c_2 \exp\left(\frac{1 - \sqrt{61}}{2}x\right)$$

Problem 3.3.16

The C-Eq is

$$9r^2 - 12r + 4 = 0$$
$$(3r - 2)^2 = 0$$

whose roots are

$$r_{1,2} = \frac{2}{3}, \frac{2}{3}$$

Thus, the GS is

$$y(x) = c_1 \exp\left(\frac{2x}{3}\right) + c_2\, x \exp\left(\frac{2x}{3}\right)$$

Problem 3.3.17

(1) This is a kth order c-coeff DE with C-Eq $(r - r_1)^{k_1} = 0$. Since $r = r_1$ has k_1 repeated roots, we have

$$y(x) = \left(C_1 + C_2 x + C_3 x^2 + \cdots + C_{k_1} x^{k_1 - 1}\right)\exp(r_1 x)$$

(2) The C-Eq for this DE is

$$(r - r_1)^{k_1}(r - r_2)^{k_2} \ldots (r - r_n)^{k_n} = 0$$

Thus, root $r = r_1$ repeats k_1 times, root $r = r_2$ repeats k_2 times,..., root $r = r_n$ repeats k_n times.

$$y(x) = (C_{11} + C_{12}x + \cdots + C_{1k}x^{k_1 - 1})\exp(r_1 x)$$
$$+ \left(C_{21} + C_{22}x + \cdots + C_{2k_2}x^{k_2 - 1}\right)\exp(r_2 x) + \cdots$$
$$+ \left(C_{n1} + C_{n2}x + \cdots + C_{nk_n}x^{k_n - 1}\right)\exp(r_n x)$$

where C_{ij} are constants.

Problem 3.3.18

Let's find the GS to the corresponding Homo DE

$$\left(x\frac{d}{dx} - \alpha\right)^n y(x) = 0$$

Let $x = \exp(t)$, we have

$$\left(\exp(t)\frac{d}{\exp(t)\,dt} - \alpha\right)^n y = 0$$

$$\left(\frac{d}{dt} - \alpha\right)^n y = 0$$

The C-Eq is $(r - \alpha)^n = 0$. There are n identical real roots.

So, the GS to the homogeneous DE is

$$y_C(t) = (C_1 + C_2 t + C_3 t^2 + \cdots + C_n t^{n-1})\exp(\alpha t)$$

$$= \left(\sum_{k=1}^{n} C_k t^{k-1}\right)\exp(\alpha t)$$

Then, plugging x back, we have

$$y_C(x) = \left(\sum_{k=1}^{n} C_k (\ln x)^{k-1}\right) x^\alpha$$

Finding the PS requires considering three cases:

Case 1 ($\alpha = 0$):

The original DE becomes

$$y^{(n)} = \exp(t)$$

whose PS is

$$y_P = \exp(t)$$

The GS is

$$y(t) = (C_1 + C_2 t + C_3 t^2 + \cdots + C_n t^{n-1}) + \exp(t)$$
$$y(x) = (C_1 + C_2 \ln x + C_3 (\ln x)^2 + \cdots + C_n (\ln x)^{n-1}) + x$$

Case 2 ($\alpha = 1$):

We select the TS

$$y_P = At^n \exp(t)$$

Thus,

$$\text{LHS} = \left(\frac{d}{dt} - 1\right)^n (At^n \exp(t))$$

$$= \left(\frac{d}{dt} - 1\right)^{n-1} \left(\frac{d}{dt} - 1\right)(At^n \exp(t))$$

$$= \left(\frac{d}{dt} - 1\right)^{n-1} \left(\frac{d}{dt}(At^n \exp(t)) - At^n \exp(t)\right)$$

$$= \left(\frac{d}{dt} - 1\right)^{n-1} (Ant^{n-1}\exp(t))$$

$$= \left(\frac{d}{dt} - 1\right)^{n-2} \left(\frac{d}{dt} - 1\right) Ant^{n-1}\exp(t)$$

$$= \left(\frac{d}{dt} - 1\right)^{n-2} \left(\frac{d}{dt}(Ant^{n-1}\exp(t)) - Ant^{n-1}\exp(t)\right)$$

$$= \left(\frac{d}{dt} - 1\right)^{n-2} An(n-1)t^{n-2}\exp(t)$$

$$= \cdots = An!\exp(t)$$

Plugging into the original DE, we get

$$\left(\frac{d}{dt} - 1\right)^{n} (At^n\exp(t)) = \exp(t)$$

$$A(n!)\exp(t) = \exp(t)$$

Thus,

$$A = \frac{1}{n!}$$

The GS for this case is

$$y(t) = \exp(t)\,(C_1 + C_2 t + C_3 t^2 + \cdots + C_n t^{n-1}) + \frac{1}{n!}t^n\exp(t)$$

$$y(x) = x(C_1 + C_2\ln x + C_3(\ln x)^2 + \cdots + C_n(\ln x)^{n-1}) + \frac{1}{n!}(\ln x)^n x$$

Case 3 ($\alpha \neq 1$):
We select the TS

$$y_P = A\exp(t)$$

Thus,

$$\text{LHS} = \left(\frac{d}{dt} - \alpha\right)^{n-1} \left(\frac{d}{dt} - \alpha\right) A\exp(t)$$

$$= \left(\frac{d}{dt} - \alpha\right)^{n-1} (1-\alpha)A\exp(t)$$

$$= \left(\frac{d}{dt} - \alpha\right)^{n-2} \left(\frac{d}{dt} - \alpha\right)(1-\alpha)A\exp(t)$$

$$= \left(\frac{d}{dt} - \alpha\right)^{n-2} (1-\alpha)^2 A\exp(t)$$

$$= \cdots = (1-\alpha)^n A\exp(t)$$

Thus,

$$\left(\frac{d}{dt} - \alpha\right)^{n} A\exp(t) = \exp(t)$$

$$A(1-\alpha)^n\exp(t) = \exp(t)$$

i.e.,

$$A = \frac{1}{(1-\alpha)^n}$$

The GS is

$$y(t) = \exp(\alpha t) \left(C_1 + C_2 t + C_3 t^2 + \cdots + C_n t^{n-1}\right) + \frac{1}{(1-\alpha)^n} \exp(t)$$

$$y(x) = x^\alpha \left(C_1 + C_2 \ln x + C_3 (\ln x)^2 + \cdots + C_n (\ln x)^{n-1}\right) + \frac{1}{(1-\alpha)^n} x$$

Apparently, this case covers the simpler case of $\alpha = 0$.

Problem 3.3.19

The original DE can be written in the following form

$$ax^2 \frac{d^2 y}{dx^2} + bx \frac{dy}{dx} + cy = 0$$

Let $t = \ln x$. Then,

$$\frac{dt}{dx} = \frac{1}{x}$$

Thus, by the chain rule,

$$\frac{dy}{dx} = \frac{dy}{dt}\frac{dt}{dx} = \frac{dy}{dt}\frac{1}{x}$$

$$\frac{d^2 y}{dx^2} = \frac{d}{dx}\left(\frac{dy}{dx}\right)$$

$$= \frac{d}{dx}\left(\frac{dy}{dt}\frac{1}{x}\right)$$

$$= -\frac{1}{x^2}\frac{dy}{dt} + \frac{1}{x}\frac{d}{dx}\left(\frac{dy}{dt}\right)$$

$$= -\frac{1}{x^2}\frac{dy}{dt} + \frac{1}{x}\frac{d}{dt}\left(\frac{dy}{dt}\right)\frac{dt}{dx}$$

$$= -\frac{1}{x^2}\frac{dy}{dt} + \frac{1}{x^2}\frac{d^2 y}{dt^2}$$

$$= \frac{1}{x^2}\left(\frac{d^2 y}{dt^2} - \frac{dy}{dt}\right)$$

Finally,

$$x^2 \frac{d^2 y}{dx^2} = \frac{d^2 y}{dt^2} - \frac{dy}{dt}$$

Plugging these back into the DE, we have

$$a\left(\frac{d^2 y}{dt^2} - \frac{dy}{dt}\right) + b\frac{dy}{dt} + cy = 0$$

$$a\frac{d^2 y}{dt^2} + (b-a)\frac{dy}{dt} + cy = 0$$

a linear DE wrt t. The C-Eq is

$$ar^2 + (b-a)r + c = 0$$

$$r_{1,2} = \frac{a - b \pm \sqrt{(b-a)^2 - 4ac}}{2a}$$

If $(b-a)^2 - 4ac > 0$, we have two distinct real roots.

$$y(t) = C_1 \exp(r_1 t) + C_2 \exp(r_2 t)$$

Using $t = \ln x$, we have

$$y(x) = C_1 x^{r_1} + C_2 x^{r_2}$$

If $(b-a)^2 - 4ac = 0$, we have a repeated root r, thus

$$y(t) = (C_1 + C_2 t) \exp(rt)$$

Using $t = \ln x$, we have

$$y(x) = (C_1 + C_2 \ln x) x^r$$

If $(b-a)^2 - 4ac < 0$, we have two complex conjugated roots

$$r_{1,2} = r \exp(\pm i\theta)$$

Thus,

$$y(t) = (C_1 \sin \theta t + C_2 \cos \theta t) \exp(rt)$$

Using $t = \ln x$, we have

$$y(x) = (C_1 \sin \theta \ln x + C_2 \cos \theta \ln x) x^r$$

Problem 3.3.20

Let $t = \ln x$. Then,

$$x \frac{dy}{dx} = \frac{dy}{dt}$$

$$x^2 \frac{d^2 y}{dx^2} = \frac{d^2 y}{dt^2} - \frac{dy}{dt}$$

Thus, the original DE becomes

$$\frac{d^2 y}{dt^2} - 9y = 0$$

The C-Eq is

$$r^2 - 9 = 0$$

$$r_{1,2} = \pm 3$$

$$y_1 = \exp(3t) = x^3$$

$$y_2 = \exp(-3t) = x^{-3}$$

$$y_C(x) = C_1 x^3 + C_2 x^{-3}$$

Problem 3.3.21

Let $t = \ln x$. Then,

$$x \frac{dy}{dx} = \frac{dy}{dt}$$

$$x^2 \frac{d^2 y}{dx^2} = \frac{d^2 y}{dt^2} - \frac{dy}{dt}$$

Thus,

$$\frac{d^2y}{dt^2} + (b-1)\frac{dy}{dt} + cy = 0$$

The function y now depends on variable v instead of x.

C-Eq:

$$r^2 + (b-1)r + c = 0$$

$$r_{1,2} = \frac{-(b-1) \pm \sqrt{(b-1)^2 - 4c}}{2}$$

$$y_{1,2} = \exp\left(\frac{-(b-1) \pm \sqrt{(b-1)^2 - 4c}}{2}v\right)$$

Problem 3.3.22

Let $t = \ln x$. Then,

$$\frac{dy}{dx} = \exp(-t)\frac{dy}{dt}$$

$$\frac{d^2y}{dx^2} = \exp(-2t)\left(\frac{d^2y}{dt^2} - \frac{dy}{dt}\right)$$

Thus, the DE becomes

$$\frac{d^2y}{dt^2} - 3\frac{dy}{dt} + 2y = 0$$

Its C-Eq is

$$r^2 - 3r + 2 = 0$$

$$r_{1,2} = 1, 2$$

Thus,

$$y(t) = C_1 \exp(t) + C_2 \exp(2t)$$
$$y(x) = C_1 x + C_2 x^2$$

Applying the IC, we get

$$\begin{cases} C_1 + C_2 = 3 \\ C_1 + 2C_2 = 1 \end{cases}$$

This gives

$$\begin{cases} C_1 = 5 \\ C_2 = -2 \end{cases}$$

Thus, the PS is

$$y(x) = 5x - 2x^2$$

Problem 3.3.23

Let $y = (x+2)^\lambda$. Then,

$$y' = \lambda(x+2)^{\lambda-1}$$

$$y'' = (\lambda - 1)\lambda(x + 2)^{\lambda-2}$$

Substituting them into the DE, we get

$$(x + 2)^{\lambda}(\lambda^2 - 2\lambda + 1) = 0$$
$$(\lambda - 1)^2 = 0$$
$$\lambda_{1,2} = 1, 1$$

The GS is

$$y = C_1(x + 2) + C_2(x + 2)\ln|x + 2|$$

Problem 3.3.24

Let $y = x^{\lambda}$. Then, $y' = \lambda x^{\lambda-1}, y'' = \lambda(\lambda - 1)x^{\lambda-2}$

Thus,

$$x^2\lambda(\lambda - 1)x^{\lambda-2} - 2x\lambda x^{\lambda-1} - 10x^{\lambda} = 0$$
$$x^{\lambda}(\lambda(\lambda - 1) - 2\lambda - 10) = 0$$
$$\lambda^2 - 3\lambda - 10 = 0$$

There are two real and distinct roots $\lambda_{1,2} = 5, -2$

So, the GS is

$$y = C_1 x^5 + C_2 x^{-2}$$

Problem 3.3.25

Let $y = x^r$, we have

$$y' = rx^{r-1}$$
$$y'' = r(r - 1)x^{r-2}$$
$$y''' = r(r - 1)(r - 2)x^{r-3}$$

Then,

$$r(r - 1)(r - 2) + r(r - 1) - r + 1 = 0$$
$$r^3 - 2r^2 + 1 = 0$$
$$(r - 1)(r^2 - r - 1) = 0$$

which has three distinct roots, resulting in three LI solutions.

$$r_{1,2,3} = 1, \frac{1 - \sqrt{5}}{2}, \frac{1 + \sqrt{5}}{2}$$

The GS is

$$y(x) = c_1 x^1 + c_2 x^{\frac{1-\sqrt{5}}{2}} + c_3 x^{\frac{1+\sqrt{5}}{2}}$$

457

3.4 Inhomogeneous Linear DEs

Problem 3.4.1

This is a 2nd.O c-coeff DE whose corresponding C-Eq is

$$r^2 - 2r - 8 = (r + 2)(r - 4) = 0$$
$$r_{1,2} = -2, 4$$

Thus, $y_C(x) = C_1 \exp(-2x) + C_2 \exp(4x)$
$$f(x) = \exp(4x)$$

The TS is

$$y_P(x) = Ax^s \exp(4x)$$

Comparing terms in y_C and y_P, we notice $s = 1$.

$$y_P(x) = Ax \exp(4x) \Longrightarrow \begin{cases} y_P' = 4Ax \exp(4x) + A \exp(4x) \\ y_P'' = 8A \exp(4x) + 16Ax \exp(4x) \end{cases}$$

Plugging into the original DE, we get

$$8A \exp(4x) + 16Ax \exp(4x) - 2(4Ax \exp(4x) + A \exp(4x)) - 8Ax \exp(4x)$$
$$= \exp(4x)$$
$$6A \exp(4x) = \exp(4x)$$
$$A = \frac{1}{6} \Longrightarrow y_p = \frac{x}{6} \exp(4x)$$

So, the GS is

$$y(x) = C_1 \exp(-2x) + C_2 \exp(4x) + \frac{x}{6} \exp(4x)$$

Problem 3.4.2

C-Eq for Homo portion of the DE is

$$r^4 - 1 = 0$$
$$r_{1,2,3,4} = 1, -1, i, -i$$
$$y_C(x) = c_1 \exp(x) + c_2 \exp(-x) + c_3 \cos x + c_4 \sin x$$

The TS is

$$y_P = A$$

and

$$y_P' = 0 = y_P'' = y_P''' = y_P''''$$

After substitution, we get

$$0 - A = 1$$
$$A = -1$$
$$y_P = -1$$

The GS is

$$y = y_C + y_P$$
$$= c_1 \exp(x) + c_2 \exp(-x) + c_3 \cos x + c_4 \sin x - 1$$

Problem 3.4.3

We first consider the Homo DE

$$y'''' - y = 0$$

The C-Eq is

$$r^4 - 1 = (r^2 + 1)(r^2 - 1)$$
$$= (r + i)(r - i)(r - 1)(r + 1)$$
$$= 0$$

whose roots are $r_{1,2,3,4} = i, -i, 1, -1$.

Then, the complementary GS is

$$y_C = C_1 \cos x + C_2 \sin x + C_3 \exp(x) + C_4 \exp(-x)$$

The TS is

$$y_P = Ax \exp(x)$$

Its derivatives are

$$y_P' = A \exp(x) + Ax \exp(x)$$
$$y_P'' = 2A \exp(x) + Ax \exp(x)$$
$$y_P''' = 3A \exp(x) + Ax \exp(x)$$
$$y_P'''' = 4A \exp(x) + Ax \exp(x)$$

Inserting them to the original InHomo DE, we have

$$4A \exp(x) = 4 \exp(x)$$

Solving this equation gives us $A = 1$. Thus,

$$y_P = x \exp(x)$$

Finally, the GS to the original DE is

$$y = y_C + y_P$$
$$= C_1 \cos x + C_2 \sin x + C_3 \exp(x) + C_4 \exp(-x) + x \exp(x)$$

Problem 3.4.4

Solve the corresponding Homo DE. The C-Eq is

$$r^4 - r^3 - r^2 - r - 2 = 0$$
$$(r + 1)(r^2 + 1)(r - 2) = 0$$
$$r_{1,2,3,4} = -1, 2, i, -i$$

The GS to the Homo DE is

$$y_C = C_1 \exp(-x) + C_2 \exp(2x) + C_3 \sin x + C_4 \cos x$$

Since $f(x) = 18x^5$, we select the TS

$$y_P = x^s(A_0 + A_1 x + A_2 x^2 + A_3 x^3 + A_4 x^4 + A_5 x^5)$$

Since there is no polynomial in the Homo solution, we let $s = 0$. Thus,

$$y_P = A_0 + A_1 x + A_2 x^2 + A_3 x^3 + A_4 x^4 + A_5 x^5$$

and

$$y_P' = A_1 + 2A_2 x + 3A_3 x^2 + 4A_4 x^3 + 5A_5 x^4$$
$$y_P'' = 2A_2 + 6A_3 x + 12A_4 x^2 + 20A_5 x^3$$
$$y_P''' = 6A_3 + 24A_4 x + 60A_5 x^2$$
$$y_P'''' = 24A_4 + 120A_5 x$$

Inserting these to the DE, we have

$$y_P'''' - y_P''' - y_P'' - y_P' - 2y_P = 18x^5$$

Equating the terms, we have

$$\begin{cases} 24A_4 - 6A_3 - 2A_2 - A_1 - 2A_0 = 0 \\ 120A_5 - 24A_4 - 6A_3 - 2A_2 - 2A_1 = 0 \\ -60A_5 - 12A_4 - 3A_3 - 2A_2 = 0 \\ -20A_5 - 4A_4 - 2A_3 = 0 \\ -5A_5 - 2A_4 = 0 \\ -2A_5 = 18 \end{cases}$$

Solving them gives

$$\begin{cases} A_0 = \dfrac{2295}{4} \\ A_1 = -\dfrac{2025}{2} \\ A_2 = \dfrac{135}{2} \\ A_3 = 45 \\ A_4 = \dfrac{45}{2} \\ A_5 = -9 \end{cases}$$

Thus, the PS is

$$y_P = \frac{2295}{4} - \frac{2025}{2}x + \frac{135}{2}x^2 + 45x^3 + \frac{45}{2}x^4 - 9x^5$$

Finally, the GS is

$$y = C_1 \exp(-x) + C_2 \exp(2x) + C_3 \sin x + C_4 \cos x + \frac{2295}{4} - \frac{2025}{2}x$$
$$+ \frac{135}{2}x^2 + 45x^3 + \frac{45}{2}x^4 - 9x^5$$

Problem 3.4.5

We select the TS

$$y_P = x^s(A_0 + A_1 x + A_2 x^2 + A_3 x^3)$$

To determine s, we have to find the GS to the Homo DE. The C-Eq is

$$r^3 + r^2 + r + 1 = 0$$
$$(r+1)(r^2+1) = 0$$
$$r_{1,2,3} = -1, i, -i$$

This means the polynomials are not part of the y_c and, thus, $s = 0$. We have

$$y_P = A_0 + A_1 x + A_2 x^2 + A_3 x^3$$

and

$$y'_P = A_1 + 2A_2 x + 3A_3 x^2$$
$$y''_P = 2A_2 + 6A_3 x$$
$$y'''_P = 6A_3$$

Inserting these to the DE, we get

$$y'''_P + y''_P + y'_P + y_P = x^3 + x^2$$

Equating the terms, we have

$$\begin{cases} A_0 + A_1 + 2A_2 + 6A_3 = 0 \\ A_1 + 2A_2 + 6A_3 = 0 \\ A_2 + 3A_3 = 1 \\ A_3 = 1 \end{cases}$$

which gives

$$\begin{cases} A_0 = 0 \\ A_1 = -2 \\ A_2 = -2 \\ A_3 = 1 \end{cases}$$

Thus, we have a PS to the DE

$$y_P = -2x - 2x^2 + x^3$$

Problem 3.4.6

We select the TS

$$y_P = A \exp(3x) + B \exp(2x) + C \exp(x) + D$$
$$y'_P = 3A \exp(3x) + 2B \exp(2x) + C \exp(x)$$
$$y''_P = 9A \exp(3x) + 4B \exp(2x) + C \exp(x)$$
$$y'''_P = 27A \exp(3x) + 8B \exp(2x) + C \exp(x)$$

Inserting these to the DE and equating the terms, we get

$$\begin{cases} 27A + 9A + 3A + A = 1 \\ 8B + 4B + 2B + B = 1 \\ C + C + C + C = 1 \\ D = 1 \end{cases}$$

This gives

$$\begin{cases} A = \dfrac{1}{40} \\ B = \dfrac{1}{15} \\ C = \dfrac{1}{4} \\ D = 1 \end{cases}$$

Thus, the PS is

$$y_P = \frac{1}{40}\exp(3x) + \frac{1}{15}\exp(2x) + \frac{1}{4}\exp(x) + 1$$

Problem 3.4.7

The C-Eq of the Homo DE is

$$r^4 + \omega^2 r^2 = 0$$

$$r_{1,2,3,4} = 0, 0, \omega i, -\omega i$$

Thus, the GS for the Homo DE is

$$y_C = A_1 + A_2 x + A_3 \sin \omega x + A_4 \cos \omega x$$

Since $\sin \omega x$ and $\cos \omega x$ are part of the GS, the TS is chosen as

$$y_P = C_1 x^2 \sin(\omega x) + C_2 x^2 \cos \omega x + C_3 x \sin \omega x + C_4 x \cos \omega x$$

$$y_P'' = (2C_1 - 2\omega C_4 - 4\omega x C_2 + \omega^2 x C_3 - \omega^2 x^2 C_1) \sin \omega x$$
$$+ (2C_2 + 2\omega C_3 + 4\omega x C_1 - \omega^2 x C_4 - \omega^2 x^2 C_2) \cos \omega x$$

$$y_P^{(4)} = (4C_4 \omega^3 - 12C_1 \omega^2 + C_4 \omega^4 x + 8C_2 \omega^3 x + C_1 \omega^4 x^2) \sin \omega x$$
$$+ (-4C_3 \omega^3 - 12C_2 \omega^2 + C_3 \omega^4 x - 8C_1 \omega^3 x$$
$$+ C_2 \omega^4 x^2) \cos \omega x$$

Inserting these to the DE and equating the terms, we get

$$\begin{cases} C_1 = 0 \\ C_2 = \dfrac{1}{4\omega^3} \\ C_3 = -\dfrac{5}{4\omega^4} \\ C_4 = \dfrac{1}{\omega^3} \end{cases}$$

Finally, we have $y = y_C + y_P$.

Problem 3.4.8

The complementary GS is

$$y_C = c_1 \cos \omega t + c_2 \sin \omega t$$

If $\omega = \omega_0$, we assume the following TS:

$$y_P = Bx \exp(i\omega_0 x)$$

and find y_P''.

Plugging y_P, y_P'' into $y'' + \omega^2 y = \exp(i\omega_0 x)$, we find

$$B = \frac{1}{2i\omega_0}$$

Thus, the PS is

$$y_P = \frac{1}{2i\omega_0} x \exp(i\omega_0 x)$$

and the GS is

$$y = y_C + y_P = c_1 \cos(\omega t) + c_2 \sin(\omega t) + \frac{1}{2i\omega_0} x \exp(i\omega_0 x)$$

If $\omega \neq \omega_0$, we use the following TS:

$$y_P = A \exp(i\omega_0 x)$$

and find y_P''.

Plugging y_P, y_P'' to $y'' + \omega^2 y = \exp(i\omega_0 x)$, we find

$$A = \frac{1}{\omega^2 - \omega_0^2}$$

Thus, the PS is

$$y_P = \frac{1}{\omega^2 - \omega_0^2} \exp(i\omega_0 x)$$

and the GS is

$$y = y_C + y_P = c_1 \cos(\omega t) + c_2 \sin(\omega t) + \frac{1}{\omega^2 - \omega_0^2} \exp(i\omega_0 x)$$

Problem 3.4.9

The C-Eq of the Homo DE is

$$r^2 + 9 = 0$$
$$r_{1,2} = \pm 3i$$

Thus, the GS for the Homo DE is

$$y_C = A_1 \cos 3x + A_2 \sin 3x$$

Since $\sin 3x$ and $\cos 3x$ are part of the GS, we select the TS

$$y_P = x(C_1 \sin 3x + C_2 \cos 3x)$$

Thus,

$$y_P' = C_1 \sin 3x + C_2 \cos 3x$$
$$+ 3xC_1 \cos 3x - 3xC_2 \sin 3x$$

and

$$y_P'' = (6C_1 - 9xC_2) \cos 3x + (-6C_2 - 9xC_1) \sin 3x$$

Putting these back and equating the terms, we get

$$C_1 = \frac{1}{6}, \qquad C_2 = 0$$

Thus, the PS is

$$y_P = \frac{1}{6} x \sin 3x$$

and the GS is

$$y = A_1 \cos 3x + A_2 \sin 3x + \frac{1}{6}x \sin 3x$$

Problem 3.4.10

Find the complementary solution, *i.e.*, the GS to
$$y'' + 2y' + y = 0$$
The C-Eq is $r^2 + 2r + 1 = 0$ whose two identical roots $r_{1,2} = -1, -1$. Thus,
$$y_C(x) = c_1 \exp(-x) + c_2 x \exp(-x)$$
Next, we find the PS $y_P(x)$. Because the InHomo term is a $\sin x$ and $\cos x$, the MUC requires us to select the TS y_P as
$$y_P = A \cos x + B \sin x$$
$$y_P' = -A \sin x + B \cos x, \quad y_P'' = -A \cos x - B \sin x$$
Substituting these values into the DE, we obtain that
$$-A \cos x - B \sin x - 2A \sin x + 2B \cos x + A \cos x + B \sin x$$
$$= 5 \sin x + 5 \cos x$$
Thus, we find that
$$-2A \sin x + 2B \cos x = 5 \sin x + 5 \cos x$$
Therefore,
$$A = -\frac{5}{2}, \quad B = \frac{5}{2}$$
The GS is
$$y(x) = y_P(x) + y_C(x)$$
$$= \frac{5}{2}(\sin x - \cos x) + c_1 \exp(-x) + c_2 x \exp(-x)$$

Problem 3.4.11

The C-Eq is
$$r^3 + r^2 + r + 1 = 0$$
$$r_{1,2,3} = -1, i, -i$$
$$y_C = C_1 \exp(-x) + C_2 \sin x + C_3 \cos x$$
The TS is
$$y_P = A + x(B \cos x + C \sin x) + D \cos 2x$$
$$+ E \sin 2x + xF \exp(-x)$$
$$y_P' = B \cos x + C \sin x + x(-B \sin x + C \cos x) - 2D \sin 2x + 2E \cos 2x$$
$$+ F \exp(-x) - xF \exp(-x)$$
$$y_P'' = -2B \sin x + 2C \cos x - x(B \cos x + C \sin x) - 4D \cos 2x - 4E \sin 2x$$
$$- 2F \exp(-x) + xF \exp(-x)$$
$$y_P''' = -3(B \cos x + C \sin x) + x(B \sin x - C \cos x) + 8D \sin 2x - 8E \cos 2x$$
$$+ (3 - x)F \exp(-x)$$

Inserting these to the DE and comparing the terms, we have

$$\begin{cases} A = 1 \\ 2C - 2B = 1 \\ -2B - 2C = 0 \\ -3D - 6E = 0 \\ -3E + 6D = 1 \\ 2F = 1 \end{cases}$$

Solving this gives

$$\begin{cases} A = 1 \\ B = -\dfrac{1}{4} \\ C = \dfrac{1}{4} \\ D = \dfrac{2}{15} \\ E = -\dfrac{1}{15} \\ F = \dfrac{1}{2} \end{cases}$$

The PS is

$$y_P = 1 - \frac{1}{4}x\cos x + \frac{1}{4}x\sin x + \frac{2}{15}\cos 2x$$
$$- \frac{1}{15}\sin 2x + \frac{1}{2}x\exp(-x)$$

Problem 3.4.12

Let $x = \exp(t)$. Then, $t = \ln x$

$$x\frac{dy}{dx} = \frac{dy}{dt}$$
$$x^2 y'' = \frac{d^2 y}{dt^2} - \frac{dy}{dt}$$
$$\frac{d^2 y}{dt^2} - \frac{dy}{dt} - 4\frac{dy}{dt} + 6y = \exp(3t)$$
$$\frac{d^2 y}{dt^2} - 5\frac{dy}{dt} + 6y = \exp(3t)$$

C-Eq of the Homo portion:

$$r^2 - 5r + 6 = 0$$
$$(r - 3)(r - 2) = 0$$
$$r_{1,2} = 3, 2$$

Thus, the complementary GS is

$$y_C(t) = C_1 \exp(3t) + C_2 \exp(2t)$$

The TS is

$$y_P(t) = At \exp(3t)$$
$$y_P'(t) = A \exp(3t) + 3At \exp(3t)$$
$$y_P''(t) = 6A \exp(3t) + 9At \exp(3t)$$

Inserting these to the DE with variables (y, t), we have
$$A \exp(3t) = \exp(3t)$$

Thus,
$$A = 1$$
$$y_P(t) = t \exp(3t)$$

The GS is
$$y(t) = C_1 \exp(3t) + C_2 \exp(2t) + t \exp(3t)$$
$$y(x) = C_1 x^3 + C_2 x^2 + x^3 \ln x$$

Problem 3.4.13

Use the trial function $y = x^\alpha$ and plug it into the Homo DE:
$$x^6(x^\alpha)'' + 2x^5(x^\alpha)' - 12x^4 x^\alpha = (\alpha(\alpha-1) + 2\alpha - 12)x^{4+\alpha}$$
$$= (\alpha^2 + \alpha - 12)x^{4+\alpha} = 0$$
$$(\alpha + 4)(\alpha - 3) = 0$$
$$\alpha = -4, 3$$

Thus,
$$y_C(x) = C_1 x^{-4} + C_2 x^3$$

The TS is
$$y_P = Ax^{-4} \ln x$$
$$x^6(Ax^{-4} \ln x)'' + 2x^5(Ax^{-4} \ln x)' - 12x^4(Ax^{-4} \ln x) = 1$$
$$x^6(20Ax^{-6} \ln x - 9Ax^{-6}) + 2x^5(-4Ax^{-5} \ln x + Ax^{-5}) - 12A \ln x = 1$$
$$(20 - 8 - 12) A \ln x + A(-9 + 2) = 1$$
$$A = -\frac{1}{7}$$

The GS is
$$y(x) = C_1 x^{-4} + C_2 x^3 - \frac{1}{7}x^{-4} \ln x$$

Problem 3.4.14

Let $x = \exp(t)$. Then, $t = \ln x$
$$x\frac{dy}{dx} = \frac{dy}{dt} = \dot{y}$$
$$x^2\frac{d^2y}{dx^2} = \frac{d^2y}{dt^2} - \frac{dy}{dt} = \ddot{y} - \dot{y}$$

Thus,

$$\ddot{y} + \dot{y} - 6y = 72 \exp(5t)$$

C-Eq:

$$r^2 + r - 6 = 0$$
$$(r + 3)(r - 2) = 0$$
$$r_{1,2} = -3, 2$$
$$y_c(t) = C_1 \exp(-3t) + C_2 \exp(2t)$$

The TS is

$$y_P = A \exp(5t)$$
$$\dot{y}_P = 5A \exp(5t)$$
$$\ddot{y}_P = 25A \exp(5t)$$

Plugging these back, we have

$$25A \exp(5t) + 5A \exp(5t) - 6A \exp(5t) = 72 \exp(5t)$$

Thus,

$$A = 3$$
$$y_P = 3 \exp(5t)$$

The GS is

$$y(t) = C_1 \exp(-3t) + C_2 \exp(2t) + 3 \exp(5t)$$
$$y(x) = C_1 x^{-3} + C_2 x^2 + 3x^5$$

Problem 3.4.15

Using the substitution $t = \ln x$, we have

$$xy' = \dot{y}$$
$$x^2 y'' = \ddot{y} - \dot{y}$$

The DE becomes

$$4\ddot{y} - 4\dot{y} - 4\dot{y} + 3y = 8 \exp\left(\frac{4}{3}t\right)$$

$$4\ddot{y} - 8\dot{y} + 3y = 8 \exp\left(\frac{4}{3}t\right)$$

which is a c-coeff linear InHomo DE wrt t. The C-Eq to the associated Homo DE is

$$4r^2 - 8r + 3 = 0$$
$$r_{1,2} = \frac{3}{2}, \frac{1}{2}$$
$$y_c(t) = C_1 \exp\left(\frac{3}{2}t\right) + C_2 \exp\left(\frac{1}{2}t\right)$$

The TS is $y_P(t) = A \exp\left(\frac{4}{3}t\right)$. Then,

$$\frac{64}{9} A \exp\left(\frac{4}{3}t\right) - \frac{32}{3} A \exp\left(\frac{4}{3}t\right) + 3A \exp\left(\frac{4}{3}t\right) = 8 \exp\left(\frac{4}{3}t\right)$$

467

$$A = -\frac{72}{5}$$

Thus,

$$y(t) = C_1 \exp\left(\frac{3}{2}t\right) + C_2 \exp\left(\frac{1}{2}t\right) - \frac{72}{5}\exp\left(\frac{4}{3}t\right)$$

That is

$$y(x) = C_1 x^{\frac{3}{2}} + C_2 x^{\frac{1}{2}} - \frac{72}{5}x^{\frac{4}{3}}$$

Problem 3.4.16

Using substitution $x = \exp(t)$, we get

$$\frac{d^2y}{dt^2} - \frac{dy}{dt} + \frac{dy}{dt} + y = t$$

$$\frac{d^2y}{dt^2} + y = t$$

The Homo portion of the solution

$$y_C(t) = C_1 \cos t + C_2 \sin t$$

The TS is

$$y_P(t) = At + B$$

We get $A = 1, B = 0$. The GS is

$$y(t) = C_1 \cos t + C_2 \sin t + t$$
$$y(x) = C_1 \cos(\ln x) + C_2 \sin(\ln x) + \ln x$$

Problem 3.4.17

Let $t = \ln x$. Then,

$$xy' = \dot{y}$$
$$x^2 y'' = \ddot{y} - \dot{y}$$
$$x^3 y''' = 2\dot{y} - 3\ddot{y} + \dddot{y}$$

The DE reduces to

$$\dddot{y} - 2\ddot{y} + 2\dot{y} - y = 1 + \exp(t) + \exp(2t)$$

The C-Eq for the Homo DE is

$$r^3 - 2r^2 + 2r - 1 = 0$$
$$(r - 1)(r^2 - r + 1) = 0$$
$$r_{1,2,3} = 1, \frac{1 + i\sqrt{3}}{2}, \frac{1 - i\sqrt{3}}{2}$$

Thus, the GS to the Homo DE is

$$y_C(t) = C_1 \exp(t) + \left(C_2 \cos\frac{\sqrt{3}}{2}t + C_3 \sin\frac{\sqrt{3}}{2}t\right)\exp\left(\frac{1}{2}t\right)$$

Since $\exp(t)$ is part of the GS, we select the TS

$$y_P(t) = A_1 t \exp(t) + A_2 \exp(2t) + A_3$$

Thus,

$$\dot{y}_P = A_1(t+1)\exp(t) + 2A_2\exp(2t)$$
$$\ddot{y}_P = A_1(t+2)\exp(t) + 4A_2\exp(2t)$$
$$\dddot{y}_P = A_1(t+3)\exp(t) + 8A_2\exp(2t)$$

Plugging these into the DE, we have

$$A_1\exp(t) + 3A_2\exp(2t) - A_3 = \exp(t) + \exp(2t) + 1$$

This gives

$$A_{1,2,3} = 1, \frac{1}{3}, -1$$

Thus,

$$y_P = t\exp(t) + \frac{1}{3}\exp(2t) - 1$$

and

$$y(t) = t\exp(t) + \frac{1}{3}\exp(2t) - 1 + C_1\exp(t)$$
$$+ \left(C_2\cos\frac{\sqrt{3}}{2}t + C_3\sin\frac{\sqrt{3}}{2}t\right)\exp\left(\frac{1}{2}t\right)$$

Finally, we have

$$y(x) = x\ln x + \frac{1}{3}x^2 - 1 + C_1 x + \left(C_2\cos\left(\frac{\sqrt{3}}{2}\ln x\right) + C_3\sin\left(\frac{\sqrt{3}}{2}\ln x\right)\right)\sqrt{x}$$

Problem 3.4.18

Plugging in the TS $y = x^\alpha$ to solve the Homo DE:

$$(\alpha(\alpha-1) + \alpha - 4)x^\alpha = 0$$
$$\alpha^2 - 4 = 0$$
$$\alpha = \pm 2$$

Thus, $y_C = C_1 x^2 + C_2 x^{-2}$ is the complementary GS. For the InHomo part, the TS is

$$y_P = Ax^2\ln x + Bx^{-2}\ln x$$

where A and B are constants. Plugging into the DE gives

$$x^2(Ax^2\ln x + Bx^{-2}\ln x)'' + x(Ax^2\ln x + Bx^{-2}\ln x)'$$
$$- 4(Ax^2\ln x + Bx^{-2}\ln x)$$
$$= x^2\big(A(2\ln x + 3) + Bx^{-4}(6\ln x - 5)\big)$$
$$+ x\big(Ax(2\ln x + 1) + Bx^{-3}(-2\ln x + 1)\big)$$
$$- 4(Ax^2\ln x + Bx^{-2}\ln x)$$
$$= 4Ax^2 - 4Bx^{-2}$$

$$A = \frac{1}{4}, \quad B = -\frac{1}{4}$$

The GS is

$$y = C_1 x^2 + C_2 x^{-2} + \frac{1}{4}(x^2 - x^{-2})\ln x$$

Problem 3.4.19

Given two LI solutions $y_1(x)$ and $y_2(x)$, which imply

$$\begin{cases} y_1'' + \alpha(x)y_1' + \beta(x)y_1 = 0 & \text{(A.9)} \\ y_2'' + \alpha(x)y_2' + \beta(x)y_2 = 0 & \text{(A.10)} \end{cases}$$

(A.9) $\times y_2 -$ (A.10) $\times y_1$ gives

$$\alpha(\tau)(y_1' y_2 - y_1 y_2') = y_1'' y_2 - y_1 y_2''$$

From the Wronskian

$$W[\tau] = \begin{vmatrix} y_1 & y_2 \\ y_1' & y_2' \end{vmatrix} = y_1 y_2' - y_1' y_2$$

we get

$$\frac{dW}{d\tau} = y_1 y_2'' - y_1'' y_2$$

The above two give

$$\frac{dW}{d\tau} = -\alpha(\tau)W[\tau]$$

Thus,

$$\frac{dW}{W} = -\alpha(\tau)d\tau$$

$$\ln W = -\int \alpha(\tau)d\tau$$

$$W[t] = \exp\left(-\int^t \alpha(\tau)d\tau\right)$$

This is the required relation between the Wronskian and $\alpha(x)$.
The VOP formula gives:

$$y_P(x) = \int^x \frac{y_2(x)y_1(t) - y_1(x)y_2(t)}{W[t]}f(t)dt$$

Inserting $W[t]$ to the above formula finds a PS

$$y_P(x) = \int^x \frac{y_2(x)y_1(t) - y_1(x)y_2(t)}{\exp\left(-\int^t \alpha(\tau)d\tau\right)}f(t)dt$$

Since y_1 and y_2 are two LI solutions of the Homo portion of $y'' + \alpha(x)y' + \beta(x)y = f(x)$, the GS is

$$y_C(x) = C_1 y_1(x) + C_2 y_2(x)$$

$$y(x) = y_C(x) + y_P(x)$$

$$= C_1 y_1(x) + C_2 y_2(x) + \int^x \frac{y_2(x)y_1(t) - y_1(x)y_2(t)}{\exp\left(-\int^t \alpha(\tau)d\tau\right)}f(t)dt$$

Problem 3.4.20

C-Eq of the Homo portion

$$x'' + \omega^2 x = 0$$
$$r^2 + \omega^2 = 0 \Rightarrow r_{1,2} = \pm i\omega$$

The GS is

$$x_C = c_1 \cos(\omega t) + c_2 \sin(\omega t)$$

(1) For a general function $f(t)$, we get

$$x_P(t) = \int K(t, \tau) f(\tau) d\tau$$

$$x(t) = x_C + x_P$$
$$= c_1 \cos(\omega t) + c_2 \sin(\omega t) + \int_{\tau_0}^{\tau} K(t, \tau) \int f(\tau) \, d\tau$$

where

$$K(t, \tau) = \frac{1}{\omega} (\cos \omega t \sin \omega \tau - \cos \omega \tau \sin \omega t)$$

(2) For a specific function $f(t) = \exp(i\omega t)$, we found that it is a solution to the Homo DE and thus, the TS is

$$x_P(t) = At \exp(i\omega t)$$

Thus,

$$x'_P(t) = A \exp(i\omega t) + At i\omega \exp(i\omega t)$$
$$x''_P(t) = -A t\omega^2 \exp(i\omega t) + 2Ai\omega \exp(i\omega t)$$

Thus, we have

$$-A t\omega^2 \exp(i\omega t) + 2Ai\omega \exp(i\omega t) + At\omega^2 \exp(i\omega t) = \exp(i\omega t)$$

From this, we get

$$A = \frac{1}{2i\omega}$$

Thus, the PS should be

$$x_P(t) = \frac{1}{2i\omega} \exp(i\omega t)$$

$$x(t) = x_C + x_P = c_1 \cos(\omega t) + c_2 \sin(\omega t) + \frac{1}{2i\omega} \exp(i\omega t)$$

(3) Function $f(t)$ is a specific function $f(t) = \exp(\omega t)$.
For $\omega \neq 0$

$$x_P = A \exp(\omega t)$$
$$x'_P = A\omega \exp(\omega t)$$
$$x''_P = A\omega^2 \exp(\omega t)$$

Since $x'' + \omega^2 x = \exp(\omega t)$, we get

$$A\omega^2 \exp(\omega t) + A\omega^2 \exp(\omega t) = \exp(\omega t)$$

$$A = \frac{1}{\omega^2}$$

$$x_P = \frac{1}{\omega^2} \exp(\omega t)$$

$$x(t) = x_C + x_P$$

$$= c_1 \cos(\omega t) + c_2 \sin(\omega t) + \frac{1}{\omega^2} \exp(\omega t)$$

Remarks

If $\omega = 0$

$$x'' = 1$$
$$x'' - 1 = 0$$
$$(r - 1)(r + 1) = 0$$
$$r_{1,2} = \pm 1$$
$$x(t) = x_C = c_1 \exp(t) + c_2 \exp(-t)$$

Problem 3.4.21

This is a 2nd.O c-coeff DE whose C-Eq is

$$r^2 - (r_1 + r_2)r + r_1 r_2 = 0$$

whose roots are r_1 and r_2 and the corresponding solutions are

$$y_1(x) = \exp(r_1 x)$$
$$y_2(x) = \exp(r_2 x)$$

Let $y_P(x) = u_1(x)y_1(x) + u_2(x)y_2(x)$ where

$$u_1(x) = -\int^x \frac{y_2(t)}{W[t]} f(t)dt = -\int^x \frac{\exp(r_2 t)}{W[t]} f(t)dt$$

$$u_2(x) = \int^x \frac{y_1(t)}{W[t]} f(t)dt = \int^x \frac{\exp(r_1 t)}{W[t]} f(t)dt$$

where

$$W[t] = \begin{vmatrix} \exp(r_1 t) & \exp(r_2 t) \\ r_1 \exp(r_1 t) & r_2 \exp(r_2 t) \end{vmatrix}$$
$$= (r_2 - r_1)\exp((r_1 + r_2)t)$$

The PS is

$$y_P(x) = u_1(x)y_1(x) + u_2(x)y_2(x) + u_2 y_2$$

$$= \frac{1}{r_2 - r_1}\int^x \left(\exp(r_2(x - t)) - \exp(r_1(x - t))\right)f(t)dt$$

Thus, the GS for the whole DE is

$$y = C_1 \exp(r_1 x) + C_2 \exp(r_2 x)$$

$$+ \frac{1}{r_2 - r_1}\int^x \left(\exp(r_2(x - t)) - \exp(r_1(x - t))\right)f(t)dt$$

Problem 3.4.22

First, we take all necessary derivatives of $y_2(x)$,
$$y_2'(x) = Zy_1' + Z'y_1$$
$$y_2''(x) = Zy_1'' + 2Z'y_1' + Z''y_1$$
Plugging these values into the original DE, we obtain
$$Zy_1'' + 2Z'y_1' + Z''y_1 + P(x)(Zy_1' + Z'y_1) + Q(x)Zy_1 = 0$$
Now note that we can collect terms to get
$$Z(y_1'' + P(x)y_1' + Q(x)y_1) + Z''y_1 + Z'(2y_1' + P(x)y_1) = 0$$
Since y_1 is a solution
$$Z(y_1'' + P(x)y_1' + Q(x)y_1) = 0$$
we get
$$Z''y_1 + Z'(2y_1' + P(x)y_1) = 0$$
Now, the following substitutions cast the DE into a 1st.O linear DE
$$u = Z' => u' = Z''$$
We obtain
$$u' + u\left(2\frac{y_1'}{y_1} + P(x)\right) = 0$$
Using the IF method, we find that
$$u(x) = \exp\left(-\int\left(2\frac{y_1'}{y_1} + P(x)\right)dx\right)$$
$$= \exp\left(-2\ln y_1 - \int P(x)dx\right)$$
Thus,
$$u(x) = \frac{\exp(-\int P(x)dx)}{y_1^2(x)}$$
Now, we must solve for Z.
$$Z(x) = \int \frac{\exp(-\int P(x)dx)}{y_1^2(x)}dx$$
The second LI solution is
$$y_2(x) = y_1(x)\int \frac{\exp(-\int P(x)dx)}{y_1^2(x)}dx$$

Problem 3.4.23

The corresponding Homo DE is
$$y'' + y = 0$$
The C-Eq is
$$r^2 + 1 = 0$$
$$r_{1,2} = \pm i$$

So, the GS to the Homo DE is

$$y_C(x)) = C_1 + C_2 \cos x$$
$$= C_1 \sin x + C_2 \cos x$$

By VOP, denote

$$y_1(x) = \sin x, \ y_2(x) = \cos x, \ f(x) = \cot x$$

Then,

$$W(y_1, y_2) = \begin{vmatrix} y_1 & y_2 \\ y_1' & y_2' \end{vmatrix}$$
$$= \begin{vmatrix} \sin x & \cos x \\ \cos x & -\sin x \end{vmatrix}$$
$$= -1$$

and the PS is

$$y_P(x) = -y_1(x) \int \frac{y_2(t)f(t)}{W[t]} dt + y_2(x) \int \frac{y_1(t)f(t)}{W[t]} dt$$
$$= \sin x \int \cos t \cot t \, dt - \cos x \int \sin t \cot t \, dt$$
$$= \sin x \int \frac{1}{\sin t} dt = \sin x \ln|\csc x - \cot x|$$

Problem 3.4.24

The Homo DE is

$$y''' + y'' + y' + y = 0$$

The C-Eq is

$$r^3 + r^2 + r + 1 = 0$$
$$(r + 1)(r^2 + 1) = 0$$
$$r_{1,2,3} = -1, i, -i$$

We have

$$y_1 = \exp(-x), \qquad y_2 = \sin x, \qquad y_3 = \cos x$$

Let

$$y_P = u_1 y_1 + u_2 y_2 + u_3 y_3$$

Then,

$$y_P' = (u_1 y_1' + u_2 y_2' + u_3 y_3') + (u_1' y_1 + u_2' y_2 + u_3' y_3)$$

Since y_p is not unique, we can assume

$$u_1' y_1 + u_2' y_2 + u_3' y_3 = 0$$

Thus,

$$y_P' = u_1 y_1' + u_2 y_2' + u_3 y_3'$$

Similarly, we have

$$y_P'' = (u_1 y_1'' + u_2 y_2'' + u_3 y_3'') + (u_1' y_1' + u_2' y_2' + u_3' y_3')$$

and we can also assume

$$u_1' y_1' + u_2' y_2' + u_3' y_3' = 0$$

Thus,

$$y_P'' = u_1 y_1'' + u_2 y_2'' + u_3 y_3''$$

Finally,

$$y_P''' = (u_1 y_1''' + u_2 y_2''' + u_3 y_3''') + (u_1' y_1'' + u_2' y_2'' + u_3' y_3'')$$

Since y_1, y_2 and y_3 are solutions of the Homo DE,

$$\begin{cases} y_1''' = -y_1'' - y_1' - y_1 \\ y_2''' = -y_2'' - y_2' - y_2 \\ y_3''' = -y_3'' - y_3' - y_3 \end{cases}$$

Plugging these into the equation for y_P''', we get

$$y_P''' = -y_P'' - y_P' - y_P + (u_1' y_1'' + u_2' y_2'' + u_3' y_3'')$$

Therefore, from the original DE, we have

$$u_1' y_1'' + u_2' y_2'' + u_3' y_3'' = f(x)$$

Now, we have

$$\begin{cases} u_1' y_1 + u_2' y_2 + u_3' y_3 = 0 \\ u_1' y_1' + u_2' y_2' + u_3' y_3' = 0 \\ u_1' y_1'' + u_2' y_2'' + u_3' y_3'' = f(x) \end{cases}$$

or in matrix format

$$\begin{pmatrix} y_1 & y_2 & y_3 \\ y_1' & y_2' & y_3' \\ y_1'' & y_2'' & y_3'' \end{pmatrix} \begin{pmatrix} u_1' \\ u_2' \\ u_3' \end{pmatrix} = \begin{pmatrix} 0 \\ 0 \\ f(x) \end{pmatrix}$$

Solving this, we get

$$u_1 = \int \frac{W(y_2, y_3) f(x)}{W(y_1, y_2, y_3)} dx$$

$$u_2 = \int \frac{W(y_3, y_1) f(x)}{W(y_1, y_2, y_3)} dx$$

$$u_3 = \int \frac{W(y_1, y_2) f(x)}{W(y_1, y_2, y_3)} dx$$

Problem 3.4.25

The Homo DE is

$$y'' + 9y = 0$$

whose C-Eq is

$$r^2 + 9 = 0$$
$$r_{1,2} = \pm 3i$$

Thus, we have

$$y_1(x) = \cos 3x, \qquad y_2(x) = \sin 3x$$

The TS is

$$y_P(x) = u_1(x) y_1(x) + u_2(x) y_2(x)$$

where u_1 and u_2 satisfy

$$\begin{cases} u_1'y_1 + u_2'y_2 = 0 \\ u_1'y_1' + u_2'y_2' = f(x) \end{cases}$$

where

$$f(x) = \sin x \tan x$$

and the Wronskian

$$W[y_1(x), y_2(x)] = \begin{vmatrix} \cos 3x & \sin 3x \\ -3\sin 3x & 3\cos 3x \end{vmatrix}$$
$$= 3$$

We get

$$u_1(x) = -\int^x \frac{y_2(t)f(t)}{W[t]}\,dt$$
$$= -\frac{1}{3}\int \sin 3t\, f(t)\,dt$$
$$u_2(x) = \int^x \frac{y_1(t)f(t)}{W[t]}\,dt$$
$$= \frac{1}{3}\int \cos 3t\, f(t)\,dt$$

The PS is

$$y_P(x) = u_1(x)y_1(x) + u_2(x)y_2(x)$$
$$= \frac{1}{3}\int \sin 3(x-t)\,f(t)\,dt$$

where we used the trigonometric identity

$$\sin 3(x-t) = \sin 3x \cos 3t - \cos 3x \sin 3t$$

Problem 3.4.26

If $\omega = 0$, we have $y'' = 0$ where $y = x$ is a PS.

Now, let's consider $\omega \neq 0$. It is easy to get the GS for the Homo DE:

$$y_C = C_1 \cos \omega x + C_2 \sin \omega x$$

Thus, we have

$$y_P = \int \frac{y_2(x)y_1(t) - y_1(x)y_2(t)}{W(y_1(t), y_2(t))} f(t)\,dt$$

where $y_1(x) = \cos \omega x$, $y_2(x) = \sin \omega x$, $f(t) = \sin \omega t \tan \omega t$ and the Wronskian $W = \omega$. Thus, we get

$$y_P = \frac{1}{2\omega^2}(\omega x \sin \omega x + 2\cos \omega x \ln \cos \omega x)$$

Problem 3.4.27

The C-Eq of the Homo DE is
$$r^2 + a^2 = 0$$
which gives
$$r_{1,2} = \pm ia$$
We get $y_1(x) = \cos(ax)$ and $y_2(x) = \sin(ax)$ and GS for the Homo DE
$$y_C(x) = C_1 \cos(ax) + C_2 \sin(ax)$$
Thus, we have the Wronskian
$$W[y_1(t), y_2(t)] = \begin{vmatrix} y_1 & y_2 \\ y_1' & y_2' \end{vmatrix} = a$$
Recognizing $f(t) = \tan(bt)$, we can compute the P
$$
\begin{aligned}
y_P(x) &= \int \frac{y_2(x)y_1(t) - y_1(x)y_2(t)}{W[y_1(t), y_2(t)]} f(t)\,dt \\
&= \int \frac{\sin(ax)\cos(at) - \cos(ax)\sin(at)}{a} \tan(bt)\,dt \\
&= \frac{1}{a} \int \sin(a(x - t)) \tan(bt)\,dt
\end{aligned}
$$

Problem 3.4.28

(1) The associated Homo DE is $y'' + y = 0$ and its C-Eq is
$$r^2 + 1 = 0$$
The two LI solutions to the Homo DE are
$$y_1(x) = \sin x, \quad y_2(x) = \cos x$$
and $f(x) = 2\sin x$.

(2) The Wronskian is
$$W(x) = \begin{vmatrix} y_1 & y_2 \\ y_1' & y_2' \end{vmatrix} = \begin{vmatrix} \sin x & \cos x \\ \cos x & -\sin x \end{vmatrix} = -1$$
$$u_1(x) = -\int \frac{y_2(t)f(t)}{W[t]}\,dt = -\int \frac{\cos t \cdot 2\sin t}{-1}\,dt = -\frac{\cos 2x}{2} + B_1$$
$$u_2(x) = \int \frac{y_1(t)f(t)}{W[t]}\,dt = \int \frac{\sin t \cdot 2\sin t}{-1}\,dx = \frac{\sin 2x}{2} - x + B_2$$
where B_1 and B_2 are arbitrary constants and setting $B_1 = B_2 = 0$ is a convenience and freedom we may enjoy. Thus, the PS is
$$
\begin{aligned}
y_P(x) &= y_1(x)u_1(x) + y_2(x)u_2(x) \\
&= \sin x \left(-\frac{\cos 2x}{2} \right) + \cos x \left(\frac{\sin 2x}{2} - x \right) = \frac{1}{2}\sin x - x\cos x
\end{aligned}
$$
(3) The GS to the DE is
$$
\begin{aligned}
y(x) &= y_C(x) + y_P(x) \\
&= (C_1 \sin x + C_2 \cos x) + \left(\frac{1}{2}\sin x - x\cos x \right)
\end{aligned}
$$

Problem 3.4.29

With $y_1 = x$, $y_2 = 1/x$, and $f(x) = 72x^3$, the VOP takes the form

$$\begin{cases} xu_1' + \dfrac{u_2'}{x} = 0 \\[2mm] u_1' - \dfrac{u_2'}{x^2} = 72x^3 \end{cases}$$

Multiplying the second DE by x and adding the two DEs, we get

$$u_1' = 36x^3$$
$$u_1 = \int 36x^3 dx = 9x^4$$

and

$$u_2' = -x^2 u_1' = -36x^5$$
$$u_2 = \int (-36x^5) dx = -6x^6$$

Then, the PS is

$$y_P = u_1 y_1 + u_2 y_2$$
$$= x(9x^4) - \frac{1}{x}(6x^6)$$
$$= 3x^5$$

Problem 3.4.30

The homogeneous portion is
$$x'' - 3x' - 4x = 0$$

C-Eq:
$$r^2 - 3r - 4 = 0$$
$$r_{1,2} = 4, -1$$
$$x_C(t) = C_1 \exp(4t) + C_2 \exp(-t)$$

Method 1: The MUC.

Since the InHomo terms are contained in $x_C(t)$, we select the following TS
$$x_P(t) = tA \exp(4t) + tB \exp(-t)$$

Thus,
$$x_P'(t) = (1 + 4t)A \exp(4t) + (1 - t)B \exp(-t)$$
$$x_P''(t) = 8(1 + 2t)A \exp(4t) + (t - 2)B \exp(-t)$$

Substituting and equating the coefficients
$$x_P''(t) - 3x_P'(t) - 4x_P(t) = 15 \exp(4t) + 5 \exp(-t)$$
yields $A = 3$, $B = -1$
$$x(t) = x_C(t) + x_P(t)$$
$$= C_1 \exp(4t) + C_2 \exp(-t) + 3t \exp(4t) - t \exp(-t)$$

Method 2: The VOP method.

From $x_c(t)$, we have

$$x_1(t) = \exp(4t)$$
$$x_2(t) = \exp(-t)$$

Thus,

$$x_1'(t) = 4\exp(4t)$$
$$x_2'(t) = -\exp(-t)$$

The Wronskian of x_1, x_2 is

$$W(t) = \begin{vmatrix} x_1 & x_2 \\ x_1' & x_2' \end{vmatrix} = \begin{vmatrix} \exp(4t) & \exp(-t) \\ 4\exp(4t) & -\exp(-t) \end{vmatrix} = -5\exp(3t)$$

Thus,

$$x_P = \int^t \frac{x_2(t)x_1(s) - x_1(t)x_2(s)}{W(s)} f(s)ds$$

where

$$x_1(s) = \exp(4s), x_2(s) = \exp(-s), W(s) = -5\exp(3s)$$

We get

$$x_P = 3t\exp(4t) + \frac{3}{5}\exp(4t) - t\exp(-t) - \frac{1}{5}\exp(-t)$$

$$x(t) = x_C(t) + x_P(t)$$

$$= C_1\exp(4t) + C_2\exp(-t) + 3t\exp(4t) + \frac{3}{5}\exp(4t) - t\exp(-t)$$

$$- \frac{1}{5}\exp(-t)$$

$$= C_3\exp(4t) + C_4\exp(-t) + 3t\exp(4t) - t\exp(-t)$$

where $C_3 = C_1 + \frac{3}{5}, C_4 = C_2 - \frac{1}{5}$

Problem 3.4.31

The Wronskian is

$$W(x_1, x_2, x_3) = \begin{vmatrix} x_1 & x_2 & x_3 \\ x_1' & x_2' & x_3' \\ x_1'' & x_2'' & x_3'' \end{vmatrix}$$

Taking the derivative, we have

$$\frac{dW}{dt} = \frac{d}{dt}\begin{vmatrix} x_1 & x_2 & x_3 \\ x_1' & x_2' & x_3' \\ x_1'' & x_2'' & x_3'' \end{vmatrix}$$

$$= \begin{vmatrix} x_1' & x_2' & x_3' \\ x_1' & x_2' & x_3' \\ x_1'' & x_2'' & x_3'' \end{vmatrix} + \begin{vmatrix} x_1 & x_2 & x_3 \\ x_1'' & x_2'' & x_3'' \\ x_1'' & x_2'' & x_3'' \end{vmatrix} + \begin{vmatrix} x_1 & x_2 & x_3 \\ x_1' & x_2' & x_3' \\ x_1''' & x_2''' & x_3''' \end{vmatrix}$$

$$= \begin{vmatrix} x_1 & x_2 & x_3 \\ x_1' & x_2' & x_3' \\ x_1''' & x_2''' & x_3''' \end{vmatrix}$$

x_1, x_2 and x_3 are solutions of the DE

$$x_1''' + p_1 x_1 + p_2 x_1 + p_3 x_1 = 0$$
$$x_2''' + p_1 x_2 + p_2 x_2 + p_3 x_2 = 0$$
$$x_3''' + p_1 x_3 + p_2 x_3 + p_3 x_3 = 0$$

Thus,

$$x_1''' = -p_1 x_1'' - p_2 x_1' - p_3 x_1 = a$$
$$x_2''' = -p_1 x_2'' - p_2 x_2' - p_3 x_2 = b$$
$$x_3''' = -p_1 x_3'' - p_2 x_3' - p_3 x_3 = c$$

$$\frac{dW}{dt} = \begin{vmatrix} x_1 & x_2 & x_3 \\ x_1' & x_2' & x_3' \\ a & b & c \end{vmatrix}$$

$$= \begin{vmatrix} x_1 & x_2 & x_3 \\ x_1' & x_2' & x_3' \\ -p_1 x_1'' & -p_1 x_2'' & -p_1 x_3'' \end{vmatrix} + \begin{vmatrix} x_1 & x_2 & x_3 \\ x_1' & x_2' & x_3' \\ -p_2 x_1' & -p_2 x_2' & -p_2 x_3' \end{vmatrix}$$

$$+ \begin{vmatrix} x_1 & x_2 & x_3 \\ x_1' & x_2' & x_3' \\ -p_3 x_1 & -p_3 x_2 & -p_3 x_3 \end{vmatrix}$$

$$= -p_1 \begin{vmatrix} x_1 & x_2 & x_3 \\ x_1' & x_2' & x_3' \\ x_1'' & x_2'' & x_3'' \end{vmatrix} = -p_1 W$$

$$W = K \exp\left(\int -p_1(t) dt \right)$$

$$W(x_1, x_2) = \begin{vmatrix} x_1 & x_2 \\ x_1' & x_2' \end{vmatrix} = x_1 x_2' - x_2 x_1'$$

Similarly,

$$W(x_2, x_3) = x_2 x_3' - x_3 x_2', \qquad W(x_3, x_1) = x_3 x_1' - x_1 x_3'$$

$$u_1(t) = \int \frac{W(x_2, x_3) f(t)}{W(x_1, x_2, x_3)} dt$$

$$\Rightarrow \quad u_2(t) = \int \frac{W(x_3, x_1) f(t)}{W(x_1, x_2, x_3)} dt$$

$$u_3(t) = \int \frac{W(x_1, x_2) f(t)}{W(x_1, x_2, x_3)} dt$$

Finally, we get the PS by substituting $u_1(t), u_2(t), u_3(t), x_1(t), x_2(t), x_3(t)$ into

$$x_P = u_1(t) x_1(t) + u_2(t) x_2(t) + u_3(t) x_3(t)$$

Chapter 4 Systems of Linear DEs

4.1 Basics of System of DEs

Problem 4.1.1

Converted to matrix format, the original DE.Syst can be written as

$$\begin{pmatrix} x_1'(t) \\ x_2'(t) \end{pmatrix} = \begin{pmatrix} 4 & -7 \\ 2 & -5 \end{pmatrix} \begin{pmatrix} x_1(t) \\ x_2(t) \end{pmatrix}$$

Problem 4.1.2

Converted to matrix format, the original DE.Syst can be written as

$$\begin{pmatrix} x_1'(t) \\ x_2'(t) \\ x_3'(t) \end{pmatrix} = \begin{pmatrix} -1 & 5 & 3 \\ 0 & 1 & 1 \\ 0 & -2 & -2 \end{pmatrix} \begin{pmatrix} x_1(t) \\ x_2(t) \\ x_3(t) \end{pmatrix} + \begin{pmatrix} \sin t \\ 0 \\ -t \end{pmatrix}$$

4.2 First-Order Systems and Applications

Problem 4.2.1

(1) Consider the forces on the first object: the spring k_1 is stretched x_1 units and spring k_2 by $x_2 - x_1$ units. Stretched force is pointed away from the object. By Newton's law, we have

$$mx_1'' = k_2(x_2 - x_1) - k_1 x_1 = -(k_1 + k_2)x_1 + k_2 x_2$$

For the second object, the spring k_2 is stretched by $(x_2 - x_1)$ units, and the spring k_3 is compressed by x_2 units. Again, by Newton's law, we have

$$mx_2'' = -k_2(x_2 - x_1) - k_3 x_2$$
$$= k_2 x_1 - (k_2 + k_3)x_2$$

(2) If the spring k_2 is broken, then, $k_2 = 0$, the DEs are

$$\begin{cases} mx_1'' = -k_1 x_1 \\ mx_2'' = -k_3 x_2 \end{cases}$$

(3) If $k_1 = k_2 = k_3 = k$

$$\begin{cases} mx_1'' = -2kx_1 + kx_2 \Rightarrow x_2 = \dfrac{mx_1''}{k + 2x_1} \\ mx_2'' = kx_1 - 2kx_2 \end{cases}$$

$$\frac{mx_1^{(4)}}{k + 2x_1''} = kx_1 - 2k\left(\frac{mx_1''}{k + 2x_1}\right)$$

$$\frac{mx_1^{(4)}}{k + 2x_1''} = kx_1 - 2mx_1'' - 4kx_1$$

$$x_1^{(4)} + \frac{2k + 2km}{m}x_1'' + \frac{3k^2}{m}x_1 = 0$$

which is a 4th-order c-coeff DE. After solving the C-Eq, and finding the characteristic roots, we can easily get the GS $x_1(t)$. Function $x_2(t)$ can be found by a similar method.

Problem 4.2.2

The force is shown by Figure A.7.

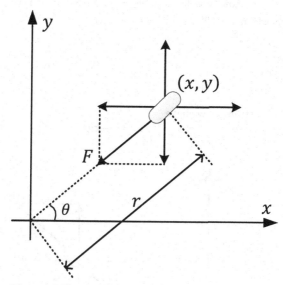

Figure A.7 The object moves under the influence of force F.

EoM in x direction,
$$F_x = -F \cos \theta = mx''$$
EoM in y direction,
$$F_y = -F \sin \theta = my''$$
We know that
$$r = \sqrt{x^2 + y^2}$$
$$F = \frac{k}{x^2 + y^2} = \frac{k}{r^2}$$
$$\sin \theta = \frac{y}{r}, \qquad \cos \theta = \frac{x}{r}$$
By substituting $\sin \theta$, $\cos \theta$, and F, we have
$$\begin{cases} mx'' = -\dfrac{kx}{r^3} \\ my'' = -\dfrac{ky}{r^3} \end{cases}$$

Problem 4.2.3

Forces on this object can be shown in Figure A.8. By Newton's law, we get
EoM in x-direction: $F_x = -F \cos \theta = mx''$

EoM in y-direction: $F_y = -F \sin \theta - F_G = my''$

Also, from the given condition, we have

$$\cos \theta = \frac{v_x}{v} = \frac{x'}{v}$$

$$\sin \theta = \frac{v_y}{v} = \frac{y'}{v}$$

$$F_G = mg, \qquad F = kv^2$$

By substitution, we have

$$\begin{cases} mx'' = -kvx' \\ my'' = -kvy' - mg \end{cases}$$

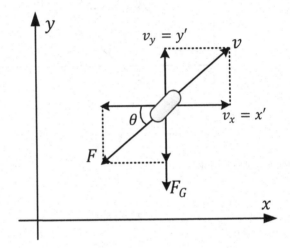

Figure A.8 A projectile model moving with speed v.

Problem 4.2.4

Assuming the displacements of blocks from the natural positions $x_1(t)$ and $x_2(t)$, we get the EoMs for the two blocks

$$\begin{cases} m_1 x_1'' = -k_1 x_1 + k_2(x_2 - x_1) + m_1 g \\ m_2 x_2'' = -k_2(x_2 - x_1) + m_2 g \end{cases}$$

By substitution, we get

$$m_1 m_2 x_2^{(4)} + \big((k_1 + k_2)m_2 + k_2 m_1\big)x_2'' - k_1 k_2 x_2$$
$$= (k_1 + k_2)m_2 g$$

This is a c-coeff InHomo DE that can be solved using the C-Eq method.

$$x_2(t) = c_1 \cos \omega_1 t + c_2 \sin \omega_1 t + c_3 \cos \omega_2 t + c_4 \sin \omega_2 t + x_{2P}(t)$$

where ω_1 and ω_2 are two characteristic frequencies. Back substitution can generate the solution for $x_1(t)$.

Problem 4.2.5

Let $x_i(t) \ \forall \ i = 1, 2, 3$ be the displacement of block-i from its initial position. The DEs for the system are

$$\begin{cases} m_1 x_1'' = -k_1 x_1 + k_2(x_2 - x_1) + m_1 g \\ m_2 x_2'' = -k_2(x_2 - x_1) + k_3(x_3 - x_2 - x_1) + m_2 g \\ m_3 x_3'' = -k_3(x_3 - x_2 - x_1) + m_3 g \end{cases}$$

and the ICs are

$$x_1(0) = x_1'(0) = x_2(0) = x_2'(0) = x_3(0) = x_3'(0) = 0$$

The above DE.Syst together with the ICs describes the motion of the three blocks.

Let

$$w_1^2 = \frac{k_1}{m_1}, \ w_2^2 = \frac{k_2}{m_2}, \ w_3^2 = \frac{k_3}{m_3}, \ w_{12}^2 = \frac{k_2}{m_1}, \ w_{23}^2 = \frac{k_3}{m_2}$$

We have

$$\begin{cases} x_1'' = -(w_1^2 + w_{12}^2)x_1 + w_{12}^2 x_2 + g \\ x_2'' = -(w_2^2 + w_{23}^2)x_2 - w_{23}^2 x_1 + w_{23}^2 x_3 + g \\ x_3'' = -w_3^2 x_3 + w_3^2 x_1 + w_3^2 x_2 + g \\ x_1(0) = x_1'(0) = x_2(0) = x_2'(0) = x_3(0) = x_3'(0) = 0 \end{cases}$$

Problem 4.2.6

On the ball, the force can be decomposed into two components, one along the direction of the spring x and another θ along which we can establish the following two DEs.

$$\begin{cases} mx'' = mg - kx & \text{(A.11)} \\ m\big(\theta(x + L_0)\big)'' = -mg\theta & \text{(A.12)} \end{cases}$$

Along the extension of the spring direction, the gravity is

$$mg \cos \theta \approx mg \times 1$$

Because the angle θ is sufficiently small, the approximation given above is not unreasonable. Eq. (A.11) is trivial to solve, one can write the solution as

$$x(t) = \sin(\omega t) + \frac{g}{\omega^2}$$

and $\frac{g}{\omega^2}$ can be absorbed to L_0 where $\omega^2 = \frac{k}{m}$. Inserting such solution to Eq. (A.12) results in

$$\theta''(x + L_0) + 2\theta' x' + \theta x'' = -g\theta$$

$$\theta''(\sin \omega t + L_0) + 2\theta'\omega \cos \omega t + \theta \omega^2 \sin \omega t = -g\theta$$

Now, we combine with IC to get

$$\begin{cases} \theta''(\sin \omega t + L_0) + 2\theta'\omega \cos \omega t + \theta \omega^2 \sin \omega t = -g\theta \\ \theta(t = 0) = \theta_0, \quad \theta'(t = 0) = 0 \end{cases}$$

We can now perform LT on the above equation to solve it or we can use other methods.

4.3 Substitution Method

Problem 4.3.1

Summing up two DEs, we get

$$x'' = \exp(-t)$$

whose solution is

$$x(t) = \exp(-t) + C_1 t + C_2$$

Substituting into the second DE, we get

$$\exp(-t) + y'' - \exp(-t) - C_1 t - C_2 = 0$$
$$y'' = C_1 t + C_2$$
$$y(t) = \frac{C_1}{6} t^3 + \frac{1}{2} C_2 t^2 + C_3 t + C_4$$

Problem 4.3.2

From the first DE $x' = -y$, we have $y = -x'$. Substituting this into the second DE, we get

$$(-x')' = 10x + 7x'$$

or

$$x'' + 7x' + 10x = 0$$

a 2nd.O c-coeff DE whose C-Eq is

$$r^2 + 7r + 10 = 0$$
$$r_{1,2} = -5, -2$$

Therefore,

$$x = C_1 \exp(-5t) + C_2 \exp(-2t)$$

and

$$y = -x'$$
$$= 5C_1 \exp(-5t) + 2C_2 \exp(-2t)$$

Applying the ICs

$$\begin{cases} C_1 + C_2 = 2 \\ 5C_1 + 2C_2 = -7 \end{cases}$$

we get $C_1 = -\frac{11}{3}$ and $C_2 = \frac{17}{3}$ and the PS's to the DEs are

$$\begin{cases} x = -\dfrac{11}{3} \exp(-5t) + \dfrac{17}{3} \exp(-2t) \\ y = -\dfrac{55}{3} \exp(-5t) + \dfrac{34}{3} \exp(-2t) \end{cases}$$

Problem 4.3.3

From the first DE

$$y = \frac{x' - x}{2}$$

$$y' = \frac{x'' - x'}{2}$$

Plugging into the second DE, we have

$$\frac{x'' - x'}{2} = 2x - 2\frac{x' - x}{2}$$

$$x'' + x' - 6x = 0$$

This is a 2nd.O c-coeff DE whose C-Eq is

$$r^2 + r - 6 = (r + 3)(r - 2) = 0$$

$$r_{1,2} = -3, 2$$

$$x(t) = C_1 \exp(-3t) + C_2 \exp(2t)$$

$$y(t) = \frac{x' - x}{2}$$

$$= -2C_1 \exp(-3t) + \frac{1}{2}C_2 \exp(2t)$$

Applying the IC's, we get $C_1 = -3/5$ and $C_2 = 8/5$ and the PS

$$\begin{cases} x(t) = \frac{1}{5}(8\exp(2t) - 3\exp(-3t)) \\ y(t) = \frac{2}{5}(2\exp(2t) + 3\exp(-3t)) \end{cases}$$

Problem 4.3.4

By the given information, we have

$$\begin{cases} m_1 x_1'' = -(k_1 + k_2)x_1 + k_2 x_2 \\ m_2 x_2'' = k_2 x_1 - (k_2 + k_3)x_2 \end{cases}$$

Given $m_1 = m_2 = 1$, $k_1 = 1$, $k_2 = 4$ and $k_3 = 1$, we have

$$\begin{cases} x_1'' = -5x_1 + 4x_2 \\ x_2'' = 4x_1 - 5x_2 \end{cases}$$

From the second DE, we have

$$x_1 = \frac{x_2''}{4} + \frac{5x_2}{4}$$

Putting it back to the first DE, we get

$$\left(\frac{x_2''}{4} + \frac{5x_2}{4}\right)'' = -5\left(\frac{x_2''}{4} + \frac{5x_2}{4}\right) + 4x_2$$

$$x_2^{(4)} + 10x_2'' + 9x_2 = 0$$

The C-Eq is

$$r^4 + 10r^2 + 9 = 0$$

$$r_{1,2,3,4} = \pm 3i, \pm i$$

The GS for x_2:

$$x_2 = C_1 \sin 3t + C_2 \cos 3t + C_3 \sin t + C_4 \cos t$$

and back substitution to find x_1:

$$x_1 = \frac{x_2''}{4} + \frac{5}{4}x_2$$

$$= -C_1 \sin 3t - C_2 \cos 3t + C_3 \sin t + C_4 \cos t$$

Problem 4.3.5

From

$$x_2' = 2x_1 + x_2 + t^2$$

we have

$$x_1 = \frac{x_2' - x_2 - t^2}{2}, \qquad x_1' = \frac{x_2'' - x_2' - 2t}{2}$$

Substituting into

$$x_1' = x_1 + 2x_2 + t$$

we have

$$x_2'' - 2x_2' - 3x_2 = -t^2 + 4t$$

C-Eq:

$$r^2 - 2r - 3 = 0$$

$$r_{1,2} = 3, -1$$

$$x_{2C} = c_1 \exp(3t) + c_2 \exp(-t)$$

Then, we select the TS

$$x_{2P} = A_2 t^2 + A_1 t + A_0$$

$$x_{2P}' = 2A_2 t + A_1$$

$$x_{2P}'' = 2A_2$$

Substituting into the original DE

$$2A_2 - 2(2A_2 t + A_1) - 3(A_2 t^2 + A_1 t + A_0) = -t^2 + 4t$$

we get

$$A_0 = \frac{38}{27}, A_1 = -\frac{16}{9}, A_2 = \frac{1}{3}$$

So, the GS is

$$x_2 = c_1 \exp(3t) + c_2 \exp(-t) + \frac{1}{3}t^2 - \frac{16}{9}t + \frac{38}{27}$$

$$x_2' = 3c_1 \exp(3t) - c_2 \exp(-t) + \frac{2}{3}t - \frac{16}{9}$$

Substituting into

$$x_1 = \frac{x_2' - x_2 - t^2}{2}$$

we have

$$x_1 = c_1 \exp(3t) - c_2 \exp(-t) - \frac{2}{3}t^2 + \frac{11}{9}t - \frac{43}{27}$$

So, the GS's are

$$x_1 = c_1 \exp(3t) - c_2 \exp(-t) - \frac{2}{3}t^2 + \frac{11}{9}t - \frac{43}{27}$$

$$x_2 = c_1 \exp(3t) + c_2 \exp(-t) + \frac{1}{3}t^2 - \frac{16}{9}t + \frac{38}{27}$$

Problem 4.3.6

$$x' = y$$
$$x'' = y' = -9x + 6x'$$
$$x'' - 6x' + 9x = 0$$
$$r^2 - 6r + 9 = 0$$
$$r_{1,2} = 3, 3$$

Thus,

$$x(t) = (A + Bt) \exp(3t)$$
$$y(t) = (3A + B + 3Bt) \exp(3t)$$

Problem 4.3.7

$$\begin{cases} x' = x - 2y & \text{(A.13)} \\ y' = x - y & \text{(A.14)} \end{cases}$$

From Eq. (A.14)

$$x = y' + y$$
$$x' = y'' + y'$$

Plugging this into Eq. (A.13)

$$x' = x - 2y$$
$$y'' + y' = y' + y - 2y$$
$$y'' + y = 0$$

The C-Eq is

$$r^2 + 1 = 0$$
$$r_{1,2} = \pm i$$
$$y(t) = C_1 \cos t + C_2 \sin t$$

Back substituting, we get

$$x(t) = (C_2 - C_1) \sin t + (C_1 + C_2) \cos t$$

Plugging in the IC, we have

$$\begin{cases} 1 = C_1 + C_2 \\ 2 = C_1 \end{cases}$$
$$\begin{cases} C_1 = 2 \\ C_2 = -1 \end{cases}$$

Thus,

$$\begin{cases} x(t) = \cos t - 3\sin t \\ y(t) = 2\cos t - \sin t \end{cases}$$

Problem 4.3.8

$$\begin{cases} x' = 3x + 4y & \text{(A.15)} \\ y' = 3x + 2y & \text{(A.16)} \end{cases}$$

From Eq. (A.16)

$$x = \frac{y' - 2y}{3}$$

$$x' = \frac{y'' - 2y'}{3}$$

Plugging this into Eq. (A.15)

$$\frac{y'' - 2y'}{3} = y' - 2y + 4y$$

$$y'' - 5y' - 6y = 0$$

The C-Eq is

$$r^2 - 5r - 6 = 0$$

$$(r + 1)(r - 6) = 0$$

$$r_{1,2} = -1, 6$$

$$y(t) = C_1 \exp(-t) + C_2 \exp(6t)$$

$$y'(t) = -C_1 \exp(-t) + 6C_2 \exp(6t)$$

Put this back, and

$$x(t) = -C_1 \exp(-t) + \frac{4}{3} C_2 \exp(6t)$$

Applying the ICs,

$$\begin{cases} 1 = -C_1 + \dfrac{4}{3} C_2 \\ 1 = C_1 + C_2 \end{cases}$$

we get $C_1 = \frac{1}{7}$ and $C_2 = \frac{6}{7}$ and the PS

$$\begin{cases} x(t) = -\dfrac{1}{7} \exp(-t) + \dfrac{8}{7} \exp(6t) \\ y(t) = \dfrac{1}{7} \exp(-t) + \dfrac{6}{7} \exp(6t) \end{cases}$$

Problem 4.3.9

$$\begin{cases} x' = 2x & \text{(A.17)} \\ y' = 3x + 3y & \text{(A.18)} \\ z' = 4x + 4y + 4z & \text{(A.19)} \end{cases}$$

Using Eq. (A.17), we find the solution

$$x(t) = a_1 \exp(2t)$$

Plugging the above into Eq. (A.18), we get

$$y'(t) = 3x + 3y$$
$$= 3(a_1 \exp(2t)) + 3y$$

whose solution is

$$y(t) = -3a_1 \exp(2t) + b_1 \exp(3t)$$

Similarly, we have

$$z'(t) = 4x + 4y + 4z$$
$$= 4(a_1 \exp(2t)) + 4(-3a_1 \exp(2t) + b_1 \exp(3t)) + 4z$$

whose solution is

$$z(t) = 4a_1 \exp(2t) - 4b_1 \exp(3t) + c_1 \exp(4t)$$

Now, we can express all three solutions as

$$x(t) = a_1 \exp(2t)$$
$$y(t) = -3a_1 \exp(2t) + b_1 \exp(3t)$$
$$z(t) = 4a_1 \exp(2t) - 4b_1 \exp(3t) + c_1 \exp(4t)$$

The solution can be expressed in a matrix form

$$\begin{pmatrix} x(t) \\ y(t) \\ z(t) \end{pmatrix} = a_1 \begin{pmatrix} 1 \\ -3 \\ 4 \end{pmatrix} \exp(2t) + b_1 \begin{pmatrix} 0 \\ 1 \\ -4 \end{pmatrix} \exp(3t) + c_1 \begin{pmatrix} 0 \\ 0 \\ 1 \end{pmatrix} \exp(4t)$$

4.4 The Operator Method

Problem 4.4.1

(a) Substitute $L_1 \equiv D^2 + D + 1$ and $L_2 \equiv D^2 + 2D$ into $L_1 L_2 x(t)$ and $L_2 L_1 x(t)$,

$$\begin{aligned}
L_1 L_2 x(t) &= (D^2 + D + 1)(D^2 + 2D)x(t) \\
&= (D^2 + D + 1)\big(x''(t) + 2x'(t)\big) \\
&= x''''(t) + 3x'''(t) + 3x''(t) + 2x'(t) \\
L_2 L_1 x(t) &= (D^2 + 2D)\big((D^2 + D + 1)x(t)\big) \\
&= (D^2 + 2D)\big(x''(t) + x'(t) + x(t)\big) \\
&= x''''(t) + 3x'''(t) + 3x''(t) + 2x'(t)
\end{aligned}$$

Therefore,

$$L_1 L_2 x(t) = L_2 L_1 x(t)$$

(b) Substitute $L_1 x(t) \equiv Dx(t) + tx(t)$ and $L_2 x(t) \equiv tDx(t) + x(t)$ into $L_1 L_2 x(t)$ and $L_2 L_1 x(t)$,

$$\begin{aligned}
L_1 L_2 x(t) &= (D + t)\big(tDx(t) + x(t)\big) \\
&= (D + t)\big(tx'(t) + x(t)\big) \\
&= tx''(t) + (2 + t^2)x'(t) + tx(t) \\
L_2 L_1 x(t) &= (tD + 1)(D + t)x(t) \\
&= (tD + 1)\big(tx'(t) + tx(t)\big) \\
&= tx''(t) + (1 + t^2)x'(t) + 2tx(t)
\end{aligned}$$

Therefore,

$$L_1 L_2 x(t) \neq L_2 L_1 x(t)$$

Problem 4.4.2

The original Homo system

$$\begin{cases}
(D - 1)x - 2y - z = 0 \\
-6x + (D + 1)y = 0 \\
x + 2y + (D + 1)z = 0
\end{cases}$$

can be written in terms of operator $D = \frac{d}{dt}$

$$\begin{pmatrix} D - 1 & -2 & -1 \\ -6 & D + 1 & 0 \\ 1 & 2 & D + 1 \end{pmatrix} \begin{pmatrix} x \\ y \\ z \end{pmatrix} = \begin{pmatrix} 0 \\ 0 \\ 0 \end{pmatrix}$$

The operational determinant is:

$$\begin{vmatrix} D - 1 & -2 & -1 \\ -6 & D + 1 & 0 \\ 1 & 2 & D + 1 \end{vmatrix} = D(D + 4)(D - 3)$$

So x, y, z all satisfy a 3rd-order Homo DE with C-Eq

$$r(r + 4)(r - 3) = 0$$

and the corresponding GS's are

$$\begin{cases} x(t) = a_1 + a_2 \exp(-4t) + a_3 \exp(3t) \\ y(t) = b_1 + b_2 \exp(-4t) + b_3 \exp(3t) \\ z(t) = c_1 + c_2 \exp(-4t) + c_3 \exp(3t) \end{cases}$$

Substituting $x(t)$ and $y(t)$ into the second DE $y' = 6x - y$ and collecting the coefficients of like terms, we have:

$$-4b_2 \exp(-4t) + 3b_3 \exp(3t)$$
$$= 6a_1 + 6a_2 \exp(-4t) + 6a_3 \exp(3t) - b_1 - b_2 \exp(-4t) - b_3 \exp(3t)$$

This gives

$$b_1 = 6a_1, b_2 = -2a_2, b_3 = \frac{3}{2}a_3$$

Similarly, we can substitute $x(t)$, $y(t)$ and $z(t)$ in the first or third DE, and compare the coefficients of the like terms. Then, we have

$$c_1 = -12a_1, c_2 = -a_2, c_3 = -a_3$$

and the GS

$$\begin{cases} x(t) = a_1 + a_2 \exp(-4t) + a_3 \exp(3t) \\ y(t) = 6a_1 - 2a_2 \exp(-4t) + \frac{3}{2}a_3 \exp(3t) \\ z(t) = -12a_1 - a_2 \exp(-4t) - a_3 \exp(3t) \end{cases}$$

Problem 4.4.3

$$\begin{cases} (D^2 + D)x + D^2 y = 2\exp(-t) & \text{(A.20)} \\ (D^2 - 1)x + (D^2 - D)y = 0 & \text{(A.21)} \end{cases}$$

By the following operations

$$(D^2 - D) \text{ Eq. (A.20)} - D^2 \text{ Eq. (A.21)}$$

we have:

$$(D^2 - D)(D^2 + D)x - D^2(D^2 - 1)x = (D^2 - D)2\exp(-t)$$

which gives

$$0 = 4\exp(-t)$$

This is impossible and the DEs have no solution.

Problem 4.4.4

$$(D - 3)x - 9y = 0$$
$$-2x + (D - 2)y = 0$$
$$\Rightarrow \begin{vmatrix} D - 3 & -9 \\ -2 & D - 2 \end{vmatrix} x = 0$$
$$\Rightarrow (D^2 - 5D - 12)x = 0$$

The C-Eq is

$$r^2 - 5r - 12 = 0$$
$$r_{1,2} = \frac{5 \pm \sqrt{73}}{2}$$

$$x(t) = A_1 \exp(r_1 t) + B_1 \exp(r_2 t)$$
$$y(t) = A_2 \exp(r_1 t) + B_2 \exp(r_2 t)$$

Apply the IC $x(0) = y(0) = 2$ gives

$$A_1 + B_1 = A_2 + B_2 = 2$$

Problem 4.4.5

The original DEs can be written as

$$\begin{pmatrix} D-1 & -2 \\ -2 & D-1 \end{pmatrix} \begin{pmatrix} x_1 \\ x_2 \end{pmatrix} = \begin{pmatrix} t \\ t^2 \end{pmatrix}$$

$$\begin{pmatrix} D-1 & -2 & t \\ -2 & D-1 & t^2 \end{pmatrix} \Rightarrow \begin{pmatrix} -2 & D-1 & t^2 \\ D-1 & -2 & t \end{pmatrix}$$

$$\Rightarrow \begin{pmatrix} -2 & D-1 & t^2 \\ 2(D-1) & -4 & 2t \end{pmatrix}$$

$$\Rightarrow \begin{pmatrix} -2 & D-1 & t^2 \\ 0 & D^2-2D-3 & -t^2+4t \end{pmatrix}$$

So, we have

$$(D^2 - 2D - 3)x_2 = -t^2 + 4t$$

whose C-Eq is

$$r^2 - 2r - 3 = 0$$
$$r_1 = 3, r_2 = -1$$
$$x_{2C} = c_1 \exp(3t) + c_2 \exp(-t)$$

Then, we select the TS

$$x_{2P} = A_2 t^2 + A_1 t + A_0$$
$$x'_{2P} = 2A_2 t + A_1$$
$$x''_{2P} = 2A_2$$

Substituting back into the original DE

$$2A_2 - 2(2A_2 t + A_1) - 3(A_2 t^2 + A_1 t + A_0) = -t^2 + 4t$$

we get,

$$A_0 = \frac{38}{27}, \qquad A_1 = -\frac{16}{9}, \qquad A_2 = \frac{1}{3}$$

So, the GS is

$$x_2 = c_1 \exp(3t) + c_2 \exp(-t) + \frac{1}{3}t^2 - \frac{16}{9}t + \frac{38}{27}$$

From the first of the original DEs, we get

$$-2x_1 + (D-1)x_2 = t^2$$

$$x_1 = \frac{x'_2 - x_2 - t^2}{2}$$

$$= c_1 \exp(3t) - c_2 \exp(-t) - \frac{2}{3}t^2 + \frac{11}{9}t - \frac{43}{27}$$

So, the GS is

$$x_1 = c_1 \exp(3t) - c_2 \exp(-t) - \frac{2}{3}t^2 + \frac{11}{9}t - \frac{43}{27}$$

$$x_2 = c_1 \exp(3t) + c_2 \exp(-t) + \frac{1}{3}t^2 - \frac{16}{9}t + \frac{38}{27}$$

Problem 4.4.6

$$Dx - 1y - 1z = \exp(-t)$$
$$-1x + Dy - 1z = 0$$
$$-1x - 1y + Dz = 0$$

$$\begin{vmatrix} D & -1 & -1 \\ -1 & D & -1 \\ -1 & -1 & D \end{vmatrix} = D(D^2 - 1) + 1(-D - 1) - 1(1 + D)$$

$$= D^3 - D - D - 1 - 1 - D$$
$$= D^3 - 3D - 2$$
$$= (D + 1)^2(D - 2) = 0$$

$$x(t) = a_1 \exp(2t) + a_2 \exp(-t) + a_3 \exp(-t)$$
$$y(t) = b_1 \exp(2t) + b_2 \exp(-t) + b_3 \exp(-t)$$
$$z(t) = c_1 \exp(2t) + c_2 \exp(-t) + c_3 \exp(-t)$$

$$\begin{cases} x = \dfrac{1}{3}C_1 \exp(2t) - (C_2 + C_3) \exp(-t), \\[2mm] y = -\dfrac{1}{2}\exp(t) + \dfrac{1}{3}C_1 \exp(2t) + C_2 \exp(-t), \\[2mm] z = -\dfrac{1}{2}\exp(t) + \dfrac{1}{3}C_1 \exp(2t) + C_3 \exp(-t). \end{cases}$$

Problem 4.4.7

Method 1: The S-method.

By the first DE, we have

$$y = x' - x$$

and

$$y' = x'' - x'$$

Plugging it into the second DE, we have

$$x'' - x' = 6x - (x' - x)$$
$$x'' = 7x$$

The C-Eq is $r^2 = 7$ whose roots are $r_{1,2} = \pm\sqrt{7}$.

Therefore

$$\begin{cases} x(t) = c_1 \exp(\sqrt{7}t) + c_2 \exp(-\sqrt{7}t) \\ y(t) = (\sqrt{7} - 1)c_1 \exp(\sqrt{7}t) - (\sqrt{7} + 1)c_2 \exp(-\sqrt{7}t) \end{cases}$$

Method 2: The operator method.

The DEs are written as

$$\begin{cases} (D-1)x - y = 0 \\ -6x + (D+1)y = 0 \end{cases}$$

(A.22)
(A.23)

By $(D+1)(A.22) + (A.23)$, we have

$$(D+1)(D-1)x - 6x = 0$$
$$(D^2 - 7)x = 0$$

The C-Eq is

$$r^2 - 7 = 0$$
$$r_{1,2} = \pm\sqrt{7}$$

Therefore

$$\begin{cases} x(t) = c_1 \exp(\sqrt{7}t) + c_2 \exp(-\sqrt{7}t) \\ y(t) = (\sqrt{7} - 1)c_1 \exp(\sqrt{7}t) - (\sqrt{7} + 1)c_2 \exp(-\sqrt{7}t) \end{cases}$$

Problem 4.4.8

We add and subtract the two DEs respectively, and we have

$$D(x+y) = \frac{1}{2}(\exp(-2t) + \exp(3t))$$

(A.24)

$$2x - 3y = \frac{1}{2}(\exp(-2t) - \exp(3t))$$

(A.25)

From (A.25) we have $x = \frac{1}{4}(\exp(-2t) - \exp(3t)) + \frac{3}{2}y$.

We return x to (A.24) and

$$\left(\frac{1}{4}(\exp(-2t) - \exp(3t)) + \frac{5}{2}y\right)' = \frac{1}{2}(\exp(-2t) - \exp(3t)).$$

Thus,

$$y' = \frac{2}{5}\exp(-2t) + \frac{1}{2}\exp(3t)$$

and

$$y(t) = -\frac{1}{5}\exp(-2t) + \frac{1}{6}\exp(3t) + C_1$$
$$x(t) = -\frac{1}{20}\exp(-2t) + \frac{3}{2}C_1$$

The GS is

$$\begin{cases} x(t) = -\frac{1}{20}\exp(-2t) + \frac{3}{2}C_1 \\ y(t) = -\frac{1}{5}\exp(-2t) + \frac{1}{6}\exp(3t) + C_1 \end{cases}$$

Problem 4.4.9

The original InHomo system can be rewritten in the following operator form:

$$\begin{cases} (D^2 - 4)x + 13Dy = 6\sin t & \text{(A.26)} \\ 2Dx + (-D^2 + 9)y = 0 & \text{(A.27)} \end{cases}$$

Operating on the new InHomo system as

$$2D \ (\text{A.26}) - (D^2 - 4) \ (\text{A.27})$$

we get

$$(D^4 + 13D^2 + 36)y = 12\cos t$$

C-Eq:

$$r^4 + 13r^2 + 36 = 0$$

$$r_{1,2} = \pm 2i, \qquad r_{3,4} = \pm 3i$$

$$y_C(t) = C_1 \sin 2t + C_2 \cos 2t + C_3 \sin 3t + C_4 \cos 3t$$

The fact that the DE has only terms with even number of derivatives allows us to select the following TS

$$y_P(t) = A\cos t$$

$$y_P'' = -A\cos t, \qquad y_P^{(4)} = A\cos t$$

This gives

$$A = \frac{1}{2}$$

So, the GS is

$$y(t) = C_1 \sin 2t + C_2 \cos 2t + C_3 \sin 3t + C_4 \cos 3t + \frac{1}{2}\cos t$$

$$y''(t) = -4C_1 \sin 2t - 4C_2 \cos 2t - 9C_3 \sin 3t - 9C_4 \cos 3t - \frac{1}{2}\cos t$$

Plugging these into (A.27), we get

$$x' = \frac{1}{2}(y'' - 9y)$$

$$= \frac{1}{2}(-13C_1 \sin 2t - 13C_2 \cos 2t - 18C_3 \sin 3t - 18C_4 \cos 3t - 5\cos t)$$

$$x(t) = \frac{1}{2}\left(\frac{13}{2}C_1 \cos 2t - \frac{13}{2}C_2 \sin 2t + 6C_3 \cos 3t - 6C_4 \sin 3t + 5\sin t\right)$$

$$+ C_5$$

In summary, the GS for the InHomo system is

$$\begin{pmatrix} x(t) \\ y(t) \end{pmatrix} = C_1 \begin{pmatrix} \frac{13}{4}\cos 2t \\ \sin 2t \end{pmatrix} + C_2 \begin{pmatrix} -\frac{13}{4}\sin 2t \\ \cos 2t \end{pmatrix} + C_3 \begin{pmatrix} 3\cos 3t \\ \sin 3t \end{pmatrix}$$

$$+ C_4 \begin{pmatrix} -3\sin 3t \\ \cos 3t \end{pmatrix} + C_5 \begin{pmatrix} 1 \\ 0 \end{pmatrix} + \frac{1}{2}\begin{pmatrix} 5\sin t \\ \cos t \end{pmatrix}$$

Problem 4.4.10

Method 1: The S-method.

From the second DE

$$x = \frac{y' - y}{2}$$

$$x' = \frac{y'' - y'}{2}$$

Substitute these into the first DE

$$\frac{y'' - y'}{2} = 3\frac{y' - y}{2} - 2y$$

$$y'' - 4y' + 7y = 0$$

C-Eq:

$$r^2 - 4r + 7 = 0$$

$$r_1 = 2 + \sqrt{3}i, \qquad r_2 = 2 - \sqrt{3}i$$

So

$$y(t) = c_1 \exp(2t) \cos(\sqrt{3}t) + c_2 \exp(2t) \sin(\sqrt{3}t)$$

$$y'(t) = \left(2c_1 + \sqrt{3}c_2\right) \exp(2t) \cos(\sqrt{3}t) + \left(2c_2 - \sqrt{3}c_1\right) \exp(2t) \sin(\sqrt{3}t)$$

$$x(t) = \frac{y' - y}{2}$$

$$= \frac{1}{2}\left(c_1 + \sqrt{3}c_2\right) \exp(2t) \cos(\sqrt{3}t) + \frac{1}{2}\left(c_2 - \sqrt{3}c_1\right) \exp(2t) \sin(\sqrt{3}t)$$

So, the GS is

$$\begin{cases} x(t) = \frac{1}{2}\left(c_1 + \sqrt{3}c_2\right) \exp(2t) \cos(\sqrt{3}t) + \frac{1}{2}\left(c_2 - \sqrt{3}c_1\right) \exp(2t) \sin(\sqrt{3}t) \\ y(t) = c_1 \exp(2t) \cos(\sqrt{3}t) + c_2 \exp(2t) \sin(\sqrt{3}t) \end{cases}$$

Method 2: The operator method.

The original DEs can be written as

$$\begin{pmatrix} D - 3 & 2 \\ -2 & D - 1 \end{pmatrix}\begin{pmatrix} x \\ y \end{pmatrix} = 0$$

$$\begin{pmatrix} D - 3 & 2 \\ -2 & D - 1 \end{pmatrix} \Rightarrow \begin{pmatrix} -2 & D - 1 \\ D - 3 & 2 \end{pmatrix}$$

$$\Rightarrow \begin{pmatrix} -2 & D - 1 \\ 2D - 6 & 4 \end{pmatrix}$$

$$\Rightarrow \begin{pmatrix} -2 & D - 1 \\ 0 & 4 + (D - 1)(D - 3) \end{pmatrix}$$

Thus,

$$\left(4 + (D - 1)(D - 3)\right)y(t) = (D^2 - 4D + 7)y(t) = 0$$

The C-Eq is $r^2 - 4r + 7 = 0$ whose roots are $r_{1,2} = 2 \pm \sqrt{3}i$. Thus, the GS is

$$y(t) = c_1 \exp(2t) \cos(\sqrt{3}t) + c_2 \exp(2t) \sin(\sqrt{3}t)$$

$$y'(t) = \left(2c_1 + \sqrt{3}c_2\right)\exp(2t)\cos(\sqrt{3}t) + \left(2c_2 - \sqrt{3}c_1\right)\exp(2t)\sin(\sqrt{3}t)$$

From the first row of the matrix, we have

$$-2x + y' - y = 0$$

$$x(t) = \frac{y' - y}{2}$$

$$= \frac{1}{2}\cos(\sqrt{3}t) + \frac{1}{2}\sin(\sqrt{3}t)$$

$$y(t) = c_1 \exp(2t)\cos(\sqrt{3}t) + c_2 \exp(2t)\sin(\sqrt{3}t)$$

4.5 The Eigen-Analysis Method

Problem 4.5.1

$$\det\begin{pmatrix} 1-\lambda & -3 \\ 3 & 7-\lambda \end{pmatrix} = (1-\lambda)(7-\lambda) + 9$$
$$= (\lambda - 4)^2$$
$$= 0$$

The system has two identical e-values

$$\lambda_1 = \lambda_2 = 4$$

The e-vectors have relation

$$V_1 = (A - \lambda)V_2$$

The e-vector V_2 is

$$(A - \lambda I)^2 V_2 = \begin{pmatrix} -3 & -3 \\ 3 & 3 \end{pmatrix}\begin{pmatrix} -3 & -3 \\ 3 & 3 \end{pmatrix} V_2 = 0$$
$$V_2 = \begin{pmatrix} 0 \\ 1 \end{pmatrix}$$

The e-vector V_1 is

$$V_1 = (A - \lambda)V_2 = \begin{pmatrix} -3 & -3 \\ 3 & 3 \end{pmatrix}\begin{pmatrix} 0 \\ 1 \end{pmatrix} = \begin{pmatrix} -3 \\ 3 \end{pmatrix}$$

The two LI solutions are

$$X_1 = \begin{pmatrix} -3 \\ 3 \end{pmatrix}\exp(4t)$$
$$X_2 = \left(\begin{pmatrix} -3 \\ 3 \end{pmatrix}t + \begin{pmatrix} 0 \\ 1 \end{pmatrix}\right)\exp(4t) = \begin{pmatrix} -3t \\ 3t+1 \end{pmatrix}\exp(4t)$$

So, finally, the GS for the original DE is

$$X = c_1 X_1 + c_2 X_2 = c_1\begin{pmatrix} -3 \\ 3 \end{pmatrix}\exp(4t) + c_2\begin{pmatrix} -3t \\ 3t+1 \end{pmatrix}\exp(4t)$$

Problem 4.5.2

$$Z(t) = \begin{pmatrix} x(t) \\ y(t) \end{pmatrix}$$
$$Z'(t) = \begin{pmatrix} 4 & -3 \\ 3 & 4 \end{pmatrix}Z, \qquad Z(0) = \begin{pmatrix} 2 \\ 3 \end{pmatrix}$$
$$\det(A - \lambda I) = \begin{vmatrix} 4-\lambda & -3 \\ 3 & 4-\lambda \end{vmatrix} = \lambda^2 - 8\lambda + 25$$

whose roots are

$$\lambda_{1,2} = 4 \pm 3i$$

For $\lambda_1 = 4 + 3i$

$$\begin{pmatrix} -3i & -3 \\ 3 & -3i \end{pmatrix}\begin{pmatrix} \eta_1 \\ \eta_2 \end{pmatrix} = \begin{pmatrix} 0 \\ 0 \end{pmatrix}$$
$$-3i\eta_1 - 3\eta_2 = 0$$

$$\eta_1 = -i\eta_2$$

$$\begin{pmatrix} \eta_1 \\ \eta_2 \end{pmatrix}' = \begin{pmatrix} -i \\ 1 \end{pmatrix}$$

For $\lambda_2 = 4 - 3i$

$$\begin{pmatrix} 3i & -3 \\ 3 & 3i \end{pmatrix} \begin{pmatrix} \eta_1 \\ \eta_2 \end{pmatrix} = \begin{pmatrix} 0 \\ 0 \end{pmatrix}$$

$$3i\eta_1 - 3\eta_2 = 0$$

$$\eta_1 = i\eta_2$$

$$\begin{pmatrix} \eta_1 \\ \eta_2 \end{pmatrix}' = \begin{pmatrix} i \\ 1 \end{pmatrix}$$

Thus,

$$Z(t) = c_1 \exp((4 + 3i)t) \begin{pmatrix} -i \\ 1 \end{pmatrix} + c_2 \exp((4 - 3i)t) \begin{pmatrix} i \\ 1 \end{pmatrix}$$

To find the constants,

$$Z(0) = \begin{pmatrix} 2 \\ 3 \end{pmatrix} = c_1 \begin{pmatrix} -i \\ 1 \end{pmatrix} + c_2 \begin{pmatrix} i \\ 1 \end{pmatrix}$$

$$2 = -ic_1 + ic_2$$

$$3 = c_1 + c_2$$

$$c_1 = \frac{3}{2} + i, \qquad c_2 = \frac{3}{2} - i$$

Plugging the constants, we now get

$$\begin{pmatrix} x(t) \\ y(t) \end{pmatrix} = \left(\frac{3}{2} + i\right) \exp((4 + 3i)t) \begin{pmatrix} -i \\ 1 \end{pmatrix} + \left(\frac{3}{2} - i\right) \exp((4 - 3i)t) \begin{pmatrix} i \\ 1 \end{pmatrix}$$

$$= \begin{pmatrix} \exp(4t) \left(2 \cos(3t) - 3 \sin(3t)\right) \\ \exp(4t) \left(3 \cos(3t) + 2 \sin(3t)\right) \end{pmatrix}$$

Problem 4.5.3

First, we solve the Homo DEs

$$\begin{pmatrix} x_1 \\ x_2 \end{pmatrix}' = \begin{pmatrix} 1 & 2 \\ 2 & 1 \end{pmatrix} \begin{pmatrix} x_1 \\ x_2 \end{pmatrix}$$

where

$$\det \begin{pmatrix} 1 - \lambda & 2 \\ 2 & 1 - \lambda \end{pmatrix} = (1 - \lambda)^2 - 4 = 0$$

So, the e-values are $\lambda_1 = 3$, $\lambda_2 = -1$

For $\lambda_1 = 3$, we get

$$\begin{pmatrix} -2 & 2 \\ 2 & -2 \end{pmatrix} V_1 = \begin{pmatrix} 0 \\ 0 \end{pmatrix}$$

We may select the e-vector:

$$V_1 = \begin{pmatrix} 1 \\ 1 \end{pmatrix}$$

For $\lambda_2 = -1$, we get

$$\begin{pmatrix} 2 & 2 \\ 2 & 2 \end{pmatrix} V_2 = \begin{pmatrix} 0 \\ 0 \end{pmatrix}$$

We can select the e-vector:

$$V_2 = \begin{pmatrix} -1 \\ 1 \end{pmatrix}$$

So, the GS to Homo DEs is

$$X_C(t) = \begin{pmatrix} x_1 \\ x_2 \end{pmatrix} = C_1 \begin{pmatrix} 1 \\ 1 \end{pmatrix} \exp(3t) + C_2 \begin{pmatrix} -1 \\ 1 \end{pmatrix} \exp(-t)$$

The InHomo term can be written as

$$F(t) = \begin{pmatrix} 3 \exp(t) \\ -t^2 \end{pmatrix} = \begin{pmatrix} 3 \\ 0 \end{pmatrix} \exp(t) + \begin{pmatrix} 0 \\ -1 \end{pmatrix} t^2$$

The TS may take the following form

$$X_P(t) = E_1 \exp(t) + B_0 + B_1 t + B_2 t^2$$

where E_1 and B_0, B_1, B_2 are 2×1 column vectors to be determined.

$$X_P' = E_1 \exp(t) + B_1 + 2B_2 t$$

Plugging them into the original InHomo system, we get

$$E_1 \exp(t) + B_1 + 2B_2 t = \begin{pmatrix} 1 & 2 \\ 2 & 1 \end{pmatrix} (E_1 \exp(t) + B_0 + B_1 t + B_2 t^2)$$

$$+ \begin{pmatrix} 3 \\ 0 \end{pmatrix} \exp(t) + \begin{pmatrix} 0 \\ -1 \end{pmatrix} t^2$$

Using

$$A = \begin{pmatrix} 1 & 2 \\ 2 & 1 \end{pmatrix}$$

and matching terms, we get

$$E_1 = AE_1 + \begin{pmatrix} 3 \\ 0 \end{pmatrix}$$

$$B_1 = AB_0$$

$$2B_2 = AB_1$$

$$0 = AB_2 + \begin{pmatrix} 0 \\ -1 \end{pmatrix}$$

Solving the above four equations for four unknowns, we get

$$B_2 = \frac{1}{3} \begin{pmatrix} 2 \\ -1 \end{pmatrix}$$

$$B_1 = \frac{2}{9} \begin{pmatrix} -4 \\ 5 \end{pmatrix}$$

$$B_0 = \frac{2}{27} \begin{pmatrix} 14 \\ -13 \end{pmatrix}$$

$$E_1 = \frac{1}{2} \begin{pmatrix} -3 \\ 0 \end{pmatrix}$$

Thus, the PS is

$$X_P(t) = \frac{1}{2} \begin{pmatrix} -3 \\ 0 \end{pmatrix} \exp(t) + \frac{2}{27} \begin{pmatrix} 14 \\ -13 \end{pmatrix} + \frac{2}{9} \begin{pmatrix} -4 \\ 5 \end{pmatrix} t + \frac{1}{3} \begin{pmatrix} 2 \\ -1 \end{pmatrix} t^2$$

$$X(t) = C_1 \begin{pmatrix} 1 \\ 1 \end{pmatrix} \exp(3t) + C_2 \begin{pmatrix} -1 \\ 1 \end{pmatrix} \exp(-t) + \frac{1}{2} \begin{pmatrix} -3 \\ 0 \end{pmatrix} \exp(t) + \frac{2}{27} \begin{pmatrix} 14 \\ -13 \end{pmatrix}$$
$$+ \frac{2}{9} \begin{pmatrix} -4 \\ 5 \end{pmatrix} t + \frac{1}{3} \begin{pmatrix} 2 \\ -1 \end{pmatrix} t^2$$

Problem 4.5.4

The E-method for the Homo portion:
$$\frac{d}{dt} \begin{pmatrix} x \\ y \end{pmatrix} = \begin{pmatrix} 1 & 3 \\ 1 & -1 \end{pmatrix} \begin{pmatrix} x \\ y \end{pmatrix}$$
$$\det \begin{pmatrix} 1 - \lambda & 3 \\ 1 & -1 - \lambda \end{pmatrix} = (1 - \lambda)(-1 - \lambda) - 3 = \lambda^2 - 4 = 0$$
whose roots, *i.e.*, the e-values, are:
$$\lambda_{1,2} = \pm 2$$

Next, we find e-vector.

For $\lambda_1 = 2$
$$\begin{pmatrix} -1 & 3 \\ 1 & -3 \end{pmatrix} V_1 = \begin{pmatrix} 0 \\ 0 \end{pmatrix}$$

we get
$$V_1 = \begin{pmatrix} 3 \\ 1 \end{pmatrix}$$

For $\lambda_2 = -2$
$$\begin{pmatrix} 3 & 3 \\ 1 & 1 \end{pmatrix} V_2 = \begin{pmatrix} 0 \\ 0 \end{pmatrix}$$

we get
$$V_2 = \begin{pmatrix} -1 \\ 1 \end{pmatrix}$$

So
$$X_C = C_1 \begin{pmatrix} 3 \\ 1 \end{pmatrix} \exp(2t) + C_2 \begin{pmatrix} -1 \\ 1 \end{pmatrix} \exp(-2t)$$

Let the TS be
$$X_P(t) = B \exp(-2t) + Et \exp(-2t)$$

we get
$$X_P' = -2B \exp(-2t) + E \exp(-2t) - 2Et \exp(-2t)$$

Plugging them into the original InHomo system, we get
$$-2B \exp(-2t) + E \exp(-2t) - 2Et \exp(-2t)$$
$$= A(B \exp(-2t) + Et \exp(-2t)) + \begin{pmatrix} 1 \\ 0 \end{pmatrix} \exp(-2t)$$

Grouping like terms, we get
$$\begin{cases} -2B + E = AB + \begin{pmatrix} 1 \\ 0 \end{pmatrix} \\ -2E = AE \end{cases}$$

Solving the 2nd equation and recognizing the fact that $\lambda_2 = -2$ is an e-value whose associated e-vector is

$$V_2 = \begin{pmatrix} -1 \\ 1 \end{pmatrix}$$

we can set the solution vector as

$$E = e \begin{pmatrix} -1 \\ 1 \end{pmatrix}$$

where e is an arbitrary constant.
Then, the first equation becomes

$$-2B + e \begin{pmatrix} -1 \\ 1 \end{pmatrix} = AB + \begin{pmatrix} 1 \\ 0 \end{pmatrix}$$

$$(A + 2)B = e \begin{pmatrix} -1 \\ 1 \end{pmatrix} - \begin{pmatrix} 1 \\ 0 \end{pmatrix}$$

$$\begin{pmatrix} 3 & 3 \\ 1 & 1 \end{pmatrix} B = \begin{pmatrix} -e - 1 \\ e \end{pmatrix}$$

Let $B = \begin{pmatrix} b_1 \\ b_2 \end{pmatrix}$, we get

$$3b_1 + 3b_2 = -e - 1$$
$$b_1 + b_2 = e$$

In order for the above system to hold, we must have

$$-\frac{1}{3}(e + 1) = e$$

$$e = -\frac{1}{4}$$

Thus,

$$E = -\frac{1}{4} \begin{pmatrix} -1 \\ 1 \end{pmatrix}$$

Also,

$$b_1 + b_2 = -\frac{1}{4}$$

We may choose

$$B = -\frac{1}{4} \begin{pmatrix} 1 \\ 0 \end{pmatrix}$$

Then,

$$X_P = -\frac{1}{4} \begin{pmatrix} 1 \\ 0 \end{pmatrix} \exp(-2t) - \frac{1}{4} \begin{pmatrix} -1 \\ 1 \end{pmatrix} t \exp(-2t)$$

Thus, the GS is

$$X = C_1 \begin{pmatrix} 3 \\ 1 \end{pmatrix} \exp(2t) + C_2 \begin{pmatrix} -1 \\ 1 \end{pmatrix} \exp(-2t) - \frac{1}{4} \begin{pmatrix} 1 \\ 0 \end{pmatrix} \exp(-2t)$$
$$-\frac{1}{4} \begin{pmatrix} -1 \\ 1 \end{pmatrix} t \exp(-2t)$$

Problem 4.5.5

First, let's find the solution to the corresponding Homo system

$$A = \begin{pmatrix} 1 & 2 \\ 4 & -1 \end{pmatrix}$$

To find the e-values, we set

$$\det(A - \lambda I) = \det\begin{pmatrix} 1-\lambda & 2 \\ 4 & -1-\lambda \end{pmatrix} = \lambda^2 - 9 = 0$$

whose roots are

$$\lambda_{1,2} = \pm 3$$

Finding the e-vectors:

for $\lambda_1 = 3$:

$$\begin{pmatrix} 1-3 & 2 \\ 4 & -1-3 \end{pmatrix}\begin{pmatrix} u_1 \\ v_1 \end{pmatrix} = \begin{pmatrix} 0 \\ 0 \end{pmatrix}$$

$$-2u_1 + 2v_1 = 0$$

$$V_1 = \begin{pmatrix} 1 \\ 1 \end{pmatrix}$$

for $\lambda_2 = -3$:

$$\begin{pmatrix} 1-(-3) & 2 \\ 4 & -1-(-3) \end{pmatrix}\begin{pmatrix} u_2 \\ v_2 \end{pmatrix} = \begin{pmatrix} 0 \\ 0 \end{pmatrix}$$

$$4u_2 + 2v_2 = 0$$

$$V_2 = \begin{pmatrix} 1 \\ -2 \end{pmatrix}$$

So, the complementary solution is:

$$X_C(t) = C_1 \begin{pmatrix} 1 \\ 1 \end{pmatrix} \exp(3t) + C_2 \begin{pmatrix} 1 \\ -2 \end{pmatrix} \exp(-3t)$$

The PS has the following form:

$$X_P = Bt + D = \begin{pmatrix} b_1 \\ b_2 \end{pmatrix} t + \begin{pmatrix} d_1 \\ d_2 \end{pmatrix}$$

$$X_P' = \begin{pmatrix} b_1 \\ b_2 \end{pmatrix}$$

Plugging back into the original system, we get:

$$\begin{pmatrix} b_1 \\ b_2 \end{pmatrix} = \begin{pmatrix} 1 & 2 \\ 4 & -1 \end{pmatrix}\left(\begin{pmatrix} b_1 \\ b_2 \end{pmatrix} t + \begin{pmatrix} d_1 \\ d_2 \end{pmatrix}\right) + \begin{pmatrix} -1 \\ 1 \end{pmatrix} t$$

$$= \begin{pmatrix} 1 & 2 \\ 4 & -1 \end{pmatrix}\begin{pmatrix} b_1 \\ b_2 \end{pmatrix} t + \begin{pmatrix} 1 & 2 \\ 4 & -1 \end{pmatrix}\begin{pmatrix} d_1 \\ d_2 \end{pmatrix} + \begin{pmatrix} -1 \\ 1 \end{pmatrix} t$$

$$= \begin{pmatrix} b_1 + 2b_2 \\ 4b_1 - b_2 \end{pmatrix} t + \begin{pmatrix} -1 \\ 1 \end{pmatrix} t + \begin{pmatrix} d_1 + 2d_2 \\ 4d_1 - d_2 \end{pmatrix}$$

$$= \begin{pmatrix} b_1 + 2b_2 - 1 \\ 4b_1 - b_2 + 1 \end{pmatrix} t + \begin{pmatrix} d_1 + 2d_2 \\ 4d_1 - d_2 \end{pmatrix}$$

We can solve first for B:

$$\begin{pmatrix} b_1 + 2b_2 - 1 \\ 4b_1 - b_2 + 1 \end{pmatrix} = \begin{pmatrix} 0 \\ 0 \end{pmatrix}$$

i.e.,

$$\begin{cases} b_1 + 2b_2 = 1 \\ 4b_1 - b_2 = -1 \end{cases}$$

Thus,

$$\begin{pmatrix} b_1 \\ b_2 \end{pmatrix} = \frac{1}{9} \begin{pmatrix} -1 \\ 5 \end{pmatrix}$$

With these, we have

$$\begin{pmatrix} d_1 + 2d_2 \\ 4d_1 - d_2 \end{pmatrix} = \begin{pmatrix} b_1 \\ b_2 \end{pmatrix} = \frac{1}{9} \begin{pmatrix} -1 \\ 5 \end{pmatrix}$$

Thus,

$$\begin{pmatrix} d_1 \\ d_2 \end{pmatrix} = \frac{1}{9} \begin{pmatrix} 1 \\ -1 \end{pmatrix}$$

The PS is

$$X_P = \frac{1}{9} \begin{pmatrix} -1 \\ 5 \end{pmatrix} t + \frac{1}{9} \begin{pmatrix} 1 \\ -1 \end{pmatrix}$$

The GS is

$$X = X_C + X_P$$
$$= C_1 \begin{pmatrix} 1 \\ 1 \end{pmatrix} \exp(3t) + C_2 \begin{pmatrix} 1 \\ -2 \end{pmatrix} \exp(-3t) + \frac{1}{9} \begin{pmatrix} -1 \\ 5 \end{pmatrix} t + \frac{1}{9} \begin{pmatrix} 1 \\ -1 \end{pmatrix}$$

Problem 4.5.6

This DE.Syst may be written as

$$\frac{d}{dt} \begin{pmatrix} x \\ y \\ z \end{pmatrix} = \begin{pmatrix} 4 & -2 & 0 \\ -4 & 4 & -2 \\ 0 & -4 & 4 \end{pmatrix} \begin{pmatrix} x \\ y \\ z \end{pmatrix}$$

Let's find the e-values and e-vectors of the coefficient matrix.

$$|A - \lambda I| = 0$$

$$\begin{vmatrix} 4 - \lambda & -2 & 0 \\ -4 & 4 - \lambda & -2 \\ 0 & -4 & 4 - \lambda \end{vmatrix} = 0$$

$$(4 - \lambda)((4 - \lambda)^2 - 8) + 2(-4(4 - \lambda)) = 0$$
$$\lambda(\lambda - 4)(\lambda - 8) = 0$$
$$\lambda_1 = 0, \lambda_2 = 4, \lambda_3 = 8$$

The corresponding e-vectors can be found from the equation $Av_i = \lambda_i v_i$, where v_i is an e-vector corresponding to λ_i. Thus,

$$v_1 = \begin{pmatrix} 1 \\ 2 \\ 2 \end{pmatrix}, \quad v_2 = \begin{pmatrix} 1 \\ 0 \\ -2 \end{pmatrix}, \quad v_3 = \begin{pmatrix} 1 \\ -2 \\ 2 \end{pmatrix}$$

The GS to the DE.Syst is given by

$$X = c_1 v_1 \exp(\lambda_1 t) + c_2 v_2 \exp(\lambda_2 t) + c_3 v_3 \exp(\lambda_3 t)$$

where c_1, c_2, c_3 are arbitrary constants and $X = \begin{pmatrix} x \\ y \\ z \end{pmatrix}$. Thus,

$$\begin{pmatrix} x \\ y \\ z \end{pmatrix} = c_1 \begin{pmatrix} 1 \\ 2 \\ 2 \end{pmatrix} \exp(0t) + c_2 \begin{pmatrix} 1 \\ 0 \\ -2 \end{pmatrix} \exp(4t) + c_3 \begin{pmatrix} 1 \\ -2 \\ 2 \end{pmatrix} \exp(8t)$$

$$\begin{pmatrix} x \\ y \\ z \end{pmatrix} = \begin{pmatrix} c_1 + c_2 \exp(4t) + c_3 \exp(8t) \\ 2c_1 - 2c_3 \exp(8t) \\ 2c_1 - 2c_2 \exp(4t) + 2c_3 \exp(8t) \end{pmatrix}$$

4.6 Examples of Systems

Problem 4.6.1

Let

$$y_1 = x$$
$$y_2 = x'$$
$$y_3 = x''$$

The DE becomes

$$\begin{cases} y_1' = y_2 \\ y_2' = y_3 \\ y_3' - 5y_3 + 9y_1 = t\sin 2t \end{cases}$$

One may convert it into matrix format.

Chapter 5 Laplace Transforms

5.2 Properties of Laplace Transforms

Problem 5.2.1

From the figure we write the square wave function as

$$g(t) = \begin{cases} 1, & 0 \le t - 2N < 1 \\ -1, & 1 \le t - 2N < 2 \end{cases} \quad N = 0,1,2,\ldots$$

Alternatively, we can write the square wave function (as given by the original problem) as

$$g(t) = 2u(t) - u(t) - 2u(t-1) + 2u(t-2) - 2u(t-3) + \cdots$$

$$= 2\sum_{n=0}^{\infty} (-1)^n u(t-n) - u(t)$$

$$= 2$$

Applying LT, we get

$$\mathcal{L}\{g(t)\} = 2\sum_{n=0}^{\infty} (-1)^n \mathcal{L}\{u(t-n)\} - \mathcal{L}\{u(t)\}$$

$$= 2\sum_{n=0}^{\infty} (-1)^n \frac{\exp(-ns)}{s} - \frac{1}{s}$$

$$= \frac{2}{s}\sum_{n=0}^{\infty} (-1)^n \exp(-ns) - \frac{1}{s}$$

$$= \frac{2}{s}\left(\frac{1}{1+\exp(-s)}\right) - \frac{1}{s}$$

$$= \frac{1}{s}\left(\frac{1-\exp(-s)}{1+\exp(-s)}\right)$$

$$= \frac{1}{s}\tanh\left(\frac{s}{2}\right)$$

Problem 5.2.2

For this calculation, we need to use three formulas:

$$\mathcal{L}\{f'(t)\} = s\,F(s) - f(0)$$

$$\mathcal{L}\{t^2 f(t)\} = (-1)^2\,F''(s)$$

$$\mathcal{L}\{\exp(\alpha t)\,f(t)\} = F(s-\alpha)$$

$$\mathcal{L}\left\{\frac{d}{dt}(t^2 \exp(\alpha t)\sin(\omega t))\right\} = s\,\mathcal{L}\{t^2 \exp(\alpha t)\sin(\omega t)\} - 0$$

$$= s(\mathcal{L}\{\exp(\alpha t)\sin(\omega t)\})''$$

$$= s \left(\frac{\omega}{(s-\alpha)^2 + \omega^2} \right)''$$

$$= -2s\omega \left(\frac{s-\alpha}{((s-\alpha)^2 + \omega^2)^2} \right)'$$

$$= 2s\omega \frac{\omega^2 - 3(s-\alpha)^2}{((s-\alpha)^2 + \omega^2)^3}$$

Problem 5.2.3

$$\mathcal{L} \left\{ \exp(t) \int_0^{\sqrt{t}} \exp(-\tau^2) \, d\tau \right\}$$

$$= \mathcal{L} \left\{ \frac{\sqrt{\pi}}{2} \frac{2 \exp(t)}{\sqrt{\pi}} \int_0^{\sqrt{t}} \exp(-\tau^2) \, d\tau \right\}$$

$$= \mathcal{L} \left\{ \frac{\exp(t)}{\sqrt{\pi}} \int_0^{\sqrt{t}} \frac{1}{\tau} \exp(-\tau^2) \, 2\tau d\tau \right\}$$

$$= \mathcal{L} \left\{ \int_0^t \frac{1}{\sqrt{\pi x}} \exp(t-x) \, dx \right\}$$

$$= \frac{\sqrt{\pi}}{2} \mathcal{L} \left\{ \exp(t) \frac{1}{\sqrt{\pi x}} \right\}$$

$$= \mathcal{L}\{\exp(t)\} \, \mathcal{L} \left\{ \frac{1}{\sqrt{\pi x}} \right\}$$

$$= \frac{1}{(s-1)\sqrt{s}}$$

$$\mathcal{L}\{f(t)\} = \frac{2A}{s^2 T} \tanh \frac{sT}{4}$$

Problem 5.2.4

Apply LT to the function, and use the property of LT to split it into two parts.

$$\mathcal{L}\{f(t)\} = \mathcal{L}\{t^2 \sin(\omega_1 t)\} + \mathcal{L}\{\exp(\alpha t) \cos(\omega_2 t)\}$$

For the first term $\mathcal{L}\{t^2 \sin(\omega_1 t)\}$, we use the property of t-multiplication

$$\mathcal{L}\{t^n f(t)\} = (-1)^n F^{(n)}(s)$$

In this case, $n = 2$, $f(t) = \sin(\omega_1 t)$. Then

$$F(s) = \frac{\omega_1}{s^2 + \omega_1^2}$$

$$\mathcal{L}\{t^2 \sin(\omega_1 t)\} = (-1)^2 F^{(n)}(s) = \frac{6\omega_1 s^2 - 2\omega_1^3}{(s^2 + \omega_1^2)^3}$$

For the second term $\mathcal{L}\{\exp(\alpha t)\cos(\omega_2 t)\}$, we use the property of translator

$$\mathcal{L}\{\exp(\alpha t)\cos(\omega_2 t)\} = F(s - \alpha)$$

In this case, $f(t) = \cos(\omega_2 t)$. Then

$$F(s) = \frac{s}{s^2 + \omega_2^2}$$

$$\mathcal{L}\{\exp(\alpha t)\cos(\omega_2 t)\} = \frac{s - \alpha}{(s - \alpha)^2 + \omega_2^2}$$

Finally, after applying LT to the function $f(t)$, we have

$$\mathcal{L}\{f(t)\} = \frac{6\omega_1 s^2 - 2\omega_1^3}{(s^2 + \omega_1^2)^3} + \frac{s - \alpha}{(s - \alpha)^2 + \omega_2^2}$$

Problem 5.2.5

1)

$$f_1(t) = \sum_{n=0}^{\infty} u(t - n)$$

By taking the LT, we get

$$\mathcal{L}\{f_1(t)\} = \mathcal{L}\left\{\sum_{n=0}^{\infty} u(t - n)\right\}$$

$$= \sum_{n=0}^{\infty} \mathcal{L}\{u(t - n)\}$$

$$= \sum_{n=0}^{\infty} \frac{\exp(-ns)}{s}$$

$$= \frac{1}{s} \sum_{n=0}^{\infty} \exp(-ns)$$

$$= \frac{1}{s(1 - \exp(-s))}$$

So, the LT is

$$\mathcal{L}\{f_1(t)\} = \frac{1}{s(1 - \exp(-s))}$$

2)

$$f_2(t) = t - \lfloor t \rfloor$$

In this case, the floor function can be expressed as

$$\lfloor t \rfloor = \sum_{n=1}^{\infty} u(t - n)$$

So, if we perform the same process of (1), then

$$\mathcal{L}\{\lfloor t\rfloor\} = \mathcal{L}\left\{\sum_{n=1}^{\infty} u(t-n)\right\}$$

$$= \sum_{n=1}^{\infty} \mathcal{L}\{u(t-n)\}$$

$$= \sum_{n=1}^{\infty} \frac{\exp(-ns)}{s}$$

$$= \frac{1}{s}\sum_{n=1}^{\infty} \exp(-ns)$$

$$= \frac{\exp(-s)}{s(1-\exp(-s))}$$

So, we obtain the following solution

$$\mathcal{L}\{f_2(t)\} = \mathcal{L}\{t\} - \mathcal{L}\{\lfloor t\rfloor\} = \frac{1}{s^2} - \frac{\exp(-s)}{s(1-\exp(-s))}$$

Problem 5.2.6

(1) $F(s) = \mathcal{L}\{t^5\} = \int_0^{\infty} t^5 \exp(-st)\,dt$

Let $u = st$. Then, $t = u/s$ and $dt = du/s$.

$$F(s) = \int_0^{\infty} \left(\frac{u}{s}\right)^5 \exp(-u)\frac{du}{s}$$

$$= \frac{1}{s^6}\int_0^{\infty} u^5 \exp(-u)\,du$$

$$= \frac{1}{s^6}\int_0^{\infty} u^{6-1} \exp(-u)\,du$$

$$= \frac{5!}{s^6}$$

Note

Gamma function is defined as

$$\Gamma(x) = \int_0^{\infty} \exp(-t)\,t^{x-1}dt$$

and for an integer n, we have

$$\Gamma(n+1) = n\Gamma(n) = n!$$

(2) $F(s) = \mathcal{L}\{\exp(at)\} = \int_0^{\infty} \exp(at)\exp(-st)\,dt = \int_0^{\infty} \exp((a-s)t)\,dt$

When $s > a$,

$$\left.\frac{\exp((a-s)t)}{a-s}\right|_0^{\infty} = \frac{1}{s-a}$$

(3) $F(s) = \mathcal{L}\{\sin(\omega t)\}$

$$= \int_0^\infty \sin(\omega t) \exp(-st)\, dt = -\frac{1}{s} \int_0^\infty \sin(\omega t)\, d(\exp(-st))$$

Using integration by parts, we get

$$F(s) = -\frac{\exp(-st)}{s} \sin(\omega t) \Big|_0^\infty + \frac{1}{s} \int_0^\infty \exp(-st)\, d(\sin(\omega t))$$

$$= 0 - 0 + \frac{\omega}{s} \int_0^\infty \cos(\omega t) \exp(-st)\, dt$$

$$= -\frac{\omega}{s^2} \int_0^\infty \cos(\omega t)\, d(\exp(-st))$$

$$= -\frac{\omega}{s^2} \cos(\omega t) \exp(-st) \Big|_0^\infty + \frac{\omega}{s^2} \int_0^\infty \exp(-st)\, d(\cos(\omega t))$$

$$= \frac{\omega}{s^2} - \frac{\omega^2}{s^2} \int_0^\infty \exp(-st) \sin(\omega t)\, dt$$

$$= \frac{\omega}{s^2} - \frac{\omega^2}{s^2} F(s)$$

Thus,

$$\left(1 + \frac{\omega^2}{s^2}\right) F(s) = \frac{\omega}{s^2}$$

Therefore,

$$F(s) = \frac{\omega}{s^2 + \omega^2}$$

(4)

$$F(s) = \mathcal{L}\{\exp(at) \sin(\omega t)\}$$

$$= \int_0^\infty \exp(at) \sin(\omega t) \exp(-st)\, dt$$

$$= \int_0^\infty \exp(-(s - a)t) \sin(\omega t)\, dt$$

Substituting $(s - a)$ with s_1, we can change the integral to the same form as the previous part. Thus,

$$F(s) = \frac{\omega}{s_1^2 + \omega^2} = \frac{\omega}{(s - a)^2 + \omega^2}$$

Problem 5.2.7

This is a periodic function (of period T) whose LT can be expressed as

$$\mathcal{L}\{f(t)\} = \frac{1}{1 - \exp(-sT)} \int_0^T f(t) \exp(-st)\, dt$$

where the integral portion can be carried out as

$$\int_0^T f(t) \exp(-st) \, dt$$

$$= \int_0^{\frac{T}{2}} 2A \left(\frac{t}{T}\right) \exp(-st) \, dt + \int_{\frac{T}{2}}^T 2A \left(1 - \frac{t}{T}\right) \exp(-st) \, dt$$

$$= \left(-\frac{A}{s} \exp\left(-\frac{T}{2}s\right) - \frac{2A}{s^2T} \exp\left(-\frac{T}{2}s\right) + \frac{2A}{s^2T}\right)$$
$$+ \left(\frac{2A}{s^2T} \exp(-Ts) + \frac{A}{s} \exp\left(-\frac{T}{2}s\right) - \frac{2A}{s^2T} \exp\left(-\frac{T}{2}s\right)\right)$$

$$= \frac{2A}{s^2T} \left(-2 \exp\left(-\frac{T}{2}s\right) + 1 + \exp(-Ts)\right)$$

$$= \frac{2A}{s^2T} \left(1 - \exp\left(-\frac{T}{2}s\right)\right)^2$$

Therefore, the LT is

$$\mathcal{L}\{f(t)\} = \frac{1}{1 - \exp(-Ts)} \frac{2A}{s^2T} \left(1 - \exp\left(-\frac{T}{2}s\right)\right)^2$$

$$= \frac{2A}{s^2T} \frac{1 - \exp\left(-\frac{T}{2}s\right)}{1 + \exp\left(-\frac{T}{2}s\right)}$$

$$= \frac{2A}{s^2T} \left(\frac{\exp\left(\frac{T}{4}s\right) - \exp\left(-\frac{T}{4}s\right)}{\exp\left(\frac{T}{4}s\right) + \exp\left(-\frac{T}{4}s\right)}\right)$$

$$= \frac{2A}{s^2T} \frac{\sinh\left(\frac{T}{4}s\right)}{\cosh\left(\frac{T}{4}s\right)} = \frac{2A}{s^2T} \tanh\left(\frac{T}{4}s\right)$$

5.3 Inverse Laplace Transforms

Problem 5.3.1

Applying LT on the RHS of the given problem, we get

$$\mathcal{L}\left\{\frac{1}{2k^3}(\sin kt - kt \cos kt)\right\}$$

$$= \frac{1}{2k^3}(\mathcal{L}\{\sin kt\} - k\mathcal{L}\{t \cos kt\})$$

$$= \frac{1}{2k^3}\left(\frac{k}{s^2 + k^2} - \frac{k(s^2 - k^2)}{(s^2 + k^2)^2}\right)$$

$$= \frac{1}{2k^3}\left(\frac{2k^3}{(s^2 + k^2)^2}\right)$$

$$= \frac{1}{(s^2 + k^2)^2}$$

$$= \mathcal{L}\{\text{LHS}\}$$

Problem 5.3.2

Using the Bessel function

$$J_0(z) = \sum_{n=0}^{\infty} \frac{1}{(n!)^2}\left(-\frac{z^2}{4}\right)^n$$

we have

$$\mathcal{L}\{J_0(2\sqrt{t})\} = \mathcal{L}\left\{\sum_{n=0}^{\infty} \frac{1}{(n!)^2}\left(-\frac{(2\sqrt{t})^2}{4}\right)^n\right\}$$

$$= \mathcal{L}\left\{\sum_{n=0}^{\infty} (-1)^n \frac{1}{(n!)^2} t^n\right\}$$

$$= \sum_{n=0}^{\infty} \frac{(-1)^n}{(n!)^2} \mathcal{L}\{t^n\}$$

$$= \sum_{n=0}^{\infty} \frac{(-1)^n}{(n!)^2}\left(\frac{n!}{s^{n+1}}\right)$$

$$= \frac{1}{s}\sum_{n=0}^{\infty} \frac{1}{n!}\left(-\frac{1}{s}\right)^n$$

$$= \frac{1}{s}\exp\left(-\frac{1}{s}\right)$$

Thus,

$$\mathcal{L}\{J_0(2\sqrt{t})\} = \frac{1}{s}\exp\left(-\frac{1}{s}\right)$$

Problem 5.3.3

We know

$$\frac{2s}{(s^2-1)^2} = -\frac{d}{ds}\left(\frac{1}{s^2-1}\right)$$

Thus,

$$\mathcal{L}^{-1}\left[\frac{2s}{(s^2-1)^2}\right] = -\mathcal{L}^{-1}\left\{\frac{d}{ds}\left(\frac{1}{s^2-1}\right)\right\}$$

$$= -\left(-t\,\mathcal{L}^{-1}\left\{\frac{1}{s^2-1}\right\}\right)$$

$$= t\,\mathcal{L}^{-1}\left\{\frac{1}{2}\left(\frac{1}{s-1}-\frac{1}{s+1}\right)\right\}$$

$$= \frac{1}{2}t(\exp(t)-\exp(-t))$$

where we have used

$$\mathcal{L}^{-1}\left\{\frac{d}{ds}F(s)\right\} = -t\,f(t)$$

Problem 5.3.4

$$\mathcal{L}^{-1}\left\{\frac{\exp(-as)}{s(1-\exp(-as))}\right\}$$

$$= \mathcal{L}^{-1}\left\{-\frac{1}{s}+\frac{1}{s}\left(\frac{1}{1-\exp(-as)}\right)\right\}$$

$$= \mathcal{L}^{-1}\left\{-\frac{1}{s}\right\}+\mathcal{L}^{-1}\left\{\frac{1}{s}\left(\frac{1}{1-\exp(-as)}\right)\right\}$$

$$= -1 + \mathcal{L}^{-1}\left\{\sum_{n=0}^{\infty}\frac{1}{s}(\exp(-as))^n\right\}$$

$$= -1 + \mathcal{L}^{-1}\left\{\sum_{n=0}^{\infty}\frac{1}{s}\exp(-ans)\right\}$$

$$= -1 + \sum_{n=0}^{\infty}u(t-an)$$

Problem 5.3.5

Performing LT on both sides of the DE, we get

$$\mathcal{L}\{x(t)\} = \mathcal{L}\{\sin(t)\} + 2\mathcal{L}\left\{\int_0^t \cos(t-\tau)\,x(\tau)d\tau\right\}$$

$$X(s) = \frac{1}{s^2 + 1} + \frac{2s}{s^2 + 1} X(s)$$

Thus, solving the above yields

$$X(s) = \frac{1}{(s-1)^2}$$

Performing inverse LT, we get

$$x(t) = \mathcal{L}^{-1}\{X(s)\} = t\exp(t)$$

Problem 5.3.6

Applying LT on both sides of the DE and using $\mathcal{L}\{x\} = X(s)$ and the definition of convolution, we get

$$X(s) = \mathcal{L}\{2\exp(3t)\} - \mathcal{L}\left\{\int_0^t \exp(2(t-\tau))x(\tau)d\tau\right\}$$

$$= \frac{2}{s-3} - \mathcal{L}\{\exp(2t)\}\mathcal{L}\{x(t)\}$$

$$= \frac{2}{s-3} - \frac{X(s)}{s-2}$$

Thus,

$$X(s) = \frac{1}{s-1} + \frac{1}{s-3}$$

Inversing LT yields

$$x(t) = \exp(t) + \exp(3t)$$

5.4 The Convolution of Two Functions

Problem 5.4.1

$$\mathcal{L}^{-1}\left\{\frac{1}{(s-1)\sqrt{s}}\right\} = \mathcal{L}^{-1}\left\{\frac{1}{(s-1)}\frac{1}{\sqrt{s}}\right\}$$

Let $F(s) = \frac{1}{s-1}$. Then, $f(t) = \mathcal{L}^{-1}\left\{\frac{1}{s-1}\right\} = \exp(t)$.

Let $G(s) = \frac{1}{\sqrt{s}}$. Then, $g(t) = \mathcal{L}^{-1}\left\{s^{-\frac{1}{2}}\right\} = \frac{t^{-\frac{1}{2}}}{\Gamma\left(\frac{1}{2}\right)} = \frac{1}{\sqrt{\pi t}}$

where we used $\Gamma\left(\frac{1}{2}\right) = \sqrt{\pi}$.

By convolution theorem, we get

$$\mathcal{L}^{-1}\{F(s)G(s)\} = \int_0^t f(t-\tau)g(\tau)\,d\tau$$

$$= \int_0^t \exp(t-\tau)\frac{1}{\sqrt{\pi t}}\,d\tau$$

$$= \frac{\exp(t)}{\sqrt{\pi}}\int_0^t \frac{\exp(-\tau)}{\sqrt{\tau}}\,d\tau$$

$$= \frac{\exp(t)}{\sqrt{\pi}}\int_0^{\sqrt{t}} \frac{\exp(-u^2)}{u}\,2u\,du$$

$$= \exp(t)\left(\frac{2}{\sqrt{\pi}}\int_0^{\sqrt{t}} \exp(-u^2)\,du\right)$$

$$= \exp(t)\,\mathrm{erf}(\sqrt{t})$$

In the above calculation, we introduced $u = \sqrt{\tau}$ and, thus, $\tau = u^2$ and $d\tau = 2u\,du$.

Problem 5.4.2

$$\mathcal{L}\{\cos(wx)\} = \frac{s}{s^2 + w^2}$$

$$\mathcal{L}\{\cos(3x)\} = \frac{s}{s^2 + 9}$$

$$\mathcal{L}\{xf(x)\} = -\frac{dF(s)}{ds}$$

$$\mathcal{L}\{x\cos 3x\} = -\frac{d}{ds}\left(\frac{s}{s^2 + 9}\right) = \frac{s^2 - 9}{(s^2 + 9)^2}$$

$$\mathcal{L}\left\{\int_0^x \exp(\tau)\,y(x-\tau)d\tau\right\} = \mathcal{L}\{\exp(x) \otimes y(x)\}$$

$$= \mathcal{L}\{\exp(x)\}\mathcal{L}\{y(x)\}$$

$$= \frac{1}{s-1} Y(s)$$

Thus,

$$Y(s) = \frac{s^2 - 9}{(s^2 + 9)^2} - \frac{1}{s-1} Y(s)$$

Solving the above equation, one gets

$$Y(s) = \left(1 - \frac{1}{s}\right) \frac{s^2 - 9}{(s^2 + 9)^2}$$

Let $G(s) = \frac{s^2 - 9}{(s^2 + 9)^2}$. Thus,

$$Y(s) = \left(1 - \frac{1}{s}\right) G(s)$$

$$= G(s) - \frac{1}{s} G(s)$$

and

$$G(s) = \frac{s^2 - 9}{(s^2 + 9)^2}$$

$$= \frac{s^2 + 9 - 18}{(s^2 + 9)^2}$$

$$= \frac{1}{s^2 + 9} - \frac{18}{(s^2 + 9)^2}$$

whose inverse $g(x)$ is

$$\mathcal{L}^{-1}\{G(s)\} = \frac{1}{3} \mathcal{L}^{-1}\left\{\frac{3}{s^2 + 3^2}\right\} - \mathcal{L}^{-1}\left\{\frac{18}{(s^2 + 9)^2}\right\}$$

$$= \frac{1}{3} \sin(3x) - \frac{18}{54}(\sin(3x) - 3x\cos(3x))$$

$$= x \cos(3x)$$

where we used

$$\mathcal{L}^{-1}\left\{\frac{1}{(s^2 + k^2)^2}\right\} = \frac{1}{2k^3}(\sin kt - kt\cos kt)$$

Therefore,

$$y(x) = g(t) - \int_0^t g(\tau)d\tau$$

$$= x \cos(3x) - \int_0^x (\tau \cos(3\tau))d\tau$$

$$= x \cos(3x) - \frac{1}{9}(3\tau \sin(3\tau) + \cos(3\tau))|_0^x$$

$$= x \cos(3x) - \frac{1}{9}(3x \sin(3x) + \cos(3x) - 1)$$

5.5 Applications

Problem 5.5.1

Applying LT to the DE with the given IC's, we get

$$s^2 X(s) + 4X(s) = \mathcal{L}\{f(t)\}$$

where $F(s)$ is the LT of $f(t)$. We got the AE for $X(s)$

$$X(s) = \mathcal{L}\{f(t)\}\frac{1}{s^2 + 4}$$

Then, using the convolution theorem, we get

$$x(t) = \mathcal{L}^{-1}\left\{\mathcal{L}\{f(t)\}\left(\frac{1}{s^2 + 4}\right)\right\}$$

$$= \mathcal{L}^{-1}\{\mathcal{L}\{f(t)\}\}\otimes\mathcal{L}^{-1}\left\{\frac{1}{s^2 + 4}\right\}$$

$$= f(t)\otimes\left(\frac{1}{2}\sin 2t\right)$$

$$= \frac{1}{2}\int_0^t f(t - \tau)\sin 2\tau\, d\tau$$

where we used

$$\mathcal{L}^{-1}\left\{\frac{1}{s^2 + 4}\right\} = \frac{1}{2}\sin 2t$$

Problem 5.5.2

Solve the IDE

$$x'(t) + 2x(t) - 4\int_0^t \exp(t - \tau)\,x(\tau)d\tau = \sin t, \quad x(0) = 0.$$

Taking the LT on both sides of the DE, we find,

$$\big(sX(s) - x(0)\big) + 2X(s) - 4\mathcal{L}\{\exp(t)\otimes x(t)\} = \frac{1}{s^2 + 1}$$

$$sX(s) + 2X(s) - 4\left(\frac{1}{s - 1}\right)X(s) = \frac{1}{s^2 + 1}$$

Solving for $X(s)$ and performing partial fraction, we obtain

$$X(s) = \frac{s - 1}{(s^2 + 1)(s^2 + s - 6)} = \frac{1}{25}\left(\frac{4 - 3s}{s^2 + 1} + \frac{1}{s - 2} + \frac{2}{s + 3}\right)$$

Applying inverse LT, we get

$$x(t) = \mathcal{L}^{-1}\{X(s)\}$$

$$= \frac{1}{25}\mathcal{L}^{-1}\left\{\frac{4 - 3s}{s^2 + 1} + \frac{1}{s - 2} + \frac{2}{s + 3}\right\}$$

$$= \frac{1}{25}\mathcal{L}^{-1}\left\{\frac{4}{s^2 + 1} - 3\frac{s}{s^2 + 1} + \frac{1}{s - 2} + \frac{2}{s + 3}\right\}$$

$$= \frac{1}{25}(4\sin t - 3\cos t + \exp(2t) + 2\exp(-3t))$$

Problem 5.5.3

Applying LT on both sides of the two DEs, we have
$$\begin{cases} \mathcal{L}\{x''\} = -4\mathcal{L}\{x\} + \mathcal{L}\{\sin t\} \\ \mathcal{L}\{y''\} = 4\mathcal{L}\{x\} - 8\mathcal{L}\{y\} \end{cases}$$

Let $\mathcal{L}\{x\} = X(s)$. We know that
$$\mathcal{L}\{x''\} = s^2 X(s) - x(0)s - x'(0)$$

Plugging into the first DE, we get
$$s^2 X(s) - x(0)s - x'(0) = -4X(s) + \frac{1}{s^2 + 1}$$

This gives
$$X(s) = \frac{x(0)s^3 + x'(0)s^2 + x(0)s + x'(0) + 1}{(s^2 + 4)(s^2 + 1)}$$

Thus, the GS is $x(t) = \mathcal{L}^{-1}\{X(s)\}$.

Similarly, for y, we have
$$\mathcal{L}\{y''\} = s^2 Y(s) - y(0)s - y'(0)$$

The second DE can be written as
$$s^2 Y(s) - y(0)s - y'(0) = 4X(s) - 8Y(s)$$

This gives
$$Y(s) = \frac{4X(s) + y(0)s + y'(0)}{s^2 + 8}$$

Plugging $X(s)$ into the above equation, we get $Y(s)$ in terms of s and the GS is $y(t) = \mathcal{L}^{-1}\{Y(s)\}$.

Problem 5.5.4

Applying LT on both sides of the DE, we have
$$\mathcal{L}\{x'''\} + \mathcal{L}\{x''\} - 6\mathcal{L}\{x'\} = 0$$

Let $\mathcal{L}\{x\} = X(s)$. Then,
$$\mathcal{L}\{x'\} = sX(s) - x(0) = sX(s)$$
$$\mathcal{L}\{x''\} = s^2 X(s) - x(0)s - x'(0) = s^2 X(s) - 1$$
$$\mathcal{L}\{x'''\} = s^3 X(s) - x(0)s^2 - x'(0)s - x''(0)$$
$$= s^3 X(s) - s - 1$$

Inserting them to the transformed DE, we get
$$s^3 X(s) - s - 1 + s^2 X(s) - 1 - 6sX(s) = 0$$
$$X(s) = \frac{s + 2}{s(s + 3)(s - 2)}$$

$$= -\frac{1}{3} \cdot \frac{1}{s} - \frac{1}{15} \cdot \frac{1}{s+3} + \frac{2}{5} \cdot \frac{1}{s-2}$$

$$x(t) = \mathcal{L}^{-1}\{X(s)\}$$

$$= -\frac{1}{3} - \frac{1}{15}\exp(-3t) + \frac{2}{5}\exp(2t)$$

Problem 5.5.5

Applying LT on the DE, using $\mathcal{L}\{\delta(t-a)\} = \exp(-as)$, and applying the ICs $x(0) = 1$ and $x'(0) = 0$, we get

$$\mathcal{L}\{x''\} + 4L\{x\} = -5\mathcal{L}\{\delta(t-3)\}$$

$$s^2 X(s) - x(0)s - x'(0) + 4X(s) = -5\exp(-3s)$$

$$s^2 X(s) - s + 4X(s) = -5\exp(-3s)$$

Thus,

$$X(s) = \frac{-5\exp(-3s)}{s^2 + 4} + \frac{s}{s^2 + 4}$$

$$= -\frac{5}{2}\left(\frac{2}{s^2 + 2^2}\right)\exp(-3s) + \frac{s}{s^2 + 2^2}$$

Therefore, the PS is

$$x(t) = -\frac{5}{2}\mathcal{L}^{-1}\left\{\left(\frac{2}{s^2 + 2^2}\right)\exp(-3s)\right\} + \mathcal{L}^{-1}\left\{\frac{s}{s^2 + 2^2}\right\}$$

$$= -\frac{5}{2}\sin 2t \otimes \delta(t-3) + \cos 2t$$

$$= -\frac{5}{2}\int_0^t \sin 2(t-\tau)\,\delta(\tau-3)d\tau + \cos 2t$$

$$= -\frac{5}{2}\sin 2(t-3) + \cos 2t$$

Problem 5.5.6

Applying LT on both sides of the DE, we have

$$\mathcal{L}\{x'' + 6x' + 8x\} = \mathcal{L}\{-\delta(t-2)\}$$

$$s^2 X(s) - s + 6sX(s) - 6 + 8X(s) = -\exp(-2s)$$

$$X(s) = \frac{s + 6 - \exp(-2s)}{s^2 + 6s + 8}$$

$$= \frac{s + 6 - \exp(-2s)}{(s+2)(s+4)}$$

$$= \frac{2}{s+2} - \frac{1}{s+4} - \frac{1}{2}\exp(-2s)\left(\frac{1}{s+2} - \frac{1}{s+4}\right)$$

Thus,

$$x(t) = \mathcal{L}^{-1}\left\{\frac{2}{s+2} - \frac{1}{s+4} - \frac{1}{2}\exp(-2s)\left(\frac{1}{s+2} - \frac{1}{s+4}\right)\right\}$$

$$= 2\exp(-2t) - \exp(-4t) - \frac{1}{2}\delta(t-2) \otimes \exp(-2t) + \frac{1}{2}\delta(t-2)$$
$$\otimes \exp(-4t)$$

Problem 5.5.7

$$x'' + 2x' + x = \delta(t) - \delta(t-2)$$
$$x(0) = 0$$
$$x'(0) = 0$$

Applying LT on both sides and using $\mathcal{L}\{\delta(t)\} = 1$ and $\mathcal{L}\{\delta(t-a)\} = \exp(-as)$, we get

$$\mathcal{L}\{x'' + 2x' + x\} = \mathcal{L}\{\delta(t) - \delta(t-2)\}$$
$$s^2 X(s) + 2sX(s) + X(s) = 1 - \exp(-2s)$$

Thus,

$$X(s) = \frac{1 - \exp(-2s)}{s^2 + 2s + 1}$$
$$= \frac{1 - \exp(-2s)}{(s+1)^2}$$
$$= \frac{1}{(s+1)^2} - \frac{\exp(-2s)}{(s+1)^2}$$

Using $\mathcal{L}\{f(t-\tau)u(t-\tau)\} = \exp(-\tau s)\, F(s)$, we get

$$x(t) = t\exp(-t) - (t-2)\exp\big(-(t-2)\big)\, u(t-2)$$

Problem 5.5.8

$$\mathcal{L}\{\delta_a(t)\} = \mathcal{L}\{\delta(t-a)\} = \exp(-as)$$
$$\mathcal{L}\{x''\} + \mathcal{L}\{\omega^2 x\} = \sum_{n=0}^{\infty} \mathcal{L}\{\delta(t-2nt_0)\}$$
$$s^2 X(s) - x(0)s - x'(0) + \omega^2 X(s) = \sum_{n=0}^{\infty} \exp(-2nt_0 s)$$
$$(s^2 + \omega^2)X(s) = \sum_{n=0}^{\infty} \exp(-2nt_0 s)$$

Thus,

$$X(s) = \sum_{n=0}^{\infty} \frac{\exp(-2nt_0 s)}{s^2 + \omega^2}$$

Using $\mathcal{L}^{-1}\{\exp(-as)\, F(s)\} = u(t-a)f(t-a)$, we get

$$x(t) = \mathcal{L}^{-1}\left\{\sum_{n=0}^{\infty} \frac{\exp(-2nt_0 s)}{s^2 + \omega^2}\right\} = \sum_{n=0}^{\infty} \mathcal{L}^{-1}\left\{\frac{\exp(-2nt_0 s)}{s^2 + \omega^2}\right\}$$

$$= \sum_{n=0}^{\infty} \mathcal{L}^{-1} \left\{ \frac{1}{\omega} \frac{\omega \exp(-2nt_0 s)}{s^2 + \omega^2} \right\}$$

$$= \frac{1}{\omega} \sum_{n=0}^{\infty} u(t - 2nt_0) \sin \omega(t - 2nt_0)$$

Problem 5.5.9

Case 1: $a > 0, b > 0$

$$\mathcal{L}\{u(t - a)\} = \frac{1}{s} \exp(-as)$$

$$\mathcal{L}\{\delta(t - b)\} = \int_0^{\infty} \delta(t - b) \exp(-st) \, dt$$

$$= \exp(-bs)$$

Applying LT to both sides of the DE, we have

$$s^2 X(s) - x(0)s - x'(0) + 2(sX(s) - x(0)) + X(s)$$

$$= \frac{1}{s} \exp(-as) + \exp(-bs)$$

Applying the ICs, we get

$$s^2 X(s) + 2sX(s) + X(s) = \frac{1}{s} \exp(-as) + \exp(-bs)$$

$$X(s) = \frac{1}{(s + 1)^2} \left(\frac{1}{s} \exp(-as) + \exp(-bs) \right)$$

Thus,

$$x(t) = \mathcal{L}^{-1} \left\{ \frac{1}{s(s + 1)^2} \exp(-as) \right\} + \mathcal{L}^{-1} \left\{ \frac{1}{(s + 1)^2} \exp(-bs) \right\}$$

$$= \mathcal{L}^{-1} \left\{ \exp(-as) \left(\frac{1}{s} - \frac{1}{s + 1} - \frac{1}{(s + 1)^2} \right) \right\} + \mathcal{L}^{-1} \left\{ \frac{1}{(s + 1)^2} \exp(-bs) \right\}$$

$$= u(t - a)\{u(t - a) - \exp(-(t - b)) - (t - a) \exp(-(t - b))\}$$

$$\qquad + u(t - b)(t - b) \exp(-(t - b))$$

$$= u(t - a)(1 - \exp(-(t - b)) - (t - a) \exp(-(t - b)))$$

$$\qquad + u(t - b)(t - b) \exp(-(t - b))$$

Case 2: $a \leq 0, b > 0$

$$\mathcal{L}\{u(t - a)\} = \frac{1}{s}$$

$$\mathcal{L}\{\delta(t - b)\} = \exp(-bs)$$

Applying LT to both sides of the DE, we have

$$s^2 X(s) + 2sX(s) + X(s) = \frac{1}{s} + \exp(-bs)$$

$$X(s) = \frac{1}{s(s + 1)^2} + \frac{\exp(-bs)}{(s + 1)^2}$$

$$x(t) = \mathcal{L}^{-1}\left\{\frac{1}{s} - \frac{1}{s + 1} - \frac{1}{(s + 1)^2}\right\} + \mathcal{L}^{-1}\left\{\frac{\exp(-bs)}{(s + 1)^2}\right\}$$

$$= u(t) - \exp(-t) - t\exp(-t) + u(t - b)(t - b)\exp\big(-(t - b)\big)$$

$$= 1 - \exp(-t) - t\exp(-t) + u(t - b)(t - b)\exp\big(-(t - b)\big)$$

Problem 5.5.10

Applying LT to both sides of both DEs, we have

$$\begin{cases} \mathcal{L}\{x'\} = \mathcal{L}\{x\} + 2\mathcal{L}\{y\} \\ \mathcal{L}\{y'\} = 2\mathcal{L}\{x\} - 2\mathcal{L}\{y\} \end{cases}$$

Given IC, this gives

$$\begin{cases} (s - 1)X(s) = 2Y(s) + 1 \\ 2X(s) = (s + 2)Y(s) \end{cases}$$

This gives

$$\begin{cases} X(s) = \frac{1}{5}\left(\frac{1}{s + 3}\right) + \frac{4}{5}\left(\frac{1}{s - 2}\right) \\ Y(s) = -\frac{2}{5}\left(\frac{1}{s + 3}\right) + \frac{2}{5}\left(\frac{1}{s - 2}\right) \end{cases}$$

Finally,

$$\begin{cases} x(t) = \frac{1}{5}(4\exp(2t) + \exp(-3t)) \\ y(t) = \frac{1}{5}(2(\exp(2t) - \exp(-3t))) \end{cases}$$

Problem 5.5.11

$$\begin{cases} \mathcal{L}\{x_1''\} = -2\mathcal{L}\{x_1\} + \mathcal{L}\{x_2\} + \exp(-\tau s) \\ \mathcal{L}\{x_2''\} = \mathcal{L}\{x_1\} - \mathcal{L}\{x_2\} + \exp(-2\tau s) \end{cases}$$

$$\begin{cases} s^2 X_1(s) = -2X_1(s) + X_2(s) + \exp(-\tau s) \\ s^2 X_2(s) = X_1(s) - X_2(s) + \exp(-2\tau s) \end{cases}$$

$$\begin{cases} X_2(s) = (s^2 + 2)X_1(s) - \exp(-\tau s) \\ (s^2 + 1)X_2(s) = X_1(s) + \exp(-2\tau s) \end{cases}$$

Inserting the first equation into the second, we get

$$(s^2 + 1)\big((s^2 + 2)X_1(s) - \exp(-\tau s)\big) = X_1(s) + \exp(-2\tau s)$$
$$(s^4 + 3s^2 + 2)X_1(s) = \exp(-\tau s)\,(\exp(-\tau s) + s^2 + 1)$$

Thus,

$$x_1(t) = \mathcal{L}^{-1}\left\{\frac{\exp(-\tau s)\,(\exp(-\tau s) + s^2 + 1)}{(s^2 + 1)(s^2 + 2)}\right\}$$

$$= u(t - \tau)\mathcal{L}^{-1}\left\{\frac{\exp(-\tau s)}{(s^2 + 1)(s^2 + 2)} + \frac{1}{s^2 + 2}\right\}$$

$$= u(t - \tau)\mathcal{L}^{-1}\left\{\exp(-\tau s)\left(\frac{1}{s^2 + 1} - \frac{1}{s^2 + 2}\right) + \frac{1}{s^2 + 2}\right\}$$

$$= u(t - \tau)\frac{1}{\sqrt{5}}\left(u(t - \tau)\left(\sin t - \frac{1}{\sqrt{2}}\sin\sqrt{2}t\right) + \frac{1}{\sqrt{2}}\sin\sqrt{2}t\right)$$

We also have

$$(s^2 + 1)X_2(s) = X_1(s) + \exp(-2\tau s)$$
$$X_2(s) = \frac{1}{s^2 + 1}(X_1 + \exp(-2\tau s))$$

Thus,

$$x_2(t) = \mathcal{L}^{-1}\left\{\frac{1}{s^2 + 1}(X_1 + \exp(-2\tau s))\right\}$$

$$= \mathcal{L}^{-1}\left\{\frac{1}{s^2 + 1}\right\} \otimes \mathcal{L}^{-1}\{X_1 + \exp(-2\tau s)\}$$

$$= \sin t \otimes \big(x_1(t) + u(t - 2\tau)\big)$$

Problem 5.5.12

Applying LT to both sides of the DE, we have

$$sX(s) + 4X(s) + 6X(s)\frac{1}{s - 1} = \mathcal{L}\{\sin \omega t\}$$

$$X(s)\left(s + 4 + \frac{6}{s - 1}\right) = \mathcal{L}\{\sin \omega t\}$$

$$X(s) = \mathcal{L}\{\sin \omega t\}\frac{s - 1}{(s + 1)(s + 2)}$$

Using partial fractions, we get

$$\frac{s - 1}{(s + 1)(s + 2)} = \frac{3}{s + 2} - \frac{2}{s + 1}$$

Then, $X(s)$ can be expressed as

$$X(s) = \mathcal{L}\{\sin \omega t\}\left(\frac{3}{s + 2} - \frac{2}{s + 1}\right)$$

Finally, applying inverse LT and using the convolution theorem

$$\mathcal{L}^{-1}\{X(s)\} = \sin \omega t \otimes \mathcal{L}^{-1}\left\{\frac{3}{s + 2}\right\} - \sin \omega t \otimes \mathcal{L}^{-1}\left\{\frac{2}{s + 1}\right\}$$

$$= 3\sin \omega t \otimes \exp(-2t) - 2\sin \omega t \otimes \exp(-t)$$

Problem 5.5.13

Applying LT to the InHomo system, we have

$$\begin{cases} x'(t) = 2x(t) + y(t) + t \\ y'(t) = x(t) + 2y(t) + \exp(t) \end{cases}$$

$$\begin{cases} sX(s) - x(0) = 2X(s) + Y(s) + \dfrac{1}{s^2} \\ sY(s) - y(0) = X(s) + 2Y(s) + \dfrac{1}{s-1} \end{cases}$$

$$X(s) = \frac{x(0)(s-2)}{(s-1)(s-3)} + \frac{s-2}{s^2(s-1)(s-3)} + \frac{y(0)}{(s-1)(s-3)}$$

$$+ \frac{1}{(s-1)^2(s-3)}$$

$$= \frac{x(0)}{2}\left(\frac{1}{s-1} + \frac{1}{s-3}\right) + \left(-\frac{5}{9s} - \frac{2}{3s^2} + \frac{1}{2(s-1)} + \frac{1}{18(s-3)}\right)$$

$$+ \frac{y(0)}{2}\left(\frac{-1}{s-1} + \frac{1}{s-3}\right)$$

$$+ \left(-\frac{1}{4(s-1)} - \frac{1}{2(s-1)^2} + \frac{1}{4(s-3)}\right)$$

Thus,

$$x(t) = \frac{x(0)}{2}(\exp t + \exp 3t) + \left(-\frac{5}{9} - \frac{2}{3}t + \frac{1}{2}\exp t + \frac{1}{18}\exp 3t\right)$$

$$+ \frac{y(0)}{2}(-\exp t + \exp 3t)$$

$$+ \left(-\frac{1}{4}\exp t - \frac{1}{2}t\exp t + \frac{1}{4}\exp 3t\right)$$

For $Y(s)$, we have

$$Y(s) = \frac{x(0)}{(s-1)(s-3)} + \frac{1}{s^2(s-1)(s-3)} + \frac{y(0)(s-2)}{(s-1)(s-3)}$$

$$+ \frac{(s-2)}{(s-1)^2(s-3)}$$

$$= \frac{x(0)}{2}\left(\frac{-1}{s-1} + \frac{1}{s-3}\right) + \left(\frac{4}{9s} + \frac{1}{3s^2} - \frac{1}{2(s-1)} + \frac{1}{18(s-3)}\right)$$

$$+ \left(\frac{y(0)}{2}\left(\frac{1}{s-1} + \frac{1}{s-3}\right)\right)$$

$$+ \left(-\frac{1}{4(s-1)} + \frac{1}{2(s-1)^2} + \frac{1}{4(s-3)}\right)$$

Thus,

$$y(t) = \frac{x(0)}{2}(-\exp t + \exp 3t) + \left(\frac{4}{9} + \frac{1}{3}t - \frac{1}{2}\exp t + \frac{1}{18}\exp 3t\right)$$
$$+ \frac{y(0)}{2}(\exp t + \exp 3t)$$
$$+ \left(-\frac{1}{4}\exp t + \frac{1}{2}t \exp t + \frac{1}{4}\exp 3t\right)$$

Problem 5.5.14

$$\mathcal{L}\{x''\} - 6\mathcal{L}\{x'\} + 8\mathcal{L}\{x\} = 2\mathcal{L}\{1\}$$

Applying the ICs, we have

$$s^2 X(s) - 6sX(s) + 8X(s) = \frac{2}{s}$$

$$X(s) = \frac{2}{s(s-2)(s-4)}$$
$$= \frac{1}{4} \cdot \frac{1}{s} - \frac{1}{2} \cdot \frac{1}{s-2} + \frac{1}{4} \cdot \frac{1}{s-4}$$

Therefore,

$$x(t) = \frac{1}{4} - \frac{1}{2}\exp(2t) + \frac{1}{4}\exp(4t)$$

Problem 5.5.15

$$\mathcal{L}\{tx''\} + 2\mathcal{L}\{(t-1)x'\} - 2\mathcal{L}\{x\} = 0$$
$$-\left(s^2 X(s)\right)' + 2\left\{-\left(sX(s)\right)' - sX(s)\right\} - 2X(s) = 0$$
$$\frac{X'(s)}{X(s)} = -\frac{4(s+1)}{s(s+2)}$$

Using the SOV method, we solve the above DE with which s is the IV and X is the DV. Thus,

$$X(s) = \frac{C}{(s^2 + 2s)^2}$$

Thus,

$$x(t) = C\mathcal{L}^{-1}\left\{\frac{1}{(s^2 + 2s)^2}\right\}$$
$$= C\mathcal{L}^{-1}\left\{\frac{1}{s^2}\frac{1}{(s+2)^2}\right\}$$
$$= C\mathcal{L}^{-1}\left\{\frac{1}{s^2}\right\} \otimes \mathcal{L}^{-1}\left\{\frac{1}{(s+2)^2}\right\}$$
$$= Ct \otimes (t\exp(-2t))$$

where C is a constant.

Problem 5.5.16

Performing LT on the DE leads to

$$\mathcal{L}\left\{t\frac{d^2x}{dt^2}\right\} + \mathcal{L}\left\{\frac{dx}{dt}\right\} + \mathcal{L}\{tx\} = 0$$

$$-\frac{d}{ds}(s^2X - as) + (sX - a) - \frac{dX}{ds} = 0$$

$$(s^2 + 1)X' + sX = 0$$

which is a separable DE for X whose solution is

$$\int \frac{dX}{X} = -\frac{sds}{s^2 + 1}$$

$$\ln X = -\frac{1}{2}\ln(s^2 + 1)$$

$$X(s) = \frac{1}{\sqrt{s^2 + 1}}$$

Thus,

$$x(t) = \mathcal{L}^{-1}\left\{\frac{1}{\sqrt{s^2 + 1}}\right\}$$

Problem 5.5.17

$$\mathcal{L}\{x''\} - 2\mathcal{L}\{x'\} + \mathcal{L}\{x\} = \mathcal{L}\{f(t)\}$$

$$s^2X(s) - 2sX(s) + X(s) = F(s)$$

$$X(s) = \frac{F(s)}{(s^2 - 2s + 1)}$$

$$= \frac{F(s)}{(s - 1)^2}$$

Thus,

$$x(t) = \mathcal{L}^{-1}\left\{\frac{F(s)}{(s - 1)^2}\right\}$$

$$= \mathcal{L}^{-1}\left\{\frac{1}{(s - 1)^2}\right\} \otimes \mathcal{L}^{-1}\{F(s)\}$$

$$= (t\exp(2t)) \otimes f(t)$$

Problem 5.5.18

$$(s^2X(s) - 2) + 4sX(s) + 13X(s) = \frac{1}{(s + 1)^2}$$

$$X(s) = \frac{2 + \frac{1}{(s + 1)^2}}{s^2 + 4s + 13} = \frac{2s^2 + 4s + 13}{(s + 1)^2(s^2 + 4s + 13)}$$

$$= \frac{1}{50}\left(-\frac{1}{s+1} + \frac{5}{(s+1)^2} + \frac{s+2}{(s+2)^2 + 9} + 32\frac{3}{(s+2)^2 + 9}\right)$$

Thus,

$$x(t) = \frac{1}{50}\left((-1 + 5t)\exp(-t) + \exp(-2t)(\cos 3t + 32\sin 3t)\right)$$

Problem 5.5.19

Method 1: LT method.

Applying LT to the DE, we get

$$\mathcal{L}\{x''\} - \mathcal{L}\{x'\} - 12\mathcal{L}\{x\} = \mathcal{L}\{\sin 4t + \exp(3t)\}$$

$$(s^2 X(s) - sx(0) - x'(0)) - (sX(s) - x(0)) - 12X(s) = \frac{4}{s^2 + 16} + \frac{1}{s-3}$$

Using the IC, we have

$$(s^2 X(s) - 2s - 1) - (sX(s) - 2) - 12X(s) = \frac{4}{s^2 + 16} + \frac{1}{s-3}$$

$$X(s) = \frac{2s^4 - 7s^3 + 36s^2 - 108s + 52}{(s-3)(s+3)(s-4)(s^2 + 16)}$$

Now, if we apply the partial fraction for the RHS, we get

$$\frac{2s^4 - 7s^3 + 36s^2 - 108s + 52}{(s-3)(s+3)(s-4)(s^2 + 16)} = \frac{A}{s-3} + \frac{B}{s+3} + \frac{C}{s-4} + \frac{Ds+E}{s^2 + 16}$$

$$\begin{cases} A + B + C + D = 2 \\ -A - 7B - 4D + E = -7 \\ 4A + 28B + 7C - 9D - 4E = 36 \\ -16A - 112B + 36D - 9E = -108 \\ -192A + 192B - 144C + 36E = 52 \end{cases} \Rightarrow \begin{cases} A = -\dfrac{1}{6} \\ B = \dfrac{1051}{1050} \\ C = \dfrac{65}{56} \\ D = \dfrac{1}{200} \\ E = -\dfrac{28}{200} \end{cases}$$

So, we get

$$X(s) = \frac{2s^4 - 7s^3 + 36s^2 - 108s + 52}{(s-3)(s+3)(s-4)(s^2 + 16)}$$

$$= -\frac{1}{6}\left(\frac{1}{s-3}\right) + \frac{1051}{1050}\left(\frac{1}{s+3}\right) + \frac{65}{56}\left(\frac{1}{s-4}\right) + \frac{1}{200}\left(\frac{s}{s^2 + 16}\right)$$

$$- \frac{7}{200}\left(\frac{4}{s^2 + 16}\right)$$

Applying inverse LT, we obtain

$$x(t) = \frac{1051}{1050}\exp(-3t) + \frac{65}{56}\exp(4t) - \frac{7}{200}\sin 4t + \frac{1}{200}\cos 4t - \frac{1}{6}\exp(3t)$$

Method 2: The C-Eq method.

For the complementary solution, we solve the C-Eq
$$r^2 - r - 12 = 0, \quad r = -3, 4$$

Thus,
$$x_C(t) = C_1 \exp(-3t) + C_2 \exp(4t)$$

For the PS, we use MUC with TS
$$x_P(t) = A \sin 4t + B \cos 4t + C \exp(3t)$$
$$x_P'(t) = 4A \cos 4t - 4B \sin 4t + 3C \exp(3t)$$
$$x_P''(t) = -16A \sin 4t - 16B \cos 4t + 9C \exp(3t)$$

$$x_P''(t) - x_P'(t) - 12x_P(t)$$
$$= (-28A + 4B) \sin 4t + (-4A - 28B) \cos 4t - 6C \exp(3t)$$
$$= \sin 4t + \exp(3t)$$

Matching the coefficients, we get
$$\begin{cases} -28A + 4B = 1 \\ -4A - 28B = 0 \\ -6C = 1 \end{cases}$$

Thus,
$$\begin{cases} A = -\dfrac{7}{200} \\ B = \dfrac{1}{200} \\ C = -\dfrac{1}{6} \end{cases}$$

So, the PS is
$$x_P(t) = -\frac{7}{200} \sin 4t + \frac{1}{200} \cos 4t - \frac{1}{6} \exp(3t)$$

The GS is
$$x(t) = x_C(t) + x_P(t)$$
$$= C_1 \exp(-3t) + C_2 \exp(4t) - \frac{7}{200} \sin 4t + \frac{1}{200} \cos 4t - \frac{1}{6} \exp(3t)$$

With the IC, we can determine the coefficients of the common solution.
$$x(0) = C_1 + C_2 + \frac{1}{200} - \frac{1}{6} = 2$$
$$x'(0) = -3C_1 + 4C_2 - \frac{28}{200} - \frac{3}{6} = 1$$

Thus,
$$\begin{cases} C_1 = \dfrac{1051}{1050} \\ C_2 = \dfrac{65}{56} \end{cases}$$

The GS is

$$x(t) = x_C(t) + x_P(t)$$
$$= \frac{1051}{1050} \exp(-3t) + \frac{65}{56} \exp(4t) - \frac{7}{200} \sin 4t + \frac{1}{200} \cos 4t - \frac{1}{6} \exp(3t)$$

Problem 5.5.20

(1) The C-Eq of the Homo portion is
$$r^2 + \omega^2 = 0$$
$$r = \pm i\omega$$
The two LI solutions are
$$x_1 = \cos \omega t, \quad x_2 = \sin \omega t$$
$$W = \begin{vmatrix} \cos \omega t & \sin \omega t \\ -\sin \omega t & \cos \omega t \end{vmatrix} = \cos^2 \omega t + \sin^2 \omega t = 1$$
The trial functions are
$$u_1 = -\int \frac{y_2(t)f(t)}{W(t)} dt = -\int \cos \omega t \, f(t) dt$$
$$u_2 = \int \frac{y_1(t)f(t)}{W(t)} dt = \int \sin \omega t \, f(t) dt$$
The GS to the original DE is
$$x = C_1 \cos \omega t + C_2 \sin \omega t + u_1 \cos \omega t + u_2 \sin \omega t$$
(2) Applying LT to both sides of the DE, we have
$$\mathcal{L}\{x''\} + \omega^2 \mathcal{L}\{x\} = \mathcal{L}\{f(t)\}$$
Let $X(s) = \mathcal{L}\{x\}$
$$\mathcal{L}\{x''\} = s^2 X(s) - x(0)s - x'(0) = s^2 X(s) - x_0 s - v_0$$
With substitution, we have
$$s^2 X(s) - x_0 s - v_0 + w^2 X(s) = \mathcal{L}\{f(t)\}$$
$$X(s) = \frac{\mathcal{L}\{f(t)\} + x_0 s + v_0}{s^2 + w^2}$$
$$= \frac{1}{w} \mathcal{L}\{f(t)\} \frac{w}{s^2 + w^2} + x_0 \frac{s}{s^2 + w^2} + \frac{v_0}{w} \frac{w}{s^2 + w^2}$$
$$x(t) = \mathcal{L}^{-1}\{X(s)\}$$
$$= \frac{1}{w} \int_0^t f(t-\tau) \sin(wt) \, dt + x_0 \cos(wt) + \frac{v_0}{w} \sin(wt)$$

Problem 5.5.21

(1) We use the VOP method. The corresponding Homo DE is $x'' + \omega_1^2 x = 0$, a 2nd.O c-coeff DE whose C-Eq is
$$r^2 + \omega_1^2 = 0 \Rightarrow r_{1,2} = \pm \omega_1 i$$
$$x_C(t) = C_1 \cos(\omega_1 t) + C_2 \sin(\omega_1 t)$$
$$\begin{cases} x_1(t) = \cos(\omega_1 t) \\ x_2(t) = \sin(\omega_1 t) \end{cases}$$

$$W(t) = \begin{vmatrix} \cos(\omega_1 t) & \sin(\omega_1 t) \\ -\omega_1 \sin(\omega_1 t) & \omega_1 \cos(\omega_1 t) \end{vmatrix} = \omega_1$$

Let $x_P(t) = u_1(t)x_1(t) + u_2(t)x_2(t)$. Then,

$$u_1(t) = -\int^t \frac{y_2(\tau)f(\tau)}{W(\tau)} d\tau = -\int^t \frac{\sin(\omega_1 \tau)\sin(\omega_2 \tau)}{\omega_1} d\tau$$

$$= \frac{1}{2\omega_1} \int \Big(\cos((\omega_1 + \omega_2)t) - \cos((\omega_1 - \omega_2)t) \Big) dt$$

$$= \frac{1}{2\omega_1} \left(\frac{\sin((\omega_1 + \omega_2)t)}{\omega_1 + \omega_2} - \frac{\sin((\omega_1 - \omega_2)t)}{\omega_1 - \omega_2} \right)$$

$$u_2(t) = -\int \frac{y_1(t)f(t)}{W(t)} dt = -\int \frac{\cos(\omega_1 t)\sin(\omega_2 t)}{\omega_1} dt$$

$$= \frac{1}{2\omega_1} \int \Big(\sin((\omega_1 + \omega_2)t) - \sin((\omega_1 - \omega_2)t) \Big) dt$$

$$= \frac{1}{2\omega_1} \left(-\frac{\cos((\omega_1 + \omega_2)t)}{\omega_1 + \omega_2} - \frac{\cos((\omega_1 - \omega_2)t)}{\omega_1 - \omega_2} \right)$$

Thus, a PS can be written as

$$x_P(x) = u_1(t)x_1(t) + u_2(t)x_2(t)$$

$$= \frac{\cos(\omega_1 t)}{2\omega_1} \left(\frac{\sin((\omega_1 + \omega_2)t)}{\omega_1 + \omega_2} - \frac{\sin((\omega_1 - \omega_2)t)}{\omega_1 - \omega_2} \right)$$

$$+ \frac{\sin(\omega_1 t)}{2\omega_1} \left(-\frac{\cos((\omega_1 + \omega_2)t)}{\omega_1 + \omega_2} - \frac{\cos((\omega_1 - \omega_2)t)}{\omega_1 - \omega_2} \right)$$

$$= \frac{\cos(\omega_1 t)\sin((\omega_1 + \omega_2)t) - \sin(\omega_1 t)\cos((\omega_1 + \omega_2)t)}{2\omega_1(\omega_1 + \omega_2)}$$

$$+ \frac{-\cos(\omega_1 t)\sin((\omega_1 - \omega_2)t) + \sin(\omega_1 t)\cos((\omega_1 - \omega_2)t)}{2\omega_1(\omega_1 - \omega_2)}$$

$$= \frac{\sin((\omega_1 + \omega_2)t - \omega_1 t)}{2\omega_1(\omega_1 + \omega_2)} + \frac{\sin(\omega_1 t - (\omega_1 - \omega_2)t)}{2\omega_1(\omega_1 - \omega_2)}$$

$$= \frac{\sin(\omega_2 t)}{\omega_1^2 - \omega_2^2}$$

Thus, the GS to the DE is

$$x(t) = x_C(t) + x_P(t)$$

$$= C_1 \cos(\omega_1 t) + C_2 \sin(\omega_1 t) - \frac{\sin(\omega_2 t)}{\omega_1^2 - \omega_2^2}$$

With the given IC $x(0) = x'(0) = 0$

$$C_1 = 0$$

$$C_2 = \frac{\omega_2}{\omega_1(\omega_1^2 - \omega_2^2)}$$

Thus,

$$x(t) = \frac{\omega_2 \sin(\omega_1 t)}{\omega_1(\omega_1^2 - \omega_2^2)} - \frac{\sin(\omega_2 t)}{(\omega_1^2 - \omega_2^2)}$$

$$= \frac{\omega_2}{\omega_1^2 - \omega_2^2}\left(\frac{\sin(\omega_1 t)}{\omega_1} - \frac{\sin(\omega_2 t)}{\omega_2}\right)$$

(2) We use the LT method. Applying LT on the DE, we have

$$\mathcal{L}\{x''\} + \omega_1^2 \mathcal{L}\{x\} = \mathcal{L}\{\sin(\omega_2 t)\}$$

Let

$$X(s) = \mathcal{L}\{x\} \Rightarrow \mathcal{L}\{x''\} = s^2 X(s) - x(0)s - x'(0) = s^2 X(s)$$

Plugging $\mathcal{L}\{x''\}$ and $\mathcal{L}\{x\}$ into the DE, we have

$$s^2 X(s) + \omega_1^2 X(s) = \frac{\omega_2}{s^2 + \omega_2^2}$$

$$X(s) = \frac{\omega_2}{(s^2 + \omega_2^2)(s^2 + \omega_1^2)}$$

$$= \frac{\omega_2}{\omega_2^2 - \omega_1^2}\left(\frac{1}{\omega_1}\frac{\omega_1}{s^2 + \omega_1^2} - \frac{1}{\omega_2}\frac{\omega_2}{s^2 + \omega_2^2}\right)$$

Thus,

$$x(t) = \mathcal{L}^{-1}\{X(s)\} = \frac{\omega_2}{\omega_2^2 - \omega_1^2}\left(\frac{\sin(\omega_1 t)}{\omega_1} - \frac{\sin(\omega_2 t)}{\omega_2}\right)$$

Problem 5.5.22

(1) Applying LT to both sides of the DE, we have

$$\mathcal{L}\{x'\} - \mathcal{L}\{x\} = \mathcal{L}\{1\} - \mathcal{L}\{(t-1)u(t-1)\}$$

$$sX(s) - X(s) = \frac{1}{s} - \frac{1}{s^2}\exp(-s)$$

During the above calculation, we used

$$\mathcal{L}\{1\} = \frac{1}{s}, \quad \mathcal{L}\{(t-1)u(t-1)\} = \frac{1}{s^2}\exp(-s)$$

which is derived by using $\mathcal{L}\{f(t-a)u(t-1)\} = F(s)\exp(-as)$
Solving the above equation, we get

$$X(s) = \frac{1}{s-1}\left(\frac{1}{s} - \frac{1}{s^2}\exp(-s)\right)$$

$$= \frac{1}{s-1} - \frac{1}{s} - \frac{\exp(-s)}{s^2(s-1)}$$

$$= \frac{1}{s-1} - \frac{1}{s} - \exp(-s)\left(\frac{1}{s-1} - \frac{1}{s} - \frac{1}{s^2}\right)$$

Thus, the PS is

$$x(t) = \mathcal{L}^{-1}\{X(s)\}$$

$$= \exp(t) - 1 - u(t-1) \otimes (\exp(t) - 1 - t)$$

(2) We divide the equation into two regions

If $t \le 1$, the DE is
$$\begin{cases} x' - x = 1 \\ x(0) = 0 \end{cases}$$
whose solution can be easily found as $x(t) = \exp(t) - 1$.

If $t > 1$, the DE is
$$\begin{cases} x' - x = 2 - t \\ x(0) = 0 \end{cases}$$
which is a 1st.O linear DE, and
$$P(t) = -1, \qquad Q(t) = 2 - t$$
$$\rho(t) = \exp\left(\int (-1) dt \right) = \exp(-t)$$

Solving this DE, we have
$$x(t) = \frac{1}{\rho(t)} \left(\int \rho Q dt + C \right)$$
$$= \exp(t) \left(\int \exp(-t) (2 - t) dt + C \right)$$
$$= \exp(t) (-2 \exp(-t) + t \exp(-t) + \exp(-t) + C)$$
$$= -1 + t + C \exp(t)$$

Applying the IC $x(0) = 0$, we get $C = 1$ and the PS
$$x(t) = \exp(t) - 1 + t$$

Combining the above two cases, we get the PS
$$x(t) = \begin{cases} \exp(t) - 1, & t \le 1 \\ \exp(t) - 1 + t, & t > 1 \end{cases}$$

Written as a step function, the PS is
$$x(t) = \exp(t) - 1 + t\, u(t - 1)$$

Problem 5.5.23

For the original DE, if $0 < t < 2\pi$, $u(t - 2\pi) = 0$

or
$$x''(t) + 4x(t) = \cos 2t$$

Thus,
$$\begin{cases} x''(t) + 4x(t) = \cos 2t \\ x(0) = 0 \\ x'(0) = 0 \end{cases}$$

Using LT method, we have
$$(s^2 + 4)X(s) = \mathcal{L}\{\cos 2t\}$$

So
$$x(t) = \frac{1}{2} \sin 2t \otimes \cos 2t$$

$$= \frac{1}{2} \int_0^t \sin 2\tau \cos 2(t - \tau) \, d\tau$$

$$= \frac{1}{4} t \sin 2t$$

For the original DE, if $t > 2\pi$, $u(t - 2\pi) = 1$
or

$$x''(t) + 4x(t) = 0$$

with starting from $t = 2\pi$.

When $t = 2\pi$, from the condition of the phase I, $0 < t < 2\pi$, we know

$$x(2\pi) = 0, x'(2\pi) = 2\pi$$

After resetting the origin to $t = 2\pi$, the IVP becomes

$$\begin{cases} x''(t) + 4x(t) = 0 \\ x(0) = 0 \\ x'(0) = 2\pi \end{cases}$$

whose PS for $t > 2\pi$ can be found using LT as follows.

$$s^2 X - 2\pi + 4X = 0$$

Thus, $X = \frac{2\pi}{s^2 + 4}$ and the PS is

$$x(t) = \mathcal{L}^{-1} \left\{ \frac{2\pi}{s^2 + 4} \right\}$$

$$= \pi \sin 2t$$

The PS may also be expressed in a convolution as

$$x(t) = \frac{1}{2} \sin 2t \otimes \left((1 - u(t - 2\pi)) \cos 2t \right)$$

Problem 5.5.24

Applying LT to both sides of DE, we have

$$s^2 X(s) - x(0)s - x'(0) + X(s) = \mathcal{L}\{(-1)^{\llbracket t \rrbracket}\}$$

Plugging the IC, we get

$$(s^2 + 1)X(s) = \mathcal{L}\{(-1)^{\llbracket t \rrbracket}\}$$

Or

$$X(s) = \frac{1}{s^2 + 1} \mathcal{L}\{(-1)^{\llbracket t \rrbracket}\}$$

Applying inverse LT, we have

$$x(t) = \mathcal{L}^{-1} \left\{ \frac{1}{s^2 + 1} \mathcal{L}\{(-1)^{\llbracket t \rrbracket}\} \right\} = \mathcal{L}^{-1} \left\{ \frac{1}{s^2 + 1} \right\} \otimes (-1)^{\llbracket t \rrbracket}$$

$$= \sin t \otimes (-1)^{\llbracket t \rrbracket}$$

The convolution can be computed directly using its definition. One may also compute the LT of the RHS.

$$\mathcal{L}\{(-1)^{[\![t]\!]}\} = \int_0^{\infty} (-1)^{[\![t]\!]} \exp(-st)\, dt$$

$$= \int_0^1 \exp(-st)\, dt - \int_1^2 \exp(-st)\, dt + \int_2^3 \exp(-st)\, dt \cdots$$

$$= \frac{1}{s} + \frac{2}{s}\left(\sum_{n=0}^{\infty} (-1)^n \exp(-ns) - 1 \right)$$

$$= -\frac{1}{s} + \frac{2}{s} \sum_{n=0}^{\infty} (-1)^n \exp(-ns)$$

Thus,

$$X(s) = \frac{1}{s^2+1}\left(-\frac{1}{s} + \frac{2}{s} \sum_{n=0}^{\infty} (-1)^n \exp(-ns) \right)$$

$$= -\frac{1}{s} + \frac{s}{s^2+1} + \frac{2}{s^2+1}\frac{1}{s} \sum_{n=0}^{\infty} (-1)^n \exp(-ns)$$

$$x(t) = -1 + \cos t + 2\sin t \otimes \sum_{n=0}^{\infty} (-1)^n u(t-n)$$

$$\sin t \otimes \sum_{n=0}^{\infty} (-1)^n u(t-n)$$

$$= \int_0^t \sin \tau \sum_{n=0}^{\infty} (-1)^n u(t-\tau-n)\, d\tau$$

$$= \sum_{n=0}^{\infty} (-1)^n \int_0^t \sin \tau\, u(t-\tau-n)\, d\tau$$

$$= \sum_{n=0}^{\infty} (-1)^n \int_0^{t-n} \sin \tau\, d\tau + \sum_{n=0}^{\infty} (-1)^n \int_{t-n}^t 0 \sin \tau\, d\tau$$

$$= \sum_{n=0}^{\infty} (-1)^n (1 - \cos(t-n))$$

Thus,

$$x(t) = -1 + \cos t + 2 \sum_{n=0}^{\infty} (-1)^n (1 - \cos(t-n))$$

Problem 5.5.25

$$x(0) = x'(0) = y(0) = y'(0) = 0$$

(1) When $\omega = \omega_0$

$$\mathcal{L}\{x''\} + \mathcal{L}\{\omega^2 x\} = \mathcal{L}\{b_0 \sin(\omega t)\}$$

$$s^2 X(s) + \omega^2 X(s) = \frac{b_0 \omega}{s^2 + \omega^2}$$

$$X(s) = \frac{b_0 \omega}{(s^2 + \omega^2)^2}$$

$$x(t) = \mathcal{L}^{-1}\left\{\frac{b_0 \omega}{(s^2 + \omega^2)^2}\right\} = \left(\frac{b_0}{\omega}\right) \mathcal{L}^{-1}\left\{\left(\frac{\omega}{s^2 + \omega^2}\right)^2\right\}$$

$$= \left(\frac{b_0}{\omega}\right)(\sin \omega t \otimes \sin \omega t)$$

$$= \left(\frac{b_0}{2\omega}\right)\left(\frac{1}{\omega}\sin \omega t - t \cos \omega t\right)$$

$$y(t) = \mathcal{L}^{-1}\left\{-\frac{b_0}{2}\left(\frac{1}{s^2 + \omega^2}\right)'\right\}$$

$$= \left(\frac{b_0}{2\omega}\right) t \sin \omega t$$

(2) When $\omega \neq \omega_0$

$$s^2 X(s) + \omega^2 X(s) = \frac{b_0 \omega_0}{s^2 + \omega^2}$$

$$X(s) = \frac{b_0 \omega_0}{(s^2 + \omega_0^2)(s^2 + \omega^2)}$$

$$= \left(\frac{b_0 \omega_0}{\omega^2 - \omega_0^2}\right)\left(\frac{1}{s^2 + \omega_0^2} - \frac{1}{s^2 + \omega^2}\right)$$

$$= \left(\frac{b_0}{\omega^2 - \omega_0^2}\right)\left(\frac{\omega_0}{\omega_0^2 + s^2} - \left(\frac{\omega_0}{\omega}\right)\left(\frac{\omega}{\omega^2 + s^2}\right)\right)$$

Thus,

$$x(t) = \left(\frac{b_0}{\omega^2 - \omega_0^2}\right) \mathcal{L}^{-1}\left\{\frac{\omega_0}{\omega_0^2 + s^2} - \left(\frac{\omega_0}{\omega}\right)\left(\frac{\omega}{\omega^2 + s^2}\right)\right\}$$

$$= \left(\frac{b_0}{\omega^2 - \omega_0^2}\right)\left(\sin(\omega_0 t) - \left(\frac{\omega_0}{\omega}\right)\sin(\omega t)\right)$$

Next,

$$Y(s) = \frac{b_0 s}{(s^2 + \omega_0^2)(s^2 + \omega^2)}$$

$$= \left(\frac{b_0}{\omega^2 - \omega_0^2}\right)\left(\frac{s}{s^2 + \omega_0^2} - \frac{s}{s^2 + \omega^2}\right)$$

Thus,

$$y(t) = \left(\frac{b_0}{\omega^2 - \omega_0^2}\right) \mathcal{L}^{-1}\left\{\frac{s}{s^2 + \omega_0^2} - \frac{s}{s^2 + \omega^2}\right\}$$

$$= \left(\frac{b_0}{\omega^2 - \omega_0^2} \right) (\cos(\omega_0 t) - \cos(\omega t))$$

Problem 5.5.26

Observe that the external force is periodic and during the first period, the external force looks like the following,

$$r(t) = \begin{cases} f & t \in [0, \frac{\tau}{2}) \\ -f & t \in [\frac{\tau}{2}, \tau) \end{cases}$$

Thus, the force as a function of time can be given as

$$r(t) = r(t + \tau)$$

We have assumed that the direction towards right are positive.

Let $x(0) = 0$ be the initial displacement and $x'(0) = 0$ be the initial velocity, where $x(t)$ represents the displacement. According to our sign convention, the force exerted by the spring on mass m is always in the negative direction. Thus, the EoM is

$$m \frac{d^2 x}{dt^2} = -kx + r(t)$$

We can write this equation as

$$\frac{d^2 x}{dt^2} + \omega^2 x = q(t)$$

where $q(t) = m^{-1} r(t)$ and $\omega = \sqrt{m^{-1} k}$.

Applying LT to both sides, we have

$$s^2 X(s) + \omega^2 X(s) = Q(s)$$

$$X(s) = \frac{Q(s)}{s^2 + \omega^2}$$

Applying inverse LT of this equation, we get

$$x(t) = \mathcal{L}^{-1} \left\{ \frac{Q(s)}{s^2 + \omega^2} \right\}$$

$$= \frac{1}{\omega} \mathcal{L}^{-1} \left\{ \frac{\omega}{s^2 + \omega^2} \right\} \otimes \mathcal{L}^{-1} \{Q(s)\}$$

$$= \frac{1}{\omega} \sin(\omega t) \otimes q(t)$$

$$= \frac{1}{m\omega} \sin(\omega t) \otimes r(t)$$

Problem 5.5.27

$$f(t) = \begin{cases} \cos 2t & t \in [0, 2\pi] \\ 0 & \text{o. w.} \end{cases}$$

$$f(t) = \left(1 - u(t - 2\pi)\right)\cos 2t$$
$$= \cos 2t - u(t - 2\pi)\cos 2(t - 2\pi)$$
$$\mathcal{L}\{f(t)\} = \mathcal{L}\{\cos 2t - u(t - 2\pi)\cos 2(t - 2\pi)\}$$
$$= \mathcal{L}\{\cos 2t\} - \mathcal{L}\{u(t - 2\pi)\cos 2(t - 2\pi)\}$$
$$= \frac{s(1 - \exp(-2\pi s))}{s^2 + 4}$$
$$\mathcal{L}\{x'' + 4x\} = \mathcal{L}\{f(t)\}$$
$$s^2 X(s) - sx(0) - x'(0) + 4X(s) = \frac{s(1 - \exp(-2\pi s))}{s^2 + 4}$$
$$X(s)(s^2 + 4) = \frac{s(1 - \exp(-2\pi s))}{s^2 + 4}$$
$$X(s) = \frac{s(1 - \exp(-2\pi s))}{(s^2 + 4)^2}$$
$$x(t) = \mathcal{L}^{-1}\{X(s)\} = \mathcal{L}^{-1}\left\{\frac{s(1 - \exp(-2\pi s))}{(s^2 + 4)^2}\right\}$$
$$\mathcal{L}^{-1}\left\{\frac{s}{(s^2 + 4)^2}\right\} = \frac{1}{4}t\sin 2t$$
$$x(t) = \frac{t}{4}\sin 2t - u(t - 2\pi)\frac{1}{4}(t - 2\pi)\sin 2(t - 2\pi)$$
$$= \begin{cases} \frac{1}{4}t\sin 2t & t \in [0, 2\pi] \\ \frac{1}{2}\pi\sin 2t & \text{o. w.} \end{cases}$$

Problem 5.5.28

By setting the natural spot at $x = 0$, we can write the EoM as
$$mx'' = F - (k_1 + k_2)x$$
Applying LTs to both sides,
$$m\left(s^2 X(s) - sx(0) - x'(0)\right) = \mathcal{L}\{F\} - (k_1 + k_2)X(s)$$
The force can be expressed as
$$F = f\left(2\sum_{n=0}^{\infty}(-1)^n u\left(t - \frac{\tau}{2}n\right) - u(t)\right)$$
$$\mathcal{L}\{F\} = \frac{f}{s}\tanh\left(\frac{\tau s}{4}\right)$$
Applying ICs $x(0) = x'(0) = 0$ and simplifying the DE, we get
$$ms^2 X(s) = \frac{f}{s}\tanh\left(\frac{\tau s}{4}\right) - (k_1 + k_2)X(s)$$
$$X(s) = \left(\frac{f}{s}\right)\frac{\tanh\left(\frac{\tau s}{4}\right)}{ms^2 + k_1 + k_2}$$

The solution is $x(t) = \mathcal{L}^{-1}\{X(s)\}$.

Problem 5.5.29

The InHomo system is

$$mx_1'' = k(x_2 - 2x_1) + f_A(t)$$
$$mx_2'' = k(x_1 - 2x_2) + f_B(t)$$

Defining $w = \sqrt{\dfrac{k}{m}}$, we get

$$x_1'' = w^2(x_2 - 2x_1) + \frac{f_A(t)}{m}$$

$$x_2'' = w^2(x_1 - 2x_2) + \frac{f_B(t)}{m}$$

Applying LT to the above yields

$$(s^2 + 2w^2)X_1 = w^2 X_2 + F_A(s)$$
$$(s^2 + 2w^2)X_2 = w^2 X_1 + F_B(s)$$

where

$$F_A(s) = \mathcal{L}\left\{\frac{f_A(t)}{m}\right\}$$

$$F_B(s) = \mathcal{L}\left\{\frac{f_B(t)}{m}\right\}$$

Solving the above DEs, we get

$$\begin{aligned}
X_1 &= \frac{F_A(s^2 + 2w^2) + F_B w^2}{s^4 + 4w^2 s^2 + 3w^2} \\
&= \frac{F_A s^2 + (2F_A + F_B)w^2}{(s^2 + w^2)(s^2 + 3w^2)} \\
&= \frac{F_A + F_B}{2}\frac{1}{s^2 + w^2} + \frac{F_A - F_B}{2}\frac{1}{s^2 + 3w^2}
\end{aligned}$$

$$\begin{aligned}
X_2 &= \frac{F_A w^2 + F_B(s^2 + 2w^2)}{s^4 + 4w^2 s^2 + 3w^2} \\
&= \frac{F_B s^2 + (2F_B + F_A)w^2}{(s^2 + w^2)(s^2 + 3w^2)} \\
&= \frac{F_A + F_B}{2}\frac{1}{s^2 + w^2} - \frac{F_A - F_B}{2}\frac{1}{s^2 + 3w^2}
\end{aligned}$$

Thus,

$$\begin{aligned}
x_1 &= \mathcal{L}^{-1}\left\{\frac{F_A + F_B}{2}\frac{1}{s^2 + w^2} + \frac{F_A - F_B}{2}\frac{1}{s^2 + 3w^2}\right\} \\
&= \frac{1}{2w}\mathcal{L}^{-1}\{F_A + F_B\} \otimes \sin wt + \frac{1}{2\sqrt{3}w}\mathcal{L}^{-1}\{F_A - F_B\} \otimes \sin\sqrt{3}\,wt \\
&= \frac{1}{2mw}\{f_A(t) + f_B(t)\} \otimes \sin wt + \frac{1}{2\sqrt{3}mw}\{f_A(t) - f_B(t)\} \otimes \sin\sqrt{3}\,wt
\end{aligned}$$

$$x_2 = \mathcal{L}^{-1} \left\{ \frac{F_A + F_B}{2} \frac{1}{s^2 + w^2} - \frac{F_A - F_B}{2} \frac{1}{s^2 + 3w^2} \right\}$$

$$= \frac{1}{2w} \mathcal{L}^{-1} \{F_A + F_B\} \otimes \sin wt - \frac{1}{2\sqrt{3}w} \mathcal{L}^{-1} \{F_A + F_B\} \otimes \sin \sqrt{3} wt$$

$$x_2 = \frac{1}{2mw} \{f_A(t) + f_B(t)\} \otimes \sin wt - \frac{1}{2\sqrt{3}mw} \{f_A(t) - f_B(t)\} \otimes \sin \sqrt{3} wt$$

The above convolutions are easy to compute.

Problem 5.5.30

Applying LT to both DEs, we get

$$ms^2 X_1(s) = -2k X_1(s) + k X_2(s) + f_0 \exp(-t_0 s)$$
$$ms^2 X_2(s) = k X_1(s) - k X_2(s) + f_0 \exp(-2t_0 s)$$

Writing these in matrix format to solve for $X_2(s)$,

$$\begin{pmatrix} s^2 + \dfrac{2k}{m} & -\dfrac{k}{m} \\ -\dfrac{k}{m} & s^2 + \dfrac{k}{m} \end{pmatrix} \begin{pmatrix} X_1(s) \\ X_2(s) \end{pmatrix} = \begin{pmatrix} \dfrac{f_0}{m} \exp(-t_0 s) \\ \dfrac{f_0}{m} \exp(-2t_0 s) \end{pmatrix}$$

$$\begin{pmatrix} s^2 + \dfrac{2k}{m} & -\dfrac{k}{m} \\ 0 & \left(s^2 + \dfrac{k}{m}\right)\left(s^2 + \dfrac{2k}{m}\right) - \dfrac{k^2}{m^2} \end{pmatrix} \begin{pmatrix} X_1(s) \\ X_2(s) \end{pmatrix}$$

$$= \begin{pmatrix} \dfrac{f_0}{m} \exp(-t_0 s) \\ \left(s^2 + \dfrac{2k}{m}\right)\dfrac{f_0}{m} \exp(-2t_0 s) + \dfrac{kf_0}{m^2} \exp(-t_0 s) \end{pmatrix}$$

$$\left(s^2 + \frac{3 - \sqrt{5}}{2} \frac{k}{m}\right)\left(s^2 + \frac{3 + \sqrt{5}}{2} \frac{k}{m}\right) X_2(s)$$

$$= \left(s^2 + \frac{2k}{m}\right)\frac{f_0}{m} \exp(-2t_0 s) + \frac{kf_0}{m^2} \exp(-t_0 s)$$

$$X_2(s) = \frac{\left((s^2 + \omega_1^2 + \omega_2^2)\dfrac{f_0}{m} \exp(-2t_0 s) + \dfrac{kf_0}{m^2} \exp(-t_0 s)\right)}{(s^2 + \omega_1^2)(s^2 + \omega_2^2)}$$

$$= \frac{\dfrac{f_0}{m} \exp(-2t_0 s)}{s^2 + \omega_2^2} + \frac{\omega_2^2 \dfrac{f_0}{m} \exp(-2t_0 s) + \dfrac{k}{m^2} f_0 \exp(-t_0 s)}{(s^2 + \omega_1^2)(s^2 + \omega_2^2)}$$

$$x_2(t) = \frac{1}{\omega_2} \frac{f_0}{m} \sin \omega_2 t \otimes \delta(t - 2t_0)$$

$$+ \frac{1}{\omega_1 \omega_2} \left(\omega_2^2 \frac{f_0}{m} \delta(t - 2t_0) + \frac{k}{m^2} f_0 \delta(t - t_0)\right)$$

$$\otimes (\sin \omega_1 t \otimes \sin \omega_2 t)$$

where we have substituted

$$\omega_{1,2}^2 = \frac{3 \mp \sqrt{5}}{2} \frac{k}{m}$$

and

$$\omega_1^2 + \omega_2^2 = \frac{3k}{m}$$

Now, we solve for $X_1(s)$.

$$\left(s^2 + \frac{2k}{m}\right) X_1(s) - \frac{k}{m} X_2(s) = \frac{f_0}{m} \exp(-t_0 s)$$

$$X_1(s) = \frac{k}{m\left(s^2 + \frac{2k}{m}\right)} \left(\frac{\frac{f_0}{m} \exp(-2t_0 s)}{s^2 + \omega_2^2} \right.$$

$$+ \frac{\omega_2^2 \frac{f_0}{m} \exp(-2t_0 s) + \frac{k}{m^2} f_0 \exp(-t_0 s)}{(s^2 + \omega_1^2)(s^2 + \omega_2^2)} + \left. \frac{f_0}{k} \exp(-t_0 s) \right)$$

$$x_1(t) = \frac{k}{m} \frac{f_0}{m} \delta(t - 2t_0) \otimes \frac{1}{\omega_3} \sin \omega_3 t \otimes \frac{1}{\omega_2} \sin \omega_2 t$$

$$+ \frac{k}{m} \left(\omega_2^2 \frac{f_0}{m} \delta(t - 2t_0) + \frac{k^2}{m^2} f_0 \delta(t - t_0) \right) \otimes \frac{1}{\omega_1} \sin \omega_1 t$$

$$\otimes \frac{1}{\omega_2} \sin \omega_2 t \otimes \frac{1}{\omega_3} \sin \omega_3 t + \frac{f_0}{m} \delta(t - t_0) \otimes \frac{1}{\omega_3} \sin \omega_3 t$$

where

$$\omega_3^2 = \frac{2k}{m} = \omega_1^2 + \omega_2^2$$

Problem 5.5.31

Two driving forces can be expressed as

$$f_A(t) = f_0 \left(u(t) - u\left(t - \frac{\tau}{2}\right) \right)$$

$$f_B(t) = f_0 \left(u\left(t - \frac{\tau}{2}\right) - u(t - \tau) \right)$$

where f_0 is a constant. Now, let's LT the DEs:

$$mx_1'' = -kx_1 + k(x_2 - x_1) + f_0 \left(u(t) - u\left(t - \frac{\tau}{2}\right) \right)$$

$$mx_2'' = -k(x_2 - x_1) + f_0 \left(u\left(t - \frac{\tau}{2}\right) - u(t - \tau) \right)$$

$$m\left(s^2 X_1(s) - sx_1(0) - x_1'(0)\right)$$

$$= -2kX_1(s) + kX_2(s) + f_0\left(\frac{1}{s} - \frac{\exp\left(-\frac{\tau s}{2}\right)}{s}\right)$$

$$m\big(s^2 X_2(s) - s x_2(0) - x_2'(0)\big)$$

$$= kX_1(s) - kX_2(s) + f_0\left(\frac{\exp\left(-\frac{\tau s}{2}\right)}{s} - \frac{\exp(-\tau s)}{s}\right)$$

Inserting ICs

$$x_1(0) = x_2(0) = 0$$
$$x_1'(0) = x_2'(0) = 0$$

$$(ms^2 + 2k)X_1(s) - kX_2(s) = f_0\left(\frac{1}{s} - \frac{\exp\left(-\frac{\tau s}{2}\right)}{s}\right)$$

$$-kX_1(s) + (ms^2 + k)X_2(s) = f_0\left(\frac{\exp\left(-\frac{\tau s}{2}\right)}{s} - \frac{\exp(-\tau s)}{s}\right)$$

$$X_1(s) = \frac{\frac{f_0}{s}\left(ms^2 + k - ms^2 \exp\left(-\frac{\tau s}{2}\right) - k\exp(-\tau s)\right)}{(ms^2 + k)(ms^2 + 2k) - k^2}$$

$$X_2(s) = \frac{\frac{f_0}{s}\left(k + (ms^2 + k)\exp\left(-\frac{\tau s}{2}\right) - (ms^2 + 2k)\exp(-\tau s)\right)}{(ms^2 + k)(ms^2 + 2k) - k^2}$$

Applying inverse LT, we get

$$x_1(t) = \mathcal{L}^{-1}\{X_1(s)\}$$
$$x_2(t) = \mathcal{L}^{-1}\{X_2(s)\}$$

Problem 5.5.32

(1) Considering the spring forces acting on the two masses, we get

$$\begin{cases} m_1 x_1'' = -(k_1 + k_2)x_1 + k_2 x_2 \\ m_2 x_2'' = k_2 x_1 - (k_2 + k_3)x_2 \end{cases}$$

(2) Plugging the parameters into the DEs, we get

$$\begin{cases} x_1'' = -3x_1 + 2x_2 \\ x_2'' = 2x_1 - 5x_2 \end{cases}$$

which can be solved by various methods, $e.g.$, the S-method or LT method. Let's use the S-method.

From the second DE, we get

$$x_1 = \frac{1}{2}(x_2'' + 5x_2)$$

Plugging into the first DE, we get

$$x_2^{(4)} + 8x_2^{(2)} + 11x_2 = 0$$

whose C-Eq is $r_2^4 + 8r_2^2 + 11 = 0$ with roots

$$b_{1,2,3,4} = \pm\sqrt{-4 \pm \sqrt{5}}$$

with which one can compose the final GS for m_1 and m_2

$$x_2 = \sum_{i=1}^{4} C_i \exp(b_i t)$$

One can find x_1 and, then, determine C_i by the IC.

Appendix B
Laplace Transforms

Selected Laplace Transforms

$f(t)$	$F(s) = \mathcal{L}\{f(t)\}$
$\delta(t - a)$	$\exp(-as)$
$\delta^n(t)$	s^n
$u(t - a)$	$\dfrac{\exp(-as)}{s}$
t	$\dfrac{1}{s^2}$
t^2	$\dfrac{2!}{s^3}$
$t^n, \ n \geq 0$	$\dfrac{n!}{s^{n+1}}$

$\sin \omega t$	$\dfrac{\omega}{s^2 + \omega^2}$
$\cos \omega t$	$\dfrac{s}{s^2 + \omega^2}$
$t \sin \omega t$	$\dfrac{2s\omega}{(s^2 + \omega^2)^2}$
$t \cos \omega t$	$\dfrac{s^2 - \omega^2}{(s^2 + \omega^2)^2}$
$\exp(-at)$	$\dfrac{1}{s + a}$
$t^n \exp(-at)$	$\dfrac{n!}{(s + a)^{n+1}}$
$\exp(-at) \sin \omega t$	$\dfrac{\omega}{(s + a)^2 + \omega^2}$
$\exp(-at) \cos \omega t$	$\dfrac{s + a}{(s + a)^2 + \omega^2}$

Selected Properties of Laplace Transforms

$f(t)$	$F(s)$
$f(t)$	$F(s)$
$cf(t)$	$cF(s)$
$f_1(t) + f_2(t)$	$F_1(s) + F_2(s)$

$\dfrac{df(t)}{dt}$	$sF(s) - f(0)$
$\dfrac{d^n f(t)}{dt^n}$	$s^n F(s) - s^{n-1} f(0)$ $-s^{n-2} f'(0) - s^{n-3} f''(0)$ $- \cdots - f^{(n-1)}(0)$
$\displaystyle\int_0^t f(\tau) d\tau$	$\dfrac{F(s)}{s}$
$\exp(-at) f(t)$	$F(s + a)$
$f(t - \tau) u(t - \tau)$	$\exp(-s\tau) F(s)$
$f(t) \otimes g(t)$	$F(s)G(s)$
$tf(t)$	$-\dfrac{dF(s)}{ds}$
$t^n f(t)$	$(-1)^n F^n(s)$
$\dfrac{f(t)}{t}$	$\displaystyle\int_s^t F(s) ds$
$f(ct), c > 0$	$\dfrac{1}{c} F\left(\dfrac{s}{c}\right)$

Remarks

(1) $\delta(t)$ is the Dirac Delta-function which is defined as

$$\delta(t) = \begin{cases} +\infty, & t = 0 \\ 0, & t \neq 0 \end{cases}$$

and also satisfies

$$\int_{-\infty}^{+\infty} \delta(t) dt = 1$$

(2) $u(t)$ is the unit step function defined as

$$u(t) = \begin{cases} 0, & t < 0 \\ 1, & t \geq 0 \end{cases}$$

(3) The sign $\otimes$ is the convolution operator which is defined as

$$f(t) \otimes g(t) = \int_0^t f(\tau)g(t - \tau)d\tau$$

Appendix C
Basic Formulas

1. $\exp(ix) = \cos x + i \sin x$

2. $\sin x = \frac{1}{2i}(\exp(ix) - \exp(-ix))$

3. $\cos x = \frac{1}{2}(\exp(ix) + \exp(-ix))$

4. $\exp(x) = \cosh x + \sinh x$

5. $\sinh x = \frac{1}{2}(\exp(x) - \exp(-x))$

6. $\cosh x = \frac{1}{2}(\exp(x) + \exp(-x))$

7. $\sin(x \pm y) = \sin x \cos y \pm \cos x \sin y$

8. $\cos(x \pm y) = \cos x \cos y \mp \sin x \sin y$

9. $\frac{d}{dx}(uv) = \frac{du}{dx}v + \frac{dv}{dx}u$

10. $\frac{d}{dx}F(g(x)) = \frac{dF}{dg}\frac{dg}{dx}$

11. $\frac{d}{dx}(x^n) = nx^{n-1}$

12. $\frac{d}{dx}(\ln x) = \frac{1}{x}$

13. $\frac{d}{dx}(\exp(x)) = \exp(x)$

14. $\frac{d}{dx}(\sin x) = \cos x$

15. $\frac{d}{dx}(\cos x) = -\sin x$

16. $\frac{d}{dx}(\tan u) = \sec^2 u$

17. $\int u\,dv = uv - \int v\,du$

18. $\int \exp(x)\,dx = \exp(x) + C$

19. $\int x^n dx = \frac{1}{n+1}x^{n+1} + C \quad (n \neq -1)$

20. $\int \frac{dx}{x} = \ln|x| + C$

21. $\int a^x dx = \frac{a^x}{\ln a} + C$

22. $\int \sin x\,dx = -\cos x + C$

23. $\int \cos x\,dx = \sin x + C$

24. $\int \tan x\,dx = -\ln|\cos x| + C$

25. $\int \sec x\,dx = \ln|\sec x + \tan x| + C$

26. $\int \csc x\,dx = \ln|\csc x - \cot x| + C$

27. $\int \frac{dx}{\sqrt{a^2-x^2}} = \sin^{-1}\frac{x}{a} + C$

28. $\int \frac{dx}{a^2+x^2} = \frac{1}{a}\tan^{-1}\frac{x}{a} + C$

29. $\int \frac{dx}{a^2-x^2} = \frac{1}{2a}\ln\left|\frac{x+a}{x-a}\right| + C$

Appendix D
Abbreviations

Abb.	Meaning
1st.O	First-order
2nd.O	Second-order
AE(s)	Algebraic Equation(s)
C	Constant, Const
BC(s)	Boundary Condition(s)
c-coeff	constant-coefficient
C-Eq	Characteristic Equation
DE(s)	Differential Equation(s)
DV(s)	Dependent variable(s)
Eq(s)	Equation(s)
E-method	Eigen-Analysis method
e-value(s)	Eigenvalue(s)

e-vector(s)	Eigenvector(s)
GS, GS's	General Solution(s)
Homo	Homogeneous, homogeneous
Homo DE(s)	Homogeneous Differential Equation(s)
Homo Systems(s)	Homogeneous System(s)
IC(s)	Initial Condition(s)
IDE(s)	Integro-Differential Equation(s)
InHomo	Inhomogeneous
InHomo DE(s)	Inhomogeneous Differential Equation(s)
InHomo System(s)	Inhomogeneous System(s)
IF	Integrating Factor
IV(s)	Independent variable(s)
IVP(s)	Initial Value Problem(s)
Linear DE(s)	Linear Differential Equation(s)
LT(s)	Laplace Transform(s)
LD	Linear Dependent, Linearly Dependent
LHS	Left-Hand Side
LI	Linear Independent, Linearly Independent
max-distance	maximum distance
max-height	maximum height
max-speed	maximum speed
MUC	Method of Undetermined Coefficients
Nonlinear DE(s)	Nonlinear DE(s), Nonlinear Differential Equation(s)

O-method	Operator method
O.W.	Otherwise
PS, PS's	Particular Solution(s)
QED	*quod erat demonstrandum* (Thus, it has been demonstrated.)
RHS	Right-Hand Side
S-method	Substitution method
SOV	Separation of Variables
SS, SS's	Singular Solution(s)
DE system(s)	System(s) of Differential Equation(s)
TS, TS's	Trial Solution(s)
VOP	Variation of Parameters
v-coeff	variable-coefficient
wrt	With Respect To

References

1. Coddington, E. A. and Levinson, N. *Theory of Ordinary Differential Equations.* New York: McGraw Hill, 1955
2. Hale, J. K. *Ordinary Differential Equations* (Unabridged Ed.). Dover Books on Mathematics. Mineola: Dover Publications, 2009; Original Publisher: New York: Wiley, 1965
3. Arnold, V. I. and Silverman, R. A. (Translator). *Ordinary Differential Equations.* Dover Books on Mathematics. Cambridge: The MIT Press, 1978
4. Tenenbaum, M. and Pollard, H. *Ordinary Differential Equations.* Dover Books on Mathematics. Mineola: Dover Publications, 1985
5. Coddington, E. A. *An Introduction to Ordinary Differential Equations* (Unabridged Ed.). Dover Books on Mathematics. Mineola: Dover Publications, 1989
6. Simmons, G. F. and Robertson, J. S. *Differential Equations with Applications and Historical Notes* (2nd Ed.). International Series in Pure and Applied Mathematics. New York: McGraw Hill, 1991
7. Hartman, P. *Ordinary Differential Equations* (2nd Ed.). Classics in Applied Mathematics (Book 38). Philadelphia: SIAM, 2002

8. Edwards, C. H. and Penney, D. E. *Elementary Differential Equations and Boundary Value Problems* (6th Ed.). New York: Pearson, 2007

9. Boyce, W. E. and DiPrima R. C. *Elementary Differential Equations* (10th Ed.). Hoboken NJ: Wiley, 2012

Index

Printed in the United States
By Bookmasters